Agronomy and Crop Production

Agronomy and Crop Production

Edited by **Alabaster Jenkins**

New York

Published by Syrawood Publishing House,
750 Third Avenue, 9th Floor,
New York, NY 10017, USA
www.syrawoodpublishinghouse.com

Agronomy and Crop Production
Edited by Alabaster Jenkins

International Standard Book Number: 978-1-68286-037-3 (Hardback)

Printed in the United States of America.

Contents

Preface

Agronomy is an important field of study in the discipline of agricultural science that primarily deals with crop production and soil management for food, fuel and other useful products. The aim of this book is to provide an understanding of the multiple aspects of agronomy with the help of concepts such as sustainable agriculture, crop rotation, plant breeding and genetics, use of fertilizers, crop yield, etc. This book, with its detailed analyses and data, will prove immensely beneficial to professionals and students engaged in this field at various levels.

After months of intensive research and writing, this book is the end result of all who devoted their time and efforts in the initiation and progress of this book. It will surely be a source of reference in enhancing the required knowledge of the new developments in the area. During the course of developing this book, certain measures such as accuracy, authenticity and research focused analytical studies were given preference in order to produce a comprehensive book in the area of study.

This book would not have been possible without the efforts of the authors and the publisher. I extend my sincere thanks to them. Secondly, I express my gratitude to my family and well-wishers. And most importantly, I thank my students for constantly expressing their willingness and curiosity in enhancing their knowledge in the field, which encourages me to take up further research projects for the advancement of the area.

Editor

Varietal Differences in Rice (*Oryza sativa* L.) Resistance to the Shield Bug, *Aspavia armigera* (Fabricius) (Hemiptera: Pentatomidae)

Abiodun O. Joda[1], Francis K. Ewete[2], B. N. Singh[3] & Olufemi O. R. Pitan[4]

[1] Department of Crop Production, Olabisi Onabanjo University, Ago-Iwoye, Nigeria

[2] Department of Crop Protection and Environmental Biology, University of Ibadan, Ibadan, Nigeria

[3] Formerly of the West African Rice Development Association, WARDA, Ibadan, Nigeria

[4] Department of Crop Protection, Federal University of Agriculture, Abeokuta, Nigeria

Correspondence: Olufemi O. R. Pitan, Department of Crop Protection, Federal University of Agriculture, Abeokuta, Nigeria. E-mail: femipitan@yahoo.com

Abstract

Sixty rice accessions were assessed in the field for resistance to *Aspavia armigera* F. attack under natural and artificial infestations. Out of these, 30 were lowland and 19 upland varieties derived from the International Institute for Tropical Agriculture (IITA) breeding programmes, while the others were obtained from other local and international organizations. From the upland varieties, the highest population density of *A. armigera* was recorded on CAN 4143 (p < 0.05), while the highest percentages of unfilled grains were obtained on ITA 132 (33.1%) and CAN 4143 (32%), and were both rated the most susceptible to *A. armigera*. CAN 6656, on which the lowest value for unfilled grains (4.7%) was found was rated resistant. Among the lowland varieties, TOX 3118-2-E2-2-1-2, on which was recorded the highest population density of *A. armigera* and highest percentage of unfilled grains (p < 0.05) was considered susceptible, while ITA 230, ITA 308, ITA 123, TOX 3100-32-2-1-3-5, TOX 3107-39-1-2-1-3, TOX 3027-43-1-E3-1-1, TOX 3441-7-1-1-1, TOX 3226-5-2-2-2, TOX 3716-15-1-1 and TOX 3561-56-2-3-2 were considered resistant. However, no-choice tests revealed that five from each of ten upland and lowland varieties initially rated as resistant were severely damaged. The other five upland ITA 315, IRAT 169, ITA 321, FAROX 41 and M 55 with consistent expression of resistance which also had low number of stylet sheath: 11, 17, 23, 24.3, and 29.3, respectively, were therefore rated resistant. Similarly, lowland ITA 230, TOX 3561-56-2-3-2, TOX 3100-2-1-3-5, TOX 3226-5-2-2-2 and TOX 3107-39-1-2-1-3 with low number of stylet sheath were rated resistant as well.

Keywords: *Aspavia armigera*, rice, resistance, damage, pod-sucking bugs, variety, pentatomid

1. Introduction

Rice (*Oryza sativa* L.) is one of the major food crops cultivated by farmers in all agro-ecological zones of Nigeria. It is an important cereal crop and food item, not only widely consumed by a large proportion of the population in Nigeria, but all around the world in various forms (Ibiam et al., 2006; Shehu et al., 2013). The global rice production is estimated at 454.6 million tonnes annually which has an average yield of 4.25 tonnes per hectare (Fazlollah et al., 2011). Land under rice cultivation in Nigeria has increased from 1,609,890 ha in 2005/2006 to 2,012,740 ha in 2009/2010, while production has also moved from 3,286,500 kg/ha in 2005/2006 to 4,080,940 kg/ha in 2009/2010 (Ibrahim, 2013).

Among the constraints to rice production in Nigeria are pests which include weeds, pathogens, insects, rodents and birds. Insect pests are serious enemies to rice production which must be protected and increased to foster human health. Over 800 insect pests attack standing and stored rice (Grist & Lever, 1969). According to Pathak and Dhaliwal (1981), these insect pests account for rice losses of 24%, while Cramer (1967) reports 35%.

Aspavia armigera is one of the insect pest complex attacking rice in Nigeria. The insect is abundant in both upland and low land ecosystems of the humid forest and Guinea savanna zones, but is more abundant in the latter (Ewete & Olagbaju, 1990). The grain-sucking bug has a wide host range. In Nigeria, the bug is a major pest of rice as well as soybean and cowpea (Ewete & Olagbaju, 1990; Pitan et al., 2007). The feeding of both the adult and nymphs of *Aspavia* species especially *A. armigera* at the dough stage results in partially filled grains of low quality. Unprocessed panicles (paddy) usually have diffused brown spots indicating areas of attack on the grains by the bugs. However, infested grains turn dirty brown to black during parboiling thus affecting their aesthetic value.

Production practices such as varietal selection impact insect populations and damage because some rice varieties are more tolerant to biotic stresses than others (Long et al., 2001). In fact, one of the ideal methods of minimizing insect damage on rice is to cultivate resistant varieties (Akinsola, 1985). However, thousands of unscreened rice lines collected from all over the world are available at the gene banks in Africa, for instance at IITA, Ibadan, and the West African Rice Development Association, WARDA (Now the African Rice Center), Ibadan, Nigeria (Ng et al., 1983). Varietal screening, therefore, is a continuous activity to increase the chances of discovering resistant qualities as few rice varieties have resistance traits (Akinsola, 1984).

Reports on rice varietal screening have been on other insects such as *Manduca sexta* (William et al., 1980), *Sitophilus graminum* (Starks & Buxton, 1977) and *Heliothis zea* (Wiseman, 1985). In Nigeria, breeding for resistance to insect pests had been on rice stem borers especially *Chilo* sp. and *Maliarpha* sp. (Akinsola, 1987; Alam, 1990; Ukwungwu & Odebiyi, 1985), and Africa Rice Gall Midge, *Orseolia oryzivora* H&G (Ukwungwu, 1985; Ukwungwu et al., 1990) but little attention was given to other insects. No doubt, there is paucity of information on rice resistance to the grain-suckers most especially, *Aspavia* sp. in Nigeria. In this study therefore, 60 lowland and upland rice varieties were evaluated for resistance to *A. armigera* on the field under natural and artificial infestation conditions. This was to select varieties as donor in the breeding programme for the integrated control of this bug.

2. Materials and Methods

2.1 Sources of Seeds

Sixty rice accessions with days to flowering ranging from 93 – 100 comprising 30 each of upland and lowland accessions were used for the study. All the lowland accessions were International Institute of Tropical Agriculture (IITA), Ibadan, Nigeria accessions: ITA 121, ITA 230, ITA 222, ITA 308, ITA 312, ITA 123, ITA 328, ITA 306, ROX 3562-65-1-2-1, TOX 3107-39-1-2-2, TOX 3098-2-2-1-2-1, TOX 3100-32-2-1-3-5, TOX 3118-2-E2-2-1-2, TOX 3118-2-E2-2-1-2, TOX 3107-39-1-2-1-3, TOX 3052-46-E2-1, TOX 3402-6-3-2-2, TOX 3440-176-1-2-1, TOX 3440-52-3-3-3, TOX 3027-43-1-E3-1-1, TOX 3562-8-2-1-2, TOX 3399-108-3-2-2, TOX 3441-22-3-3-2, TOX 3441-7-1-1-1, TOX 3226-5-1-2-2, TOX 3716-15-1-1, TOX 3561-56-2-3-2, TOX 3226-5-2-2-2, TOX 3440-133-2-3-2, and TOX 3162-11-1-2-1-1. Nineteen of the upland accessions were IITA germplasms: ITA 132, ITA 337, ITA 162, ITA 321, ITA 317, ITA 325, ITA 333, ITA 301, ITA 305, ITA 307, ITA 235, ITA 315, ITA 186, ITA 185, ITA 118, ITA 338, ITA 116, ITA 187, and ITA 182. Other varieties, FARO 11, FARO 31, and FARO 41 were from the Federal Department of Agriculture, Nigeria, IDSA 6 and IDSA 62 (Institut des Savanes, Bouake, Cote d'Ivoire), CAN 6656 and CAN 4143 (Rice Experiment Station, Chainat, Thailand), CT 6261 (Bangladesh Rice Research Institute, Chittagong, Bangladesh), IRAT 169 and IRAT 104 (Institute des Gerdart, IRAT station, Bouake), and M 55 (Macros Agricultural Research Institute, Indonesia). The field screening experiment was carried out under natural infestation as well as artificial infestation in a field cage.

2.2 Preliminary Field Screening of 60 Rice Varieties for Resistance to Aspavia armigera

The experiment was carried out at the ES-17 and F-paddy fields at WARDA in IITA, Ibadan, Nigeria. Seeds were sown, one variety per line, at a spacing of 20 cm × 20 cm using a randomized complete block design with three replicates. Total plot size was 12 m × 6 m for upland or lowland variety, and each replicate occupied 3 m × 6 m with 1 m border row between them. Plots were hand-weeded twice, and NPK fertilizer was applied to the plots at the rate of 80-60-60 kg/ha in two splits - at planting and tillering. *A. armigera* number (nymphs and adults) was counted visually from 10 plants/panicles per treatment (variety) per replicate at weekly intervals starting from flowering to harvesting period between 7.00 – 9.00 hrs local time.

2.3 Field Screening of Twenty Rice Varieties for Resistance to Aspavia armigera under Artificial Infestation

The experiment consisted of (i) mass rearing of *A. armigera* (ii) field establishment of test rice varieties and (iii) caging of the insects with the test rice plants.

2.3.1 Mass Rearing of A. armigera

A. armigera mass rearing method was adapted from Nilakhem (1976), Bowling (1979a, 1979b) and Heinrichs et al. (1985). Thirteen cages which were constructed using water-resistant wood and aluminum wire mesh were placed in a trench of about 15 cm deep outside the Plant Biology Laboratory, Olabisi Onabanjo University, Ago-Iwoye, Nigeria. Carbofuran (1.0 kg a. i/ha) was applied to the trench while permethrin powder was applied round each cage to trap ants and other crawling insects. The 13 cages were shared into 3 sets - Oviposition cages (3), Culture Maintenance cages (3), and Test cages (7). Potted rice variety ITA 257 which served as food source for the insect was planted in 30 plastic pots (20 cm diameter) filled with sieved top soil. Granules of NPK were incorporated into the soil before planting, and six seeds were sown per pot each week before thinning to two plants per pot at two weeks after planting. The potted rice plants were watered everyday during the dry season and every other day during the rainy season.

Thirty adult *A. armigera* were collected from the field, sexed and placed (1M:1F) in one of the oviposition cages containing two potted milk-stage and six potted booting-stage ITA 257 plants. After 7 days, the potted plants at booting stage containing eggs were transferred to one of the Culture Maintenance cages. At the same time, another six potted plants were placed near the already egg-laden rice plants so that the newly hatched nymphs could easily transfer to the panicle of the new food plants. This arrangement was left for 1 week. The six potted plants (now with nymphs) were thereafter transferred into Test cages where they were reared to adults. Only the newly emerged adults were used in the screen cage experiment. Food was changed in the Test cages twice weekly on Mondays and Thursdays, and all contacts with the rearing cages were in the cool hours of the day (7.00 – 9.00 hrs and 16.30 – 18.00 hrs local time). These steps were repeated throughout the screening exercise.

2.3.2 Field Establishment of Test Rice Varieties

The varieties used in this study were those considered resistant to *A. armigera* in the previous preliminary screening exercise, These were: ten upland rice: IRAT 169; FAROX 41; ITA 321; ITA 315; M 55; IDSA 52; CAN 6656; IDSA 6; IRAT 104; and ITA 301, and ten lowland rice: ITA 230; ITA 123; TOX 3100-2-1-3-5; TOX 3107-39-1-2-1-3; TOX 3441-7-1-1-1 and TOX 3716-15-1-1. This experiment was sited at the F-4 and EH15 at IITA, Ibadan, Nigeria. The 20 upland or lowland rice varieties (treatments) were in the nursery for 28 days before transplanting into the field in a randomized complete block design (RCBD) with four replicates. The rice plants were spaced at 20 cm × 20 cm giving a total of 20 hills of each rice variety per replicate. The plot was hand-weeded twice and NPK fertilizer was applied at the rate of 80-60-60 kg/ha in two splits. There was also top-dressing of rice with urea at 50 kg/ha.

2.3.3 Caging of the Insects with the Test Rice Plants

Large removable field screening cages with each ($14 \times 5 \times 1.5$ m^3) covering an area of 70 m^2 were constructed with water-resistant wood and muslin cloth for confining *A. armigera* to rice plants (both lowland and upland) on the field. Each field cage was installed on 10 well-established lowland or upland rice varieties at EH15 and F4, IITA, Ibadan, when about one-fourth of the grains reached the milk stage. The muslin cloth enveloped the cage but maneuverable entrance for the introduction of insect was created at the corner of the cage. Adult male and female *A. armigera* were collected from the mass culture into a muslin cloth bag (30 cm × 15 cm) containing fresh soft rice panicle which served as food and were transported to the field. Adult *A. armigera* were introduced into the field cage at 15 bugs/m^2 (IRRI, 1988) from a corner close to the rice panicles. The cages were checked every other day to ascertain their integrity, and were left until the rice panicles were ripened for harvest.

2.4 Data Collection and Analysis

In both experiments, at maturity ten rice panicles were randomly harvested from each replicate, placed in separate paper bag, dried and labeled. These panicles were taken to the laboratory for the evaluation of bug feeding activity by the unfilled grain seed cleaner technique. The 10 panicles harvested from each replicate were oven-dried at 50 °C for 7 days and thereafter threshed in a panicle. The filled and unfilled grains were weighed before running it through a seed cleaner. Chaffy (unfilled grains) were brown-off. The weight of filled grains was taken. The percentage of unfilled grains was taken as:

$$\% \text{ Unfilled grains} \quad = \quad \frac{X-Y}{X} \quad = \quad \frac{\text{Weight of the unfilled grains}}{\text{Total weight of filled + Unfilled grains}} \quad \times \quad 100$$

In the cage experiment, ten panicles per plot were harvested into a sleeve made of nylon mesh. This was immersed in the staining solution for one hour. After removal, they were rinsed in tap water, air-dried and later examined under a binocular dissecting microscope (Hamberg Brand) for stylet punctures to determine the percentage grain damage. Resistance was based on percentage of unfilled grains and *A. amigera* infestation level. In the cage experiment, resistance was confirmed using *A. armigera*-induced stylet grain damage.

Data on *A. armigera* population density were square root-transformed while those in percentages were arc sine-transformed before subjecting them to analysis of variance using SAS (2002). Significant means were thereafter, separated using Duncan's Multiple Range Test at 5% probability.

3. Results

Results showed that only 16.7% of the rice accessions tested was found to be resistant to *A. armigera*. Among the upland rice varieties, significantly higher population ($p < 0.05$) (5.0) of *A. armigera* was recorded on CAN 4143 compared to other varieties screened. The lowest bug population (2.0) was however, found on thirteen other varieties. The highest percentage of unfilled grains (33.1) was obtained on ITA 132, and 32% on CAN 4143, making them the most susceptible varieties to *A. armigera*. However, CAN 6656, on which the lowest value of unfilled grains (4.7) was found was considered resistant and selected for further screening (Table 1).

Table 1. Mean number of *Aspavia armigera* and percentage unfilled grains on thirty upland rice varieties under natural infestation

Variety	Mean number of *Aspavia sp*	Percentage unfilled grains	Decision
ITA 132	3.0b	33.1 ± 8.5a	S
ITA 337	2.3c	14.0 ± 3.0d	S
ITA 162	2.7b	19.6 ± 3.8d	S
ITA 321	2.0c	8.4 ± 2.7e	R
ITA 317	2.0c	17.8 ± 1.5d	S
ITA 325	2.3c	15.9 ± 8.2d	S
ITA 333	2.0c	16.6 ± 2.5d	S
ITA 310	2.3c	9.7 ± 2.1 e	R
ITA 305	2.0c	17.7 ± 4.5d	S
ITA 307	2.0c	21.8 ± 9.1c	S
ITA 235	3.0b	15.9 ± 8.6d	S
ITA 315	2.3c	5.5 ± 1.9e	R
ITA 118	2.7b	15.9 ± 8.2d	S
ITA 135	2.3c	16.2 ± 2.7d	S
ITA 118	2.3c	15.7 ± 1.1d	S
FAROX 41	2.0c	9.5 ± 1.7e	R
ITA 338	2.0c	20.8 ± 3.3c	S
IDSA 6	2.0c	9.0 ± 3.5e	R
CAN 4143	5.0a	32 ± 7.3a	S
IRAT 104	2.7b	7.5 ± 2.5e	R
M 55	2.0c	9.1 ± 0.7e	R
CT 6261	2.0c	12.8 ± 5.6d	S
IRAT 169	2.0c	9.9 ± 2.1d	R
IDSA 62	2.3c	9.8 ± 5.6d	R
CNA 6656	2.0c	4.7 ± 1.6e	R
ITA 116	2.3c	23.7 ± 4.4c	S
ITA 187	2.0c	21.9 ± 6.1c	S
ITA 182	2.7b	16 ± 2.1d	S
FARO 11(OS 6)	2.7b	14.4 ± 6.7d	S
FARO 31	2.3c	25.3 ± 5.3b	S

Note. Means followed by same letter in a column are not significantly different at 5% probability level; R = Resistant, S = Susceptible.

Similarly, among the lowland varieties, TOX 3118-2-E2-2-1-2 recorded the highest population density of *A. armigera* (4.7) which was not significantly different from values observed on ITA 121, ITA 222, TOX 3052-46-E2-1, TOX 3562-65-1-2-1; ITA 328 and TOX 3562-8-2-1-2. The lowest bug density (1.0) was however, obtained from eleven varieties. A significantly ($p < 0.05$) higher percentage of unfilled grains (21.8) was recorded on TOX 3118-2-E2-1-2, while 2.2% which was the lowest value was recorded on TOX 3716-15-1-1.

Therefore, 10 lowland varieties: ITA 230, ITA 308, ITA 123, TOX 3100-32-2-1-3-5, TOX 3107-39-1-2-1-3, TOX 3027-43-1-E3-1-1, TOX 3441-7-1-1-1, TOX 3226-5-2-2-2, TOX 3716-15-1-1 AND TOX 3561-56-2-3-2 were considered resistant and were selected for further no-choice tests (Table 2).

Table 2. Mean number of *Aspavia armigera* and percentage unfilled grains on thirty lowland rice varieties under natural infestation

Variety	Mean number of Aspavia sp	Percentage unfilled grains	Decision
ITA 121	3.7a	17.8 ±18.7b	S
ITA 230	1.0a	5.2 ± 2.3f	R
ITA 222	3.0a	16.6 ± 4.1b	S
ITA 308	1.0c	7.1 ± 1.6e	R
ITA 312	2.0b	21.0 ± 6.7a	S
ITA 123	1.0c	5.5 ± 2.7f	R
TOX 3107-39-1-2-1	10c	10.8 ± 1.9d	S
TOX 3098-39-1-2-1	2.0b	10.1 ± 1.6d	S
TOX 3100-32-2-1-3-5	1.0c	6.1 ±1.7e	R
TOX 3081-36-E1-4-3	2.0b	11.2 ± 3.1c	S
TOX 3118-2-E2-2-1-2	4.7a	21.8 ± 8.5a	S
TOX 3107-39-1-2-3	1.0c	8.4 ±2. 1e	R
TOX 3052-46-E2-1	3.0a	16.2 ± 6.2b	S
TOX 306	2.0b	20.2 ± 6.2a	S
TOX 3402-6-3-2-2	2.0b	10.0 ± 1.3d	S
TOX 3440-176-1-2-1	2.0b	12.2 ± 4.1c	S
TOX 3440-52-3-3-3	1.0c	10.7 ± 3.0d	S
TOX 3027-43-1-E3-1	1.0c	5.0 ± 0.6f	R
TOX 3562-8-2-1-2	3.0a	16.0 ± 2.7b	S
TOX 328	3.0a	13.8 ± 3.0c	S
TOX 3562-8-2-1-2	3.0a	10.0 ± 1.1d	S
TOX 3441-22-3-3-2	2.0b	13.1 ± 5.2c	S
TOX 3441-7-1-11	1.0c	7.0 ± 2.8e	R
TOX 3226-5-2-2-2	1.0c	6.5 ± 2.1e	R
TOX 3716-15-1-1	1.0c	2.2 ± 0.4g	R
TOX 3561-56-2-3-2	0.7a	3.0 ±04.7g	R
TOX 3440-133-2-3-2	1.3c	10.5 ± 1.2d	S
TOX 3226-5-3-2-2	1.3c	9.5 ± 2.4d	S
TOX 3399-108-3-2-2	2.3b	12.8 ± 3.6c	S
TOX 3162-11-1-2-1-1	2.0b	9.9 ± 1.8d	S

Note. Means followed by same letter in a column are not significantly different at 5% probability level; R = Resistant, S = Susceptible.

Table 3 shows the feeding activity of *A. armigera* on upland rice varieties screened under artificial infestation. Based on the number of stylet sheath recorded, the most susceptible variety was IRAT 104 which had 36.8 stylet sheaths statistically similar to that obtained for ITA 301. The most resistant variety was ITA 315 which had significantly lower number of stylet sheaths (11.0) compared to other upland varieties. Consequently, only five out of the ten initially considered resistant varieties from each of upland and lowland were consistent when further tested in the cage experiment. The upland varieties were: ITA 315, IRAT 169, ITA 321, FAROX 41 and M 55 with 11, 17, 23, 24.3, and 29.3 stylet sheaths, respectively. Similarly, the lowland rice varieties with significant low *A. armigera* feeding activity ($P < 0.05$) which were considered resistant were: ITA 230, TOX 3561-56-2-3-2, TOX 3100-2-1-3-5, TOX 3226-5-2-2-2 and TOX 3107-39-1-2-1-3 (Table 3).

4. Discussion

The preliminary field screening of 60 upland and lowland rice varieties based on the number of *A. armigera* and the percentage of unfilled grain recorded on each variety revealed that ten from each of upland and lowland varieties were resistant, while others were considered susceptible. The upland varieties: ITA 321, ITA 301, ITA 315, FAROX 41, IRAT 104, M 55, IRAT 169, IDSA 62, CAN 6656, and IRAT 6 were considered resistance based on higher unfilled grains and were selected for further screening, as screening under natural infestation is not enough for meaningful selection of resistant rice varieties (Akinsola, 1987). Similarly, ten lowland rice varieties: ITA 230, ITA 308, ITA 123, TOX 3100-32-1-3-5, TOX 3107-39-1-2-1-3, TOX 3027-43-1-E3-1-1, TOX 3441-7-1-1-1, TOX 3226-5-2-2-2, TOX 3176-15-1-1-1 and TOX 3561-56-2-3-2 were selected as resistant, based on the unfilled grains criterion, for further screening.

Out of these twenty upland and lowland varieties that were further screened in field cages, ten varieties: ITA 301, CAN 6656, IRAT 104, IDSA 6, IDSA 62, ITA 123, ITA 308, TOX 3716-15-1-1, TOX 3441-7-1-1-1 and 3027-43-1-E3-3-1-1, showed inconsistency in the expression of resistance and were rated susceptible after all. The expression of resistance to this bug in the remaining ten varieties were found to be reliable in the following order: upland-lowland.

Table 3. Feeding activity of adult *Aspavia armigera* on upland and lowland rice varieties under artificial infestation

Variety	Mean number of stylet	Rating
Upland		
ITA 301	36.0 ± 2.9a	S
CNA 6656	29.5 ± 2.9b	S
IRAT 104	36.8 ± 6.7a	S
ITA 321	13.3 ± 2.5c	R
IDSA 6	30.3 ± 4.6b	S
ITA 315	11.0 ± 2.1d	R
IRAT 169	7.0 ± 4.2d	R
IDSA 62	31.0 ± 5.5b	S
FAROX 41	4.3 ± 2.9e	R
M 55	9.3 ± 7.5d	R
Lowland		
ITA 230	8.3 ± 0.9c	R
ITA 123	21.8 ± 1.9a	S
ITA 308	17.8 ± 2.8b	S
TOX 3716-15-1-1	17.8 ± 2.3b	S
TOX 3561-56-2-3-2	7.8 ± 2.3d	R
TOX 3441-7-1-1-1	18.5 ± 1.5b	S
TOX 3100-32-2-1-3-5	6.8 ± 1.4d	R
TOX 3226-5-2-2-2	6.0 ± 1.2d	R
TOX 3107-39-1-2-1-3	4.3 ± 0.5e	R
TOX 3027-43-1-E3-1-1	16.3 ± 1.8b	S

Note. Means followed by same letter in column within a variety group are not significantly different at 5% probability level; ± s.e.; * Average of 1000 grains/variety; R = variety considered resistant; S = Susceptible.

FAROX 41 > ITA 321 > ITA 315 > M 55 > IRAT 169; lowland: TOX 3107-39-1-2-1-3 > TOX 3226-5-2-2-2 > TOX 3100-32-2-1-3-5 > TOX 3561-56-2-3-2 > ITA 230. This result showed that half of the 20 varieties selected from the first experiment succumbed to the feeding activity of the bug when confronted with a no-choice situation in the second experiment. This further underscores the importance of artificial bombardments of pest organisms in order to arrive at consistent resistance to pests in crops.

The mechanism by which insect resistance is conferred on crops is by non-preference, antibiosis or tolerance

(Painter, 1951; Kogan & Ortman, 1978). The choice to feed on a particular rice variety more than the other by a specific insect species may be classified as antixenosis or antibiosis modality of resistance, while tolerance has to do with the plant characteristics itself that withstand insect pest attack. Incidentally, all the rice varieties considered as resistant had fewer number of *A. armigera* on them and lower grain damage compared to others making them to exhibit either antixenosis or antibiosis type of resistance. Although previous works on rice resistance to *A. armigera* is scanty, Ewete and Olagbaju (1990) showed that ITA 257 was more susceptible to *A. armigera* than ITA 128 due to significantly higher damage and shorter developmental period recorded on the former.

That only 16.7% of the rice accessions tested was found to be resistant to *A. armigera* underscored the high level of rice susceptibility to *A. armigera* in Nigeria. It also shows that the grain-sucker is a serious economic pest of field-grown rice, which attack must be made a priority in crop improvement programmes. There is therefore, the need for continuous assessment of varieties resistant to the grain-sucker since there are large quantities of rice genotypes collected all over the world in many gene banks in Africa awaiting evaluation (Akinsola, 1984). The identified grain-sucker-resistant accessions should be made adaptable to the various rice cropping systems of the country, and be encouraged for cultivation.

Acknowledgements

The authors are grateful to the International Institute for Tropical Agriculture (IITA), Ibadan, Nigeria, and the West African Rice Development Association (WARDA) now Africa Rice Center (*AfricaRice*), Ibadan, Nigeria for providing the rice genotypes used for this study.

References

Akinsola, E. A. (1984). Insect pests of upland rice in Africa. *Overview of Upland Rice Research*. Lagona, Philippines: IRRI Los Bano.

Akinsola, E. A. (1985). Problems and prospects of rice varietal resistance in West Africa. *Insect Science and Its Application, 6*, 467-471.

Akinsola, E. A. (1987). Varietal resistance to stem borer pests of rice in West Africa: A review. *Insect Science and Its Application, 8*, 771-776.

Alam, M. S. (1990). *Checklist of rice insect pests and their parsitoids and predators in Nigeria* (p. 15). IITA, Ibadan, Nigeria.

Bowling, C. C. (1979a). Breeding for host plant resistance to rice field insects in the USA. In M. V. Harris (Ed.), *Biology and breeding for resistance to arthropods and pathogens in agricultural plants* (pp.329-340). Texas A&M, College Station USA.

Bowling, C. C. (1979b). The stylet sheath as an indicator of feeding activity of the rice stink bug. *Journal of Economic Entomology, 72*, 259-260. http://dx.doi.org/10.1093/jee/72.2.259

Cramer, H. H. (1967). Plant protection and world crop protection. *Pflanzenshutz Nachrichten bayer 20* (p. 524).

Ewete, F. K., & Olagbaju, R. A. (1990). The development of Aspavia armigera F. (Hemiptera: Pentatomidae) and its status as a pest of cowpea and rice. *Insect Science and Its Application, 11*(2), 171-177.

Fazlollah, E. C., Hoshang, B., & Abbas A. (2011). Evaluation of traditional, mechanized and chemical weed control methods in rice field. *Australian Journal of Crop Science, 5*(8), 1007-1013.

Grist, D. H., & Lever, R. J. A. W. (1969). *Pests of Rice*. Longmans, Green and Co., London, Harlow.

Heinrichs, E. A, Medrano, F. G., & Rapusas, H. R. (1985). *Genetic evaluation for insect resistance in rice*. International Rice Research Institute, Los Banos, Philippines.

Ibiam, O. F., Umechuruba, A. C. I., & Arinze, A. E. (2006). Seed-borne fungi associated with seeds of rice (*Oryzae sativa* L.) in storage and from the field in Ohaozara and Onicha Local Government areas of Ebonyi State. *World Journal of Biotechnology, 7*, 1062-1-72.

Ibrahim, U. (2012). Evaluation of Post Emergence Herbicides on Weed Control, Performance and Profitability of Rice (Oryza sativa) at Lafiagi, Kwara State of Nigeria. *Libyan Agriculture Research Center Journal International, 3*(5), 236-240.

Ibrahim, U., Rahman, S. A., & Babaji, B. A. (2011). Cost –benefit analysis of rice production under weed control methods in Lafiagi area of Kwara. *NSUK Journal of Science and Technology, 1*(1&2), 40-47.

International Rice Research Institute (IRRI). (1988). *Standard evaluation system for rice* (pp. 285-304). IRRI

Los Banos, Laguna Philippines.

Kogan, M., & Ortman, E. E. (1978). Antixenosis - A new term proposed to replace Painter's "non-preference" modality of resistance. *Bulletin of the Entomology Society of America, 24*, 175-176. http://dx.doi.org/10.1093/besa/24.2.175

Long, D. H., Correll, J. C., Lee, F. N., & TeBeest, O. O. (2001). Rice blast epidemics initiated by infested rice grain on the soil surface. *Plant Disease, 85*, 612-616. http://dx.doi.org/10.1094/PDIS.2001.85.6.612

Ng, N. Q., Jacquot, M., Abifarin, A. O., Crolik, A. J., Ghesquievea, A., & Mile, Z. (1988). Rice genetic resources collection and conservation activities in Africa and Latin America. *Rice Germplasm*. IRRI, Los Banos, Philippines.

Nilakhem, S. S. (1976). Rice lines screened resistance to rice stunk bug. *Journal of Economic Entomology, 69*, 703-705. http://dx.doi.org/10.1093/jee/69.6.703

Painter, R. H. (1951). *Insect resistance in crop plants*. The Macmillan Company, New York.

Pathak, P. K., & Dhaliwal, G. S. (1981). Trends and strategies for rice pest problems in tropical Asia. *IRRI Research Paper Series 64*.

Pitan, O. O. R., Odubiyi, S. I. I., & Olatunde, G. O. (2007). Yield response of cowpea, *Vigna unguiculata* L. (Walp.), to infestation of *Aspavia armigera* F. (Hemiptera: Pentatomidae). *Journal of Applied Entomology, 131*, 704-708. http://dx.doi.org/10.1111/j.1439-0418.2007.01209.x

SAS Institute. (2002) Statistical Analytical Systems. *SAS/STAT User's guide version, 8*(2). Cary NC: SAS Institute Inc.

Shehu, K., Abdullahi, A., Abiala, M. A., & Odebode, A. C. (2013). In vitro Bio-prospecting of Botanicals Towards Inhibition of Microbial Pathogens of Rice (*Oryza sativa* L.). *World Applied Sciences Journal, 22*(2), 227-232.

Starks, K. J., & Burton, R. H. (1977). Green bugs: Determining biotypes, culturing and screening for plant resistance with notes on rearing parasitoids. *USDAARS Tech. Bull., 1556*.

Ukwungwu, M. N., & Odebiyi, J. A. (1985). Resistance of some rice varieties to the African striped borer, *Chilo zacconious* (Bloszgush). *Insect Science and Its Application, 6*, 163-166.

Ukwungwu, M. N. (1985). Effect of Nitrogen and carbofuran on gall midge (GM) and white stem borer (SB) infestation in Nigeria. *International Rice Research Newsletter, 10*(6), 20.

Ukwungwu, M. N., Joshi, R. C., & Winslow, M. D. (1990). *Varietal resistance as a potential strategy for the management of Africa Rice Gall midge*. Paper presented at 25th (Silver Jubilee) Annual Conference of the Entomological Society of Nigeria, October 8-11, 1990, Zaria, Nigeria.

Williams, W. G., Kennedy, G. G., Yamamoto, R. T., Thacker, J. D., & Bordner, J. (1980). 2 – Tricanomic: a naturally occurring insecticide from the wild tomato, Lycopersicon hirsutum *F. Glabratus*. *Science, 207*, 888-889. http://dx.doi.org/10.1126/science.207.4433.888

Wiseman, B. R. (1985). Types and mechanisms of host plant resistance to insect attack. *Insect Science and Its Application, 6*, 239-242.

Impact of Kenaf (*Hibiscus cannabinus* L.) Leaf, Bark, and Core Extracts on Germination of Five Plant Species

Charles L. Webber III[1], Paul M. White Jr.[1], Dwight L. Myers[2], Merritt J. Taylor[3] & James W. Shrefler[4]

[1] USDA, Agriculture Research Service, Sugarcane Research Unit, Houma, LA, USA

[2] East Central University, Chemistry Department, Ada, OK, USA

[3] Oklahoma State University, Division of Agriculture Sciences and Natural Resources, Department of Agricultural Economics, Durant, OK, USA

[4] Oklahoma State University, Division of Agriculture Sciences and Natural Resources, Cooperative Extension Service, Durant, OK, USA

Correspondence: Charles L. Webber III, Research Agronomist, USDA, Agriculture Research Service, Sugarcane Research Unit, Houma, LA 70360, USA. E-mail: chuck.webber@ars.usda.gov

Abstract

The chemical interaction between plants, which is referred to as allelopathy, may result in the inhibition of plant growth and development. The objective of this research was to determine the impact of kenaf (*Hibiscus cannabinus* L.) plant extracts on the seed germination of five plant species. Four concentrations (0, 16.7, 33.3 and 66.7 g/L) of kenaf leaf, bark, and core extracts were applied to the germination medium of redroot pigweed (*Amaranthus retroflexus* L.), green bean (*Phaseolus vulgaris* L.), tomato (*Solanum lycopersicum* Mill.), cucumber (*Cucumis sativus* L.), and Italian ryegrass (*Lolium multiflorum* Lam.) seeds. The treated seeds were placed in a non-illuminated incubator at 27 °C. Germination was recorded after 7 days in the incubator. Seed germination decreased with increasing extract concentration for all the plant species tested, except for green bean. Tomato, cucumber, Italian ryegrass, and redroot pigweed followed similar trends in their responses to the extract source (kenaf bark, core, and leaves) and the impact of extract concentration. The research demonstrated that kenaf leaf extracts were allelopathic by reducing seed germination for tomato, cucumber, Italian ryegrass and redroot pigweed. Sensitivity to the allelopathic impact of the kenaf leaf extracts from highest to lowest was Italian ryegrass > tomato > redroot pigweed > cucumber > green bean, with reductions in percentage germination of 79% (Italian ryegrass), 78% (tomato), 53% (redroot pigweed), 40% (cucumber), and 0% (green bean). Future research should pursue cultural practices to utilize these natural allelopathic materials to benefit crop production and limit weed competition, assess the impact of kenaf extracts on post-germination growth, and isolate the active ingredients in the kenaf leaf extracts that are allelopathic.

Keywords: allelopathy, cucumber, green bean, Italian ryegrass, kenaf, pestiphytology, redroot pigweed, seed germination, tomato, weed control

1. Introduction

1.1 Allelopathy

"Allelopathy" as coined and defined by Molisch (1937) is the biochemical interaction between plants, whether inhibiting or stimulating plant growth and development. Many plant species are now known to produce chemicals when released into the environment that can impact the growth and development of other plants (Rice, 1984). The demand by the general public for more naturally produced crops is a positive incentive to explore the use of natural plant chemicals to either promote crop growth and production, or inhibit weed growth and development.

1.2 Kenaf

Kenaf is a warm season annual fiber crop in the same family as cotton (*Gossypium hirsutum* L., Malvaceae) and okra (*Abelmoschus esculentus* L., Malvaceae) that can be successfully produced in a various areas of the United States, particularly in the southern states (Webber & Bledsoe, 1993). The commercial use of kenaf continues to diversify from its historical role as a cordage crop (rope, twine, and sackcloth) to its various new applications including paper products, building materials, absorbents, and livestock feed (Webber & Bledsoe, 1993).

Historically, kenaf has been used as a cordage crop to produce twine, rope, and sackcloth for over six millennia (Dempsey, 1975). Kenaf was first domesticated and used in northern Africa. India has produced and used kenaf for the last 200 years, while Russia started producing kenaf in 1902 and introduced the crop to China in 1935 (Dempsey, 1975). In the United States, kenaf research and production began during World War II to supply cordage material for the war effort (Wilson et al., 1965). The war not only interrupted the foreign fiber supplies from countries such as the Philippines, but the US involvement in the war also increased the use of these fibers by the US.

Once it was determined that kenaf was suited to production in the US, research was initiated to maximize US kenaf yields. As a result, scientists successfully developed high-yielding anthracnose-resistant cultivars, cultural practices, and harvesting machinery that increased fiber yields (Nieschlag et al., 1960; White et al., 1970; Wilson et al., 1965). Then in the 1950s and early 1960s, as USDA researchers were evaluating various plant species to fulfill future fiber demands in the US, it was determined that kenaf was an excellent cellulose fiber source for a large range of paper products (newsprint, bond paper, and corrugated liner board). It was also determined that pulping kenaf required less energy and chemical inputs for processing than standard wood sources (Nelson et al., 1962). More recent research and development work in the 1990s has demonstrated the plant's suitability for use in building materials (particle boards of various densities, thicknesses, with fire and insect resistance), absorbents, textiles, livestock feed, and fibers in new and recycled plastics (injected molded and extruded) (Webber and Bledsoe 1993).

1.3 Kenaf and Allelopathy

Research by Russo et al. (1997a) comparing the impact of plastic and kenaf mulches on soil erosion and production of vegetable crops indicated that the kenaf mulch may have had an allelopathic influence on the growth of some vegetables. A kenaf mulching study further indicated that extracts from non-weathered kenaf plant material decreased germination of redroot pigweed, annual ryegrass, tomato, and cucumber, while having no impact on green bean germination (Russo et al., 1997b). In addition, whereas redroot pigweed germination continued to be detrimentally impacted across all mulching treatments (non-weather, 2 month weathered, and 4 month weathered kenaf) as the extract concentration increased, there appeared to be no ill effects on the other seedlings due to exposure to the weathered kenaf (Russo et al., 1997b). These research studies provided a clear indication that kenaf plant material had allelopathic characteristics, but the research did not isolate which portion of the kenaf plant, leaves or stems, were allelopathic. In addition, related species in the Malvaceae family have exhibited allelopathic activity. Chuah et al. (2011) reported that the aqueous extracts of pods of a related plant species, okra (*Abelmoschus esculentus* L.), inhibited goosegrass (*Eleusine indica* L. Gaertn) germination and seedling growth. Jalali et al. (2013) determined that aqueous leaf extracts of the common mallow (*Malva sylvestris*) also exhibited allelopathic activity decreasing the germination and seedling growth of blanket flower (*Gaillardia pulchella*), plumed cockscomb (*Celosia argentea*), sweet William (*Dianthus barbatus*). To maximize the use of allelopathic activity of kenaf as a positive influence in agriculture, it is important to isolate the portion of the kenaf plant that is most allelopathically active and determine the impact across numerous plant species. Such a situation obviously demanded further investigation. The purpose of this research was to determine which portion of the kenaf plant exhibited the greatest allelopathic activity and the impact across different species.

2. Material and Methods

2.1 Kenaf Plant Material Preparations

Mature, 186 day-old kenaf plants, were harvested at ground level in October prior to a killing frost. The plants were immediately separated into three major plant components, leaves, bark (stem bark), and core (stem core). The few flower and seed pods present were discarded and not included in the experiment. The separate plant portions (leaves, bark, and core) were then dried in a forced air oven for 48 h or until constant weight at 66 °C. The samples were then ground using a Thomas-Wiley Laboratory Mill with a 1 mm sieve.

2.2 Kenaf Extract Preparations

The plant materials and distilled water were added to 3000 ml flasks and placed on a Lab-Line Orbit Shakers at 100 rpm for 12 h at room temperature (22 °C). The samples were vacuum filtered through Whatman # 42 filter paper twice. The samples were then diluted as needed with distilled water to produce concentrations of 66.7 g/L (full strength), 33.3 g/L (half strength), and 16.7 g/L quarter strength solutions of kenaf leaf, bark, and core extracts. The pH for all dilutions was adjusted to 7.0 using 1M KOH and 1M HCL.

2.3 Seed Germination

Seeds of green bean (*Phaseolus vulgaris* L.) var. 10 Hystyle, tomato (*Solanum lycopersicum* Mill.), cucumber (*Cucumis sativus* L.) var. 403 Royal, annual Italian ryegrass (*Lolium multiflorus* Lam.), and redroot pigweed (*Amaranthus retroflexus* L.) were surface sterilized for 1 min using with 50% sodium hypochlorite solution. The seeds were then rinsed with water and allowed to air dry for 10 min. Twenty-five seeds of each plant species were placed in separate Petri plates which contained 2.9 cm Whatman No. 2 filter papers. To each Petri plate was added 10 ml of either kenaf plant extract (leaf, bark, or core) at each of the concentrations [66.7 g/L, 33.3 g/L, 16.7 g/L, and 0 g/L (distilled water)]. The Petri plates were covered and placed in a non-illuminated incubator at 27 °C. Seven days later the Petri dishes were removed and seed germination was measured. Seeds were considered germinated when the seed radicle was at least the length of the width of seed of the specific plant species being measured.

This germination experiment was conducted twice. Each experiment included 3 kenaf plant extracts (leaf, bark, and core), 4 concentrations [0 (control), 16.7, 33.3 and 66.7 g/L], five plant species (green bean, tomato, cucumber, annual Italian ryegrass, and redroot pigweed), and 5 replications. All data were subjected to ANOVA and mean separation using LSD with P = 0.05 (SAS Inc., SAS, Ver. 9.0, Cary, NC).

3. Results and Discussion

3.1 Statistical Analysis

Statistical analysis determined that there were significant interactions among plant species (green bean, tomato, cucumber, Italian ryegrass, and redroot pigweed), between experiments (1 and 2), among sources of kenaf extracts (bark, core, and leaf) and extract concentration (0, 16.7, 33.3 and 66.7 g/L), therefore the results will be discussed by plant species with each interaction addressed separately.

3.2 Green Bean

No significant interactions were detected between or among the experimental factors (plant parts, extract concentrations, and experiments) for green bean germination. As a result, green bean data will be discussed averaged across plant parts, extract concentration, and years (Table 1). The only differences due to main affects for green bean germination were observed when comparing the source of the plant extract. The core extracts reduced seed germination compared to the bark and leaf extract. And, although the core extract did reduce germination by 8.4% and 5.5% compared to the bark and core extracts, respectively, the decrease was only 3.7% less than the control (76.8%), 0 g/L concentration (Table 1). No differences were observed among extract concentrations or between the experiments for green bean germination. Although these results were consistent with earlier research by Russo et al. (1997b), where the green bean germination was the least detrimentally affected by kenaf extracts, the current research finding of decreased green bean germination to core, but not other extracts indicates that the kenaf core is potentially the best location for isolating detrimental allelopathic chemicals for green beans. There is a slight beneficial impact of the kenaf bark and leaf extracts on seed germination, which probably explains the trend for increasing green bean germination observed by Russo et al. (1997b) when green bean seeds were exposed to increasing rates of whole stalk kenaf (leaves, bark, and core).

Table 1. Influence of kenaf plant extracts (bark, core, and leaves) and extract concentrations (0, 16.7, 33.3, and 66.7 g/L) on green bean germination

Kenaf lant part[Z]	Germination	Extract oncentration[Y]	Germination	Experiment[X]	Germination
	(%)		(%)	#	(%)
Bark	81.5 a[W]	0 g/L	76.8 a	1	78.6 a
Core	73.1 b	16.7 g/L	76.7 a	2	76.9 a
Leaves	78.6 a	33.3 g/L	79.6 a		
		66.7 g/L	77.9 a		

[Z]Kenaf plant part means averaged across the four extract concentrations and the two experiments.

[Y]Extract concentration means averaged across kenaf plant parts and experiments.

[X]Experiment means averaged across kenaf plant parts (bark, core, and leaves) and extract concentration (0, 16.7, 33.3, and 66.7 g/L).

[W]Means in a column within a main effect (kenaf plant part, extract concentration, and experiment) followed by the same lower case letter are not significantly different at P ≤ 0.05, ANOVA.

3.3 Tomato

Two significant interactions were detected for tomato germination in relation to the source of the kenaf extract (bark, core, and leaves). The main effect plant parts interacted separately with the extract concentrations and with the experiments. Therefore, tomato germination will be discussed separated by plant parts, extract concentrations, and experiments (Table 2). Although differences did occur among the sources of the plant extracts and between experiments, the percentage of seed germination decreased as extract concentration increased for all sources of plant extracts and for both experiments. The greatest decrease was observed in Experiments 1 and 2 for the leaf extract, where germination decreased from 88% to 0% germination for Experiment 1, and 89.6% germination to 14.4% germination in Experiment 2. Although the core and bark did decrease germination as the extract concentrations increased, the decrease was only 12.8% (Experiment 1) and 9.6% (Experiment 2) for the core extract, and 14.4% (Experiment 1) and 6.4% (Experiment 2) for the bark extract, comparing the 0 g/L concentration (control) to 66.7 g/L concentrations. These results reinforce earlier research which identified tomato as being sensitive to kenaf plant extracts (Russo et al., 1997b), while the current research provides insight into which portion of the kenaf plant, the leaves, has the greatest impact on seed germination for tomato. It is interesting to note that all portions of the above ground kenaf plant were allelopathically active with tomato.

Table 2. Impact of kenaf core, bark, and leaf extracts and extract concentrations on percentage of tomato seed germination

Extract Concentration	Core[z]		Bark		Leaves	
	Experiment		Experiment		Experiment	
	1	2	1	2	1	2
	%	%	%	%	%	%
0 g/L	88.0 a[Y]	89.6 a	88.0 a	89.6 a	88.0 a	89.6 a
16.7 g/L	91.2 a	88.0 a	87.2 a	90.4 a	78.4 b	88.0 a
33.3 g/L	80.8 b	80.0 b	88.0 a	82.4 b	64.0 c	72.8 b
66.7 g/L	75.2 c	80.0 b	73.6 b	83.2 b	0.0 d	14.4 c

[Z]Extract concentration means averaged across either the kenaf core, bark, or leaf extracts.

[Y]Means in a column followed by the same lower case letter are not significantly different at $P \leq 0.05$, ANOVA.

3.4 Cucumber, Italian Ryegrass, and Redroot Pigweed Interaction Analysis

The statistical analysis did not detect interactions between the kenaf extract sources (bark, core, and leaf) or extract concentrations (0, 16.7, 33.4, and 66.7 g/L) and the experiments, therefore the results will be discussed averaged across experiments (Table 3).

3.4.1 Cucumber

The kenaf leaf extract was the only source of extract that decreased cucumber seed germination as the extract concentration increased (Table 3). This is in contrast to the kenaf bark and core extract concentrations, which actually elevated germination at the 16.7 g/L rate. This data helps explain earlier research by Russo et al. (1997b) where cucumber germination increased as whole stalk (leaves, bark, and core) extract concentrations increased. Russo et al. (1997b) were using the above ground portion of 193 day-old kenaf plants for their mulching study, which would contain disproportionate greater amounts of stalks (bark and core) than leaves, due to the naturally occurring kenaf leaf abscission as the plant matures through the growing season. The greater proportion of bark and core plant material in mulching experiments explains why they only observed a beneficial impact in reference to cucumber germination (Russo et al., 1997b) compared to our mixed response to the kenaf extracts depending of the source of the kenaf extracts.

3.4.2 Italian Ryegrass and Redroot Pigweed

Italian ryegrass and redroot pigweed response to the source of kenaf extract and extract concentrations were very similar (Table 3). Seeds from both of these plants were very sensitive to kenaf plant extracts. Seed germination for both plants decreased as the extract concentrations increased, independently of the source of the kenaf extract, but with the greatest decrease observed for kenaf leaf extracts. Italian ryegrass germination decreased by 12.2%, 14.4%, and 78.4% from the 0 g/L concentration (control) to the 66.7 g/L concentration for the bark, core, and leaf extracts, respectively. The impact on redroot pigweed germination was also dramatic, decreasing 41.2%, 24.0%, and 53.2% from the 0 g/L concentration to the 66.7 g/L concentration for the bark, core, and leaf extracts, respectively. These results are consistent with Russo et al. (1997b) where the greatest detrimental impact to kenaf extracts was observed with Italian ryegrass and redroot pigweed, while the current research further illuminated which portion of the kenaf plant created the greatest impact, the leaves.

Table 3. Impact of the kenaf extract source (core, bark, and leaf) and extract concentration on cucumber, Italian ryegrass, and redroot pigweed germination percentage

Kenaf extract		Cucumber	Italian Ryegrass	Redroot Pigweed
Bark	Concentration^Z	%	%	%
	0 g/L	71.6 c^X	84.4 a	58.0 a
	16.7 g/L	76.4 a	75.2 b	40.4 b
	33.3 g/L	73.6 b	70.4 c	26.0 c
	66.7 g/L	72.4 bc	71.2 c	16.8 d
	Experiment^Y	%	%	%
	1	73.0 a	73.4 a	39.0 a
	2	74.0 a	77.2 a	31.6 b
Core	Concentration	%	%	%
	0 g/L	71.6 b	84.4 a	58.0 a
	16.7 g/L	82.0 a	80.0 b	47.2 b
	33.3 g/L	72.8 b	71.6 c	42.8 b
	66.7 g/L	72.8 b	70.0 c	34.0 c
	Experiment	%	%	%
	1	74.4 a	78.2 a	47.6 a
	2	75.2 a	74.8 a	43.4 b
Leaves	Concentration	%	%	%
	0 g/L	71.6 a	84.4 a	58.0 a
	16.7 g/L	70.8 a	73.6 b	32.0 b
	33.3 g/L	61.6 b	48.4 c	21.2 c
	66.7 g/L	32.0 c	6.0 d	4.8 d
	Experiment	%	%	%
	1	59.8 a	54.2 a	31.4 a
	2	58.2 a	52.0 a	26.6 b

^{Z}Extract concentration means averaged across either the kenaf core, bark, or leaf extracts across 2 experiments.

^{Y}Experiment means averaged across extract concentration.

^{X}Means in a column within either extract concentration, extract source or experiment followed by the same lower case letter are not significantly different at P ≤ 0.05, ANOVA.

4. Conclusions

Seed germination decreased with increasing extract concentration for all the plant species tested, except for green bean. Tomato, cucumber, Italian ryegrass, and redroot pigweed followed similar trends in their responses to the extract source (kenaf bark, core, and leaves) and the impact of extract concentration. The research demonstrated that kenaf leaf extracts were allelopathic by reducing seed germination for tomato, cucumber, Italian ryegrass and redroot pigweed. Sensitivity to the allelopathic impact of the kenaf leaf extracts from highest to lowest was Italian ryegrass > tomato > redroot pigweed > cucumber > green bean, with reductions in percentage germination of 78.4% (Italian ryegrass), 77.6% (tomato), 53.2% (redroot pigweed), 39.6% (cucumber), and 0% (green bean). The research also indicated that kenaf core and bark extracts may provide slight beneficial impact of cucumber germination at the low extract concentration (16.7 g/L). Future research should pursue cultural practices to utilize these natural allelopathic materials to benefit crop production and limit weed competition, assess the impact of kenaf extracts on post-germination growth, and isolate the active ingredients in the kenaf leaf extracts that are allelopathic.

References

Chuah, T. S., Tiun, S. M., & Ismail, B. S. (2011). Allelopathic potential of crops on germination and growth of goosegrass (*Eleusine indica* L. Gaertn) weed. *Allelopathy Journal, 27*(1), 33-42.

Dempsey, J. M. (1975). *Fiber Crops*. The Univ. Presses of Florida, Gainesville.

Jalali, M, Gheysari, H., Azizi, M., Zahedi, S. M., & Moosavi, S. A. (2013). Allelopathic potential of common mallow (*Malva sylvestris*) on the germination and the initial growth of blanket flower, plumed cockscomb and sweet William. *Intl. J. Agri. Crop Sci., 5*(15), 1638-1641.

Molisch, H. (1937). *Der Einfluss einer pflanze auf die andere*. Allelopathic Fischer, Jena.

Nieschlag, H. J., Nelson, G. H., Wolff, I. A., & Perdue, R. E. Jr. (1960). A search for new fiber crops. *TAPPI, 43*, 193-201.

Nelson, G. H., Nieschlag, H. J., & Wolff, I. A. (1962). A search for new fiber crops, V. Pulping studies on kenaf. *TAPPI, 45*(10), 780-786.

Rice, E. L. (1984). *Allelopathy* (2nd ed., p. 422). Acedemic Press, New York.

Russo, V. M., Cartwright, C., & Webber III, C. L. (1997a). Mulching effects on erosion of soil beds and on yield of autumn and spring planted vegetables. *Biological Agriculture and Horticulture, 14*, 85-93. http://dx.doi.org/10.1080/01448765.1997.9754799

Russo, V. M., Webber III, C. L., & Myers, D. L. (1997b). Kenaf extract affects germination and post-germination development of weed, grass and vegetable seeds. *Ind. Crops Prod., 6*, 59-69. http://dx.doi.org/10.1016/S0926-6690(96)00206-3

Webber III, C. L., & Bledsoe, R. E. (1993). Kenaf: Production, harvesting, and products. In J. Janick & J. E. Simon (Eds.), *New Crops* (pp. 416-421). New York: Wiley.

White, G. A., Cummins, D. G., Whiteley, E. L., Fike, W. T., Greig, J. K., Martin, J. A., ... Clark, T. F. (1970). Cultural and harvesting methods for kenaf. *USDA Prod. Res. Rpt.* (Vol. 113). Washington, DC.

Wilson, F. D., Summers, T. E., Joyner, J. F., Fishler, D. W., & Seale, C. C. (1965). 'Everglades 41' and 'Everglades 71', two new cultivars of kenaf (*Hibiscus cannabinus* L.) for the fiber and seed. *Florida Agr. Expt. Sta. Cir. S-168*.

3

Application of Bio-Image Analysis for Classification of Different Ripening Stages of Banana

Ganesh C. Bora[1], Donqing Lin[1], Pritha Bhattacharya[1], Sukhwinder Kaur Bali[2] & Rohit Pathak[1]

[1] Agricultural and Biosystems Engineering, North Dakota State University, Fargo, ND, USA

[2] Natural Resources Management, North Dakota State University, Fargo, ND, USA

Correspondence: Ganesh C. Bora, Agricultural and Biosystems Engineering, NDSU Dept. 7620, PO Box 6050, Fargo, ND 58108-6050, USA. E-mail: ganesh.bora@ndsu.edu

Abstract

Image processing is one of the important and extensively used techniques for determining physical and chemical characteristics of fruits. These characteristics are used for defining the shelf-life, packaging systems, mode of transportation and storage technique. In this method, green, yellow, and ripened banana with black spots are used for identification and classification of different ripening stages of banana. Automatic threshold separation and Aphelion 3.2 version software have been used for capturing images and analyzing structures. Hue, Saturation, and Intensity were the color coordinates used for extracting information in an image size of 451×256 pixels. Simplified histogram analysis on different color band was carried out. Result shows green band histogram can be used for classification of various ripening stages. In blue and red band histogram, it's difficult to differentiate stages.

Keywords: image processing, banana, color, pixel, ripening stages

1. Introduction

1.1 General

In agriculture, automated imaging use high resolution photography to characterize food quality. Images of individual food item is classified and analyzed based on size, shape and physical properties. Multiple entities of items ranging from 10's to 100's of thousands of items can be measured with statistical representative distributions of item's features. Static imaging system is focused on stationary item but dynamic imaging system is capable of collecting image of items flowing through the optics. Computer vision or computer imaging is used to develop the theoretical and algorithmic techniques in which the information about the image can be extracted and analyzed. Image is taken with the camera and sends that image to the computer. In computer the other processing take place with the help of different software. As extraction of useful information from image include image acquisition, image processing and determination of correlation formula and calculation of projected area (Omid et al., 2010). Image processing techniques is having expensive kit. So for digital calculation of the product needs digital balance and caliper. Computer vision technology had been used for characterization and description of object in agriculture from many years. Soil color determination was done by Han and Hayes (1990), color vision of peach grading by Miller and Delwiche (1990) and Tarbell and Reid (1991) studied the growth and development of corn plants and developed data collection system on the basis of computer vision method.

Apples and their geometric properties are modeled by Tabatabaeefar and Rajabipour (2005). Their model shows a strong relationship between apple diameter and mass. According to Khodabandehloo (1999) surface area, mass and volume are very valuable parameter for determining sizing system. Keramat Jahromi et al. (2007) evaluated that fruits should be graded by weight rather than only size because it is more economical to provide best packing pattern on the basis of weight. But packing agricultural products by electrical mechanism on the basis of size and weight is costly and by mechanical method is full of errors. The relationship of mass, projected area and dimension developed by Tabatabaeefar et al. (2000) can be beneficial. Lorestani and Tabatabaeefar (2006) determined a mathematical model to predict the mass of kiwi fruit. In this model, dimension and projected area of the fruit were used to prefigure mass. In food technology, surface and projected area are used to ascertain the amount of chemical application, peeling time, and the microbial concentrations present in produce (Sabilov et al., 2002). Khojastehnazhand et al. (2009) found tape method as very time consuming and labor–intensive method and also human error prone. First the object is fully covered with tape, then the tape is removed and measured with area meter.

In recent past, image processing technique has been effectively applied for quality inspection and shape sorting. Development of imaging algorithm by Hahn and Sanchez (2000) made volume prediction of irregular shaped fruits much easier compared to conventional methods.

Omid et al. (2010) designed an image processing algorithm for citrus volume and mass. The algorithm was executed in Visual Basic programming language. They showed that citrus volume and mass are correlated to each other and design sizing systems. Panigrahi et al. (1995) applied computer vision technology for morphological characteristics of corn germplasm and studied the background segmentation and dimensional measurement. For dimensional measurement and morphological properties proper background segmentation is very important.

Bio-imaging is often integrated with particle sizing methods such as laser diffraction to validate ensemble based measurements. Applications of bio imaging include measurement of 1) texture 2) color 3) surface area or volume of non-spherical items 4) defects and 5) diseases. Currently bio-imaging technology has been widely used for quality evaluation of food products. Since quality is not a single attribute but collections of different optical features of food product. Bio-imaging technology has been recognized as one of the most important in situ technique for food quality assessment. Softening of texture and change of yellow coloration of peels are fruit ripeness indicator (Marriot et al., 1981). Texture specifies eating quality whereas peel color indicates shelf-life. Shelf-life is one of the important factors for retail distribution.

According to FAO (2002), above 67 million metric tons is produced worldwide. Usually, evenly green and matured bananas are harvested to increase the shelf life and prevent loss during transportation. They remain firm and green with no any significant physical (color and texture) and chemical changes. Consumers primarily evaluate bananas on skin color. Peel color corresponds to physical and chemical transformations during ripening process (Deullin, 1963; Wainwright & Hughes, 1989). Commercial grading of bananas are usually performed through eye inspection and is related to change in pigment as shown in Table 1 (Le et al., 1997). Ripeness is evaluated with the help of standardized color charts. The skin color is differentiated on the basis of seven ripening stages of the color charts (Von Loesecke, 1950; Le et al., 1997). Sometimes ripeness is also estimated by instrumental techniques (Wainwright & Hughes 1990).

Table 1. Characteristics of different grades of banana

Stage	Color
1	Green
2	Green and traces of yellow
3	More green than yellow
4	More yellow than green
5	Green tip and yellow
6	All yellow
7	Yellow, flecked with green

Source: Le et al. (1997).

During ripening, first, there's a breakdown of cell walls. Due to solublization of pectic substances, the cohesion of middle lamella reduces. Second, migration of water takes place from skin to flesh, because of osmosis (Palmer, 1971; Smith et al., 1989). In case of Musa cavendish, process of starch hydrolysis and sugar synthesis completes on full ripening. But in other species of Musa, the process continues till very ripe and senescent stage.

1.2 Objective

The project objective is to apply image analysis technique to determine the changes in banana quality in self life (*Musa domestica*) with respect to change in color features, i.e., 1) RGB and 2) HSI (Hue Saturation Intensity).

2. Material and Methods

2.1 Samples of Bananas

Mature green peel banana (*Musa cavendish) was* purchased from local market. It was then treated with ethylene. Uniform color and obvious defect free bananas were selected at random for image analysis in the Bio-Imaging and Sensing Center, Department of Agricultural and Biosystems Engineering, North Dakota State University. Six samples of banana were selected for each stage and there were 18 in total. Though seven stages are considered in self-life, only three stages were selected to make the proof of concept and then could be extrapolated for self-life prediction.

2.2 Computer Vision System

Skin color and length of the fruit is analyzed with image analysis techniques. Image processing technique was used to measure the actual projected area and surface area. In this study method (Ostu, 1979), automatic threshold separation approach has been applied to extract the banana image from plain background. Nikon camera with D3100 SLR 14.2 MP (Figure 1) and Aphelion 3.2 version, an image processing and understanding software have been used for taking images and analyzing the features.

Figure 1. Nikon camera with D3100 SLR 14.2 MP

2.3 Analysis

Sample banana was put on a white sheet as shown in Figure 1, in longitudinal and lateral orientation. Images were taken at different stages shown in Figure 2. Calibration factor, number of pixels to linear dimension was computed from an image of 8½ × 11 square inch of white sheet. The calibration factor was then used to calculate the axial dimensions, area and volume. The color image of banana study was carried out by computer vision system (Panigrahi & Mishra, 1989) using D3100 SLR 14.2 MP camera. Banana was placed on a rectangular platform with white background.

(1)   (2)   (3)

Figure 2. Changes in color and brown spot development

Hue, Saturation and Intensity (HIS) were the color coordinates used for the study of image. Each image had a size of 451 × 256 pixels. Hue and saturation were used to acquire color information whereas intensity for dimensional attributes. Histogram analysis was carried out on image analysis. The image area was measured in number of pixels. The number of pixels was multiplied by calibration factor to convert it into mm^2. Height and width of minimum bounding rectangle (MBR) in pixels was also converted into mm^2. Image of known geometry was used for the calibration of MBR. For volume measurement, calibrated area of image was multiplied with the width of other image of the same bean whose orientation was changed by 90°. So we can say that volume is product of area in longitudinal and thickness in lateral direction.

2.4 Red-Green-Blue Color System

It is hard for human eyes to detect quality differences from the minute color variations and check the quality of thousands of items. Computer imaging system has been successfully applied to acquire and process color images. The computer imaging system is composed of a color camera, capable of handling color information and color monitor that displays the information. Information is represented in term of color coordinates like Red-Green-Blue (RGB) system. The chromaticity coordinates r, g, and b are represented by following equations:

$$r = \frac{R}{R+G+B}$$
(1)

$$g = \frac{G}{R+G+B}$$
(2)

$$b = \frac{B}{R+G+B}$$
(3)

Where: R is red, B is blue and G is green.

3. Results and Discussion

Color is most important factor for image processing and object description. As per Tri-chromatic theory, RGB color model is a combination of red, green, and blue light array. This three color component contains different value, which helps to define different color spaces. Computer images are stored in the form of matrix. These elements are referred as pixel, where there are two types of information – geometric information (pixel location in the image), and surface information (pixel intensity values). During digital image processing each red, green and blue image pixel can be represented in computer memory as binary values. These images are converted to intensities and voltages which reproduces on display. The color features which affects banana can be determined by visual image an alternative color coordinate system is HIS. HUE is the dominant spectral color or the attribute of color perception by means of which an object identified as red, green or any other specific color. Intensity (I) is the relative darkness or lightness of the dominant spectral color. Saturation(S) represents the intensity of color that shows degree of deviation from gray of the same lightness.

3.1 Red

From Figure 3, it is observed that green colored banana (amplitude near about 1420 at pixel value of 198, Green) has the highest peak response than yellow colored banana (amplitude near about 1220 at pixel value of 194, Banana (1)) and ripened banana with black spots (amplitude near about 1020 at pixel value of 192, Banana (2)). Yellow colored banana peak is observed the earliest followed by ripened banana with black spots and then green banana. After reaching peak, amplitude drop-off is instant for yellow colored banana and almost similar trend in case of ripened banana with black spots. But for green banana amplitude drop-off is gradual. In case of green banana, if we follow the data population we can say that the population decreases with increase in pixel value.

For ripened banana with black spots, the density of population is almost similar throughout. However, for yellow colored banana, data population density decreases on both the sloping sides of peak point.

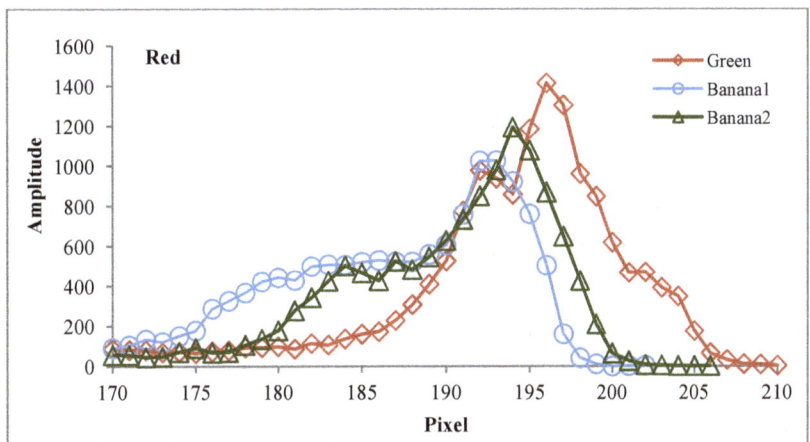

Figure 3. Pixel versus amplitude graph for three color stages of banana in RED of RGB (Banana (1) was yellow and Banana (2) was ripened banana with black spots)

We can also see that the green banana red property peak dropped off to almost 800 at 195 pixels from 1000 at 192 pixels. After that again it rises to peak point of 1420 and falls suddenly. If we follow ripened banana with black spots, we will see the graph falls sharply twice in-between 180 and 190 pixels before reaching the peak. Unlikely to green and black spot banana, yellow colored banana rises steadily to the peak without any significant rise and fall. It's plateaued at the top.

3.2 Blue

From Figure 4, it is observed that ripened banana with black spots (amplitude near about 990 at pixel value of 176, Banana (2)) has the highest peak response than yellow colored banana (amplitude near about 900 at pixel value of 184, Banana (1)) and green colored banana (Amplitude near about 700 at pixel value of 182, Green). Contrary to RED, here we can see ripened banana with black spots peak is observed the earliest followed by green colored and then yellow colored.

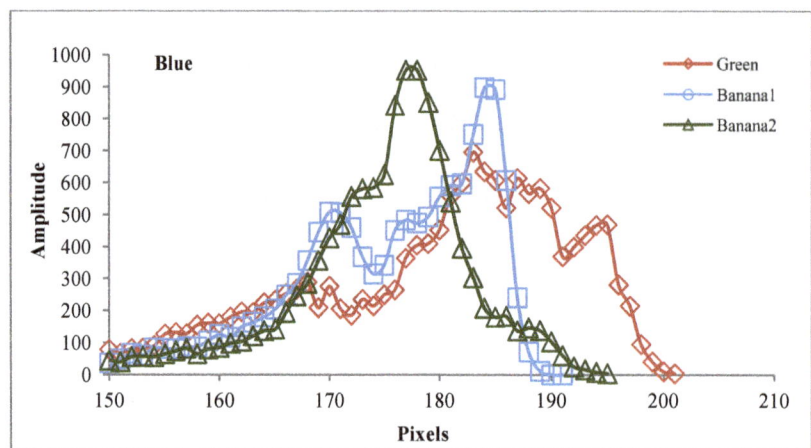

Figure 4. Pixel versus amplitude graph for three color stages of banana in BLUE of RGB (Banana (1) was yellow and Banana (2) was ripened banana with black spots)

It can also be seen from the Figure 4, green banana fluctuates wildly between 165 and 195 pixel value whereas ripened banana with black spots grows smoothly till the highest point at 176 pixel value and falls sharply. But in case of green color banana amplitude drop-off is gradual rather than sudden. Similarly, yellow colored banana shows some fluctuation at the beginning and there's a significant fall from around 500 to 300 at pixel values 170 to 175 respectively before reaching the peak and dropping suddenly. For both yellow colored and black spot banana the graph falls dramatically after it reaches the peak.

If we look at data population for ripened banana with black spots (banana-2), we can say that population is higher at beginning and end, and decreases at the rising and falling portion. Yellow colored banana shows a similar trend as banana-2. In case of green banana the density of population seems to be almost same all the way through.

3.3 Green

From Figure 5, it is observed that green banana (amplitude near about 1500 at pixel value of 195, Green) has the highest peak response than yellow colored banana (amplitude near about 1100 at pixel value of 189, Banana (1)) and ripened banana with black spots (amplitude near about 1050 at pixel value of 185, Banana (2)). Similar to BLUE, ripened banana with black spots peak is observed first. After that yellow colored and green colored peaks are observed second and third in turn. There's a marked fall of amplitude from around 700 to 400 at 185 and 190 pixels respectively. After that it reaches the highest point and falls back rapidly to almost zero. For yellow colored, at the beginning we can see an upward trend with some insignificant rise and fall. After reaching the peak it follows the similar trend as green banana. Ripened banana with black spots also shows an upward trend and rises to a peak value of around 1000 with slight fall at a pixel value of 185. The declination is gradual compared to other two bananas.

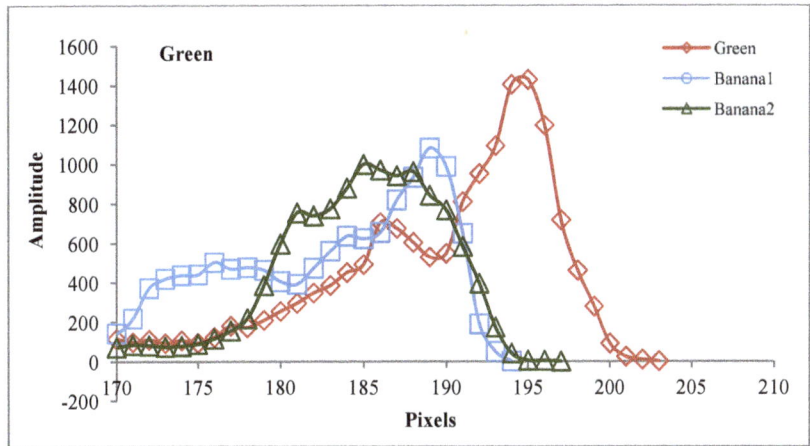

Figure 5. Pixel versus amplitude graph for three color stages of banana in GREEN of RGB (Banana (1) was yellow and Banana (2) was ripened banana with black spots)

The data population for green banana decreases with increasing pixel value. In case of ripened banana with black spots the density of population seems to be similar during the course of graph till the end. Yellow colored banana data population density decreases on both the sloping sides of peak whereas at the beginning it is denser relative to end area.

3.4 Intensity

From Figure 6, it is observed that green banana (amplitude near about 1200 at pixel value of 194) has the highest peak response than ripened banana with black spots (amplitude near about 1100 at pixel value of 187) and yellow colored banana (Amplitude near about 1050 at pixel value of 189). Ripened banana with black spots hit the highest point earlier than yellow and green colored banana. Green color banana attains the highest peak point. The second and third highest peak point is of ripened banana with black spots and yellow colored banana respectively. After reaching the peak, steep drop off is observed for all three bananas. Black spots banana shows an upward trend until it reaches the peak point and falls slightly just before reaching peak. Yellow colored banana rises smoothly till the pixel value of 173 and then falls to amplitude of around 400 at pixel value of 180. After this rises rapidly to peak point and decline sharply. Green colored banana also has upward trend till the topmost point with some fluctuation. It falls down markedly once from amplitude of 550 to 400 and almost flattened out to an amplitude of 625 in between the pixel value of 186 to 190.

In case of green banana, if we follow the data population we can say the population is reducing as the pixels are increasing and also the population density seems to be less at end area. In case of yellow colored banana, the density of population is near about same throughout graph except on the steeper sides where it seems to be reduced to some extent. But ripened banana with black spots, data population density is decreasing in the steeper areas else it's denser at the beginning and termination area.

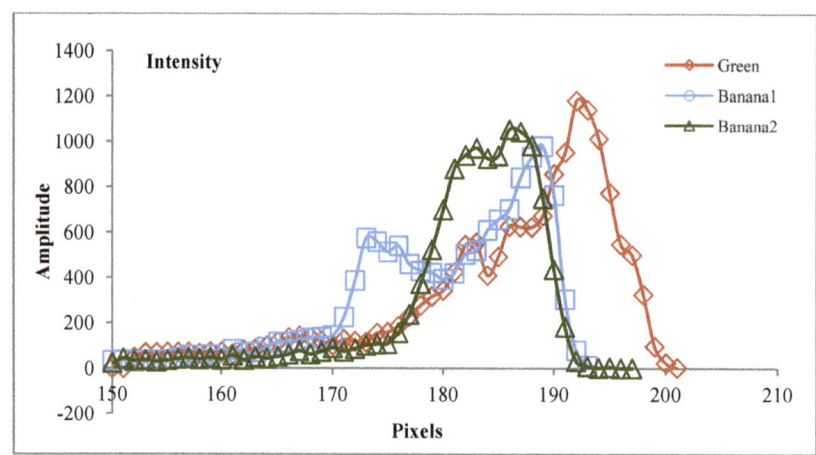

Figure 6. Pixel versus amplitude graph for three color stages of banana in INTENSITY of RGB (Banana (1) was yellow and Banana (2) was ripened banana with black spots)

4. Conclusion

Green band histogram can be chosen as distinguishable indicator in analyzing the banana at various stages. Histogram shapes are approximately similar as normal distribution but are skewed to the left. The decreasing pixel peaks can be used to show the banana quality in image processing system.

Blue band histogram is not recommended because the histograms are confusedly overlapped. Red band histogram looks like the green band but it's impossible to differentiate the ripe bananas. Intensity histogram is simply the average of three RGB histograms and it can be tested to compare with green band.

References

Deullin, R. (1963). Measurement of the color of the pulp of banana in preclimacteric phase. *Fruit, 18*, 23-26.

Food and Agriculture Organization (FAO). (2002). *FAO statistical databases*. FAO/United Nations. Retrieved October, 2002, from http://apps.fao.org

Hahn, F., & Sanchez, S. (2000). Carrot volume evaluation using imaging algorithms. *J. Agric. Eng. Res., 75*, 243-249. http://dx.doi.org/10.1006/jaer.1999.0466

Han, Y. J., & Hayes, J. C. (1990). Soil cover determination using color image analysis. *Transactions of the ASAE, 33*(4), 1402-1408. http://dx.doi.org/10.13031/2013.31486

Keramat Jahromi, M., Rafiee, S., Jafari, A., & Tabatabaeefar, A. (2007). Determination of dimension and area properties of date (Berhi) by image analysis. *International Conference on Agricultural, Food and Biological Engineering and Post-Harvest /Production Technology, KhonKaen, Thailand, 21-24 January.*

Khodabandehloo, H. (1999). *Physical properties of Iranian export apples* (M.Sc. Thesis). Tehran University, Karaj, Iran.

Khojastehnazhand, M., Omid, M., & Tabatabaeefar, A. (2009). Determination of orange volume and surface area using image processing technique. *Int. Agrophysics, 23*, 237-24.

Le, M., Slaughter, D. C., & Thompson, J. E. (1997). Optical chlorophyll sensing system for banana ripening. *Postharvest Biol. Technol., 12*(3), 273-83. http://dx.doi.org/10.1016/S0925-5214 (97) 00059-8

Lorestani, A. N., & Tabatabaeefar, A. (2006). Modeling the mass of kiwi fruit by geometrical attributes. *Int. Agrophysics, 20*, 135-139.

Marriot, J., Robinson, M., & Karikari, S. K. (1981). Starch and sugar transformation during ripening of plantains and bananas. *Trop. Sci., 32*, 1021-1026.

Miller, B., & Delwiche, M. (1990). A color vision system for peach grading. *Transactions of the ASAE, 32*(4), 1484-1490. http://dx.doi.org/10.13031/2013.31177

Omid, M., Khojasteh Nazhand, M., & Tabatabaeifar, A. (2010). Estimating volume and mass of citrus fruits by image processing technique. *J. Food Engineering, 100*(2), 315-321. http://dx.doi.org/10.1016/j.jfoodeng.2010.04.015

Palmer, J. K. (1971). The Banana. In A. C. Hulme (Ed.), *The Biochemistry of Fruit and Their Products* (Vol. 2, pp. 65-115). London: Academic Press.

Panigrahi, S., Mishra, M. K., Bern, C., & Marley, S. (1995). Background segmentation and dimensional measurement of corn germplasm. *Transactions of the ASAE, 38*(1), 291-297. http://dx.doi.org/10.13031/2013.27841

Sabilov, C. M., Boldor, D., Keener, K. M., & Farkas, B. E. (2002). Image processing method to determine surface area and volume of axi-symmetric agricultural products. *Int. J. Food Prop., 5*, 641-653. http://dx.doi.org/10.1081/JFP-120015498

Smith, N. J. S., Tucker, G. A., & Jeger, J. (1989). Softening and cell wall changes in bananas and plantains. *Aspects Appl. Biol., 20*, 57-65.

Tabatabaeefar, A., & Rajabipour, A. (2005). Modeling the mass of apples by geometrical attributes. *Sci. Hortic., 105*, 373-382. http://dx.doi.org/10.1016/j.scienta.2005.01.030

Tabatabaeefar, A., Vefagh-Nematolahee, A., & Rajabipour, A. (2000). Modeling of orange mass based on dimensions. *J. Agr. Sci. Tech., 2*, 299-305.

Tarbell, K. A., & Reid, J. F. (1991). A computer vision system for characterizing corn growth and development. *Transactions of the ASAE, 34*(5), 2245-2255. http://dx.doi.org/10.13031/2013.31864

Von Loesecke, H. W. (1950). *Bananas - Chemistry, physiology, technology*. Interscience Publishers, New York.

Wainwright, H., & Hughes, P. (1989). Objective measurement of banana pulp color. *Int. J. Food Sci. & Technol., 24*, 553-558. http://dx.doi.org/10.1111/j.1365-2621.1989.tb00679.x

Wainwright, H., & Hughes, P. (1990). Changes in banana pulp color during ripening. *Fruits, 45*(1), 25-28.

4

Simulating Soil Moisture under Different Tillage Practices, Cropping Systems and Organic Fertilizers Using CropSyst Model, in Matuu Division, Kenya

Muli M. N.[1], Onwonga R. N.[1], Karuku G. N.[1], Kathumo V. M.[1] & Nandukule M. O.[1]

[1] Department of Land Resource Management and Agricultural Technology, University of Nairobi, Nairobi, Kenya

Correspondence: Onwonga R. N., Department of Land Resource Management and Agricultural Technology, University of Nairobi, P.O. Box 29053-00625 Nairobi, Kenya. E-mail: dr.onwonga@gmail.com

Abstract

Soil moisture stress is a limiting factor in crop production particularly in arid and semi-arid lands (ASALs) as it affects many physiological and biochemical processes of plants. Research on moisture conservation measures is thus imperative. The current study used CropSyst model to simulate soil moisture under different tillage practices (oxen plough, tied ridges and furrows and ridges), cropping systems (monocropping, intercropping and crop-rotation) and organic fertilizers; farm yard manure, rock phosphate (RP) and Farmyard manure (FYM) combined with rock phosphate (RP+FYM). The study was conducted in Matuu Division, Kenya for two seasons; October 2012 to February 2013 short rain season (SRS) and March to August 2013 long rain season (LRS). The experiment was laid out in a Randomized Complete Block design with a split-split plot arrangement and replicated three times. The main plots were tillage practices whereas the split plots were cropping systems and split-split plots were organic fertilizers and a control (nothing applied). The test crops were sorghum (*Sorghum bicolor* L.) and sweet potato (*Ipomea batatas* L. *lam*) rotated and/or intercropped with dolichos (*Lablab purpureus*) and chickpea (*Cicer arietinum*). The CropSyst model was calibrated using measured soil texture, permanent wilting point, bulk density and initial soil moisture at the experimental site. Model validation was done using Root Mean Square Error (RMSE), percentage differences (PD) and willmott index (WI) of agreement. CropSyst model was reasonably validated as indicated by the low RMSE (0.5 to 1.3), PD (less than ±15) and WI index (close to 1). In the first season and second season, simulated soil moisture (101.91 and 108.3 mm) was significantly ($P < 0.05$) high in sorghum/dolichos intercrop with RP+FYM application under tied ridges and least (13.52 and 15.4 mm) in control treatment of sorghum mono crop under oxen plough. In sweet potato plots, both individual treatment and treatment interaction significantly influenced simulated soil moisture. Sweet potato-dolichos rotation (75.32 and 79.63 mm), with application of RP+FYM (75.03 and 79.39 mm) under tield ridges (95 and 100.24 mm) had highest simulated soil moisture levels under oxen plough (32.49 and 34.36 mm), sweet potato monocrop (53.46 and 55.26 mm) and control (52.52 and 55.39 mm) having the least during the first and second season, respectively. In both sorghum and sweet potato based cropping systems, soil moisture was correspondingly highest in tied ridges, intercropping and rotation systems involving dolichos and application of FYM+RP and least in control of monocropping under oxen plough. Information on effects of tillage practices, cropping systems and organic inputs could be very useful for soil water conservation purposes. Thus, using simulation models to attain the same could be the ultimate solution. A good agreement between observed and simulated soil moisture implied that CropSyst model is capable of investigating sustainable alternatives of increasing soil moisture in the ASALs.

Keywords: arid and semi-arid lands, cropping systems, CropSyst model, soil moisture

1. Introduction

In the arid and semi lands, plant production is limited by soil moisture availability and actual evapotranspiration (Biamah, 2005). The two parameters will influence the occurrence of water stress in rainfed agricultural systems. Fluctuations in soil moisture often have negative effects on crop productivity (Purcell et al., 2007). Moisture loss from the soil through evaporation and presence of erratic rainfall in the middle of the cropping season leads to crop failure. Rain water harvesting techniques could thus be used to improve soil moisture availability or reduce deficit

(Gicheru et al., 2004). Information on effects of tillage practices, cropping systems and organic fertilizers that ensure effective capture and utilization of rainfall for sustainable crop production could therefore be very useful for water conservation purposes.

Behavior of agricultural systems cannot, however, be evaluated over a long period of time using field experiments due to their complexity (Grabisch, 2003). Nonetheless, dynamics of these systems can be simulated with use of decision support tools. These tools are valuable for representing long-term productivity and environmental effects on cropping systems and extrapolating the experimental results in time and space (Dillon, 1992). Crop simulation models help researchers to ascertain the relationships among environment, management and yield variability (Sinclair & Seligman, 1996). This is in addition to predicting the effect of weather, soil properties, plant characteristics and management practices on soil moisture. Various groups use these models for decision making in agricultural systems such as efficient use of resources by providing potential plant responses to different inputs (Staggenborg et al., 2005) and hence improved efficiency of input management for cropping systems. Cropping systems simulation model represents an effort to simulate growth of single crops or crop rotations in response to weather, soil and management scenarios and provide an estimate of environmental impact (Stockle & Nelson, 1994). One such model that has been widely used is CropSyst (Stockle et al., 1994).

CropSyst is a multi-year, multi-crop, daily time step cropping systems simulation model developed to serve as an analytical tool to study effect of climate, soils and management on cropping systems productivity and environment (Stockle et al., 1994). CropSyst simulates soil water and nitrogen budgets, crop growth and development, crop yield, residue production and decomposition, soil erosion by water and salinity (Donatelli et al., 1999). The objective of the current study was to simulate soil moisture under different tillage practices, cropping systems and organic inputs in sorghum and sweet potato based cropping systems using CropSyst model.

2. Materials and Methods

2.1 Experimental Site

The study was conducted in Matuu Division (1°37′S and 1°45′S latitude and 37°15′E and 37°23′E longitude and an altitude of 700-800 metres above sea level) located in Eastern province, Kenya. It falls in agro-climatic zone IV which is classified as semi-arid land (Jaetzold & Schmidt, 2006). Rainfall patterns exhibits distinct bimodal distribution. The first rains fall between mid-March and end of May and are locally known as long rains (LR), and second rains, short rains (SR), are received between mid-October and end of December. Average seasonal rainfall is between 250-400 mm. Interseasonal rainfall variation is large with a coefficient of variation ranging between 45-58 per cent, while temperature ranges between 17 and 24 °C.

The soils are a combination of Luvisols, Lithisols, and Ferralsols according to USDA (1978) and WRB (2006) criteria. The soils are well drained, moderately to very deep, dark reddish brown to dark yellowish brown, friable to firm, sandy clay to clay and low nutrient availability (Kibunja et al., 2010). The majority of farmers in the division are small-scale mixed farmers with low income investment for agricultural production. Crop performance and yield are significantly influenced by amount of rainfall and distribution throughout the rainy season (Macharia, 2004).

2.2 Treatments and Experimental Design

To obtain data for CropSyst model calibration, field experiments were conducted for two seasons; short rain season and long rain season. Data for season one was used to calibrate the model while season two data was used for model validation. The experimental layout was a Randomized Complete Block Design with a split-split plot arrangement and replicated three times. The main plots were tillage practices; Oxen plough (OP), tied ridges (TR) and, furrows and ridges (FR) whereas the split plots were cropping systems; mono cropping, intercropping and crop rotation and split-split plots were organic fertilizers; FYM, RP and FYM+RP and a control (no organic fertilizer was applied) giving a total of 60 treatments (Table 1). The test crops were sorghum and sweet potato intercropped and/or grown in rotation with legumes; Dolichos and chickpea.

Table 1. Tillage practices, cropping systems and organic fertilizers applied during experimental period

No.	Tillage practice (Note 1)	Cropping system (Note 2)	Description	Organic fertilizer	Cropping Season	
					LRS	SRS
1	Oxen Plough	Monocropping	SOR-MONO	FYM	Sorghum	Sorghum
2				RP	Sorghum	Sorghum
3				RP+FYM	Sorghum	Sorghum
4				Control	Sorghum	Sorghum
5		Intercropping	SOR/DOL	FYM	Sorghum/Dolichos	Sorghum/Dolichos
6				RP	Sorghum/Dolichos	Sorghum/Dolichos
7				RP+FYM	Sorghum/Dolichos	Sorghum/Dolichos
8				Control	Sorghum/Dolichos	Sorghum/Dolichos
9			SOR/CP	FYM	Sorghum/Chick pea	Sorghum/Chick pea
10				RP	Sorghum/Chick pea	Sorghum/Chick pea
11				RP+FYM	Sorghum/Chick pea	Sorghum/Chick pea
12				Control	Sorghum/Chick pea	Sorghum/Chick pea
13		Crop Rotation	SOR-DOL	FYM	Dolichos	Sorghum
14				RP	Dolichos	Sorghum
15				RP+FYM	Dolichos	Sorghum
16				CONTROL	Dolichos	Sorghum
17			SOR-CP	FYM	Chickpea	Sorghum
18				RP	Chickpea	Sorghum
19				RP+FYM	Chickpea	Sorghum
20				CONTROL	Chickpea	Sorghum

SOR-MONO; Sorghum monocropping, SOR/DOL; Sorghum dolichos intercrop, SOR/CP; Sorghum chickpea intercrop, SOR-DOL; Sorghum dolichos rotation, SOR-CP; Sorghum chickpea rotation, RP; Rock phosphate, FYM; farm yard manure.

2.3 Agronomic Practices

2.3.1 Land Preparation and Planting

Land was prepared manually using oxen plough and hand hoes for OP and, TR and, FR tillage practices, respectively. Planting was done in October during the short rain season of 2012 and in April during long rain season of 2013. Sorghum and sweet potato were planted during the short rains. Sorghum seeds were sown at a spacing of 30 by 60 cm. Sweet potato cuttings were planted at a spacing of 30 by 90 cm. Weeding was done every 4 weeks after planting. Harvesting of sorghum was done by hand after 3 months when it had reached physiological maturity while sweet potato was harvested manually using hand hoe after four months.

2.3.2 Soil Analysis

Initial soil sampling was done in a zigzag manner across the field using a soil auger at 0-15, 15-30 and 30-45 cm depths and composited into one sample per depth for physical and chemical analysis before application of treatments. Thereafter soil samples were collected at 0-15 cm depth during flowering and harvesting stages of sorghum and sweet potato from each treatment. Soil was analyzed for chemical properties; pH, and mineral nitrogen and physical; soil texture, bulk density, field capacity and permanent wilting point, using standard laboratory methods (Okalebo et al., 2002). The observed soil properties were used for initial soil characterization and to prepare soil file used in calibrating CropSyst model. Soil moisture was determined by the gravimetric method (weight basis) and converted into volumetric proportion by multiplying with bulk density (Equation 1) and converted to volumetric water (mm) by multiplying by soil depth divided by 10 (Equation 2).

$$\text{Volumetric Water (\%)} = \text{Gravimetric Water (\%)} \times \text{Bulk Density (g/cm}^3) \tag{1}$$

$$\text{Volumetric Water (mm)} = \frac{\text{Volumetric \%} \times \text{Soil Test Depth (cm)}}{10} \tag{2}$$

Bulk density was determined according to Blake and Hartage (1986). Field capacity and permanent wilting point was determined using the procedure described by Klute (1986), mineral nitrogen was determined by Kjeldahl method (Bremner & Mulvaney, 1982).

2.4 CropSyst Model Description

CropSyst model is premised on assumption that actual biomass/output growth is a result of interactions involving various independent variables which include weather, soil types, management practices and crop physiology (Stockle et al. (2003), Table 1).

Table 1. Data sets required to run CropSyst model

File	Parameters Required by the Model	Parameters Used in the Model
Location	Latitude, Longitude, Altitude	Latitude: 37°15′E and 37°23′E
		Longitude: 1°37′S and 1°45′S
		Altitude: 700-800m a.s.l
Soil	pH, Permanent wilting point, Field capacity, Bulk density, Soil texture	Table 2 (determined in the field).
Crop	Growing degree days (GDD) to emergence, GDD to peak leaf area index, GDD to flowering, GDD to maximum grain filling, GDD to maturity, Base temperatures, Cut-off temperatures, maximum root depth.	GDD were observed in the experimental site. Other crop input parameters were taken as default values.
Management	**Nitrogen fertilization** (application date, amount, source-organic and inorganic-, and application mode- broadcast, incorporated, injected). **Tillage operations** (primary and secondary tillage operations).	**Organic inputs**: FYM, RP, FYM+RP, calibration was done for RP which is not currently in the model. **Tillage practices**: Tillage operations were calibrated for oxen plough, tied ridges, furrows and ridges

GDD: growing degree days; FYM: farm yard manure; RP: rock phosphate.

The model simulates soil water budget, crop canopy and root growth, dry matter production, yield, residue production and decomposition, and erosion. Management options include: cultivar selection, crop rotation, irrigation, nitrogen fertilization, tillage operations and residue management. The dates for phenological stages; emergence, flowering stage, grain filling and physiological maturity were used to calculate growing degree days ($GDD = T_{mean} - T_{base}$; where $T_{mean} = (T_{max} + T_{min})/2$). Location file was also prepared using observed weather data from nearest weather station. For each tillage practices, management files were prepared to represent each cropping systems and organic inputs. Soil moisture measurements were used for model calibration. The values of crop input parameters (maximum harvest index, maximum expected LAI, base temperature, cut-off temperature and maximum root depth) were obtained from the CropSyst manual (Stockle et al., 2003).

2.5 CropSyst model Calibration

The calibrated values (Table 2) were permanent wilting point, field capacity and mineral nitrogen. Observed mineral nitrogen was adjusted from 24 Kg N ha⁻¹ to 58.91 Kg N ha⁻¹, permanent wilting point was adjusted from 0.17 m³/m³ to 0.29 m³/m³ while field capacity was also adjusted from 0.23 m³/m³ to 0.38 m³/m³ to ensure closeness between observed and simulated soil moisture values. The values were adjusted by comparing observed soil water content with model output. Calibrated values ensured closeness between observed soil water values and simulated values. Crop growth was majorly affected by soil moisture and nitrogen content and adjustment to the required amount was accordingly done. Soil texture and bulk density were not calibrated since they were within the CropSyst required range.

Table 2. Observed and calibrated physic-chemical soil properties

Soil properties	Observed soil properties/Depth (cm)			Calibrated soil properties/Depth (cm)		
Depth (cm)	**0-10**	**10-20**	**20-30**	**0-10**	**10-20**	**20-30**
Sand (%)	49.32	49.30	49.36	49.32	49.30	49.36
Silt (%)	38.88	38.97	38.77	38.88	38.97	38.77
Clay(%	11.8	11.71	11.78	11.8	11.71	11.78
Textural class	*Sand – Clay (USDA Classification)*					
pH (H_2O)	6.5	6.7	6.8	6.5	6.7	6.8
Permanent wilting point (m^3/m^3)	0.17	0.18	0.20	0.27	0.28	0.29
Field capacity (m^3/m^3)	0.23	0.25	0.27	0.34	0.36	0.38
Bulk density (g cm^{-3})	1.503	1.508	1.67	1.503	1.508	1.67
NH_4-N (Kg N ha^{-1})	28.54	27.02	34.76	58.91	57.39	55.46
NO_3--N (Kg N ha^{-1})	24.87	29.34	25.72	52.67	51.83	50.44

2.6. Model Validation

CropSyst was validated by comparing model outputs withobserved soil moisture in different tillage practices, cropping systems and organic inputs. The agreement between model and reality was verified by means of percentage differences (PD) and root mean square error (RMSE). This is frequently used measure of the difference between values simulated by a model and those actually observed from the experiment that is being modelled (Equation 3).

$$RMSE = [n^{-1} \sum(Yield_{meas} - Yieldpred)^2] \qquad (3)$$

Additionally, Willmott index (WI) of agreement was calculated, which take a value between 0.0 and 1.0; where 1.0 means perfect fit (Willmott, 1981).

2.7 Simulations

The input files required by CropSyst model for Matuu Division, sorghum and sweet potato crops were used to run the model. Planting dates were set as 10th October, 2012 for both crops. Simulations were run from 10th, September, 2012 a month before planting and ended in 31st, March 2013 for sorghum and 31st May for sweet potato. The experiment was repeated for the second season in 2013. The starting and ending dates indicated simulation period. Sweet potato required more time to mature compared to sorghum and hence the difference in ending simulation date. Soil moisture was simulated by specifying the soil, location, crop and management practices (Table 1).

2.8 Statistical Test

Effect of different treatments on soil moisture were statistically evaluated by analysis of variance (ANOVA) as a split-split plot design (Genstat 14.0 for Windows). Least Significant Differences (LSD) at the 5% level were used to detect differences among means.

3. Results and Discussion

3.1 Validation of CropSyst Model for Soil Moisture (mm) in Sorghum and Sweet Potato Cropping Systems.

Sorghum cropping systems: Simulated soil moisture showed low values of RMSE and percentage differences (PD) compared to observed moisture in the sorghum cropping system. The PD (range −3.43 to +7.04), RMSE (0.582) and a willmott index of agreement (WI) of 0.989 between observed and simulated values in all cropping systems with application of FYM under oxen plough were indicative of good model performance. In all cropping systems with combined application of FYM and RP under furrows and ridges, the PD ranged from −3.128 to +6.203 with RMSE and WI of 0.512 and 0.974, respectively. For RP application, across all cropping systems and tillage practices, the PD ranged from −2.002 to +4.661 while the RMSE and WI were correspondingly 0.487 and

0.999. In the control, the PD ranged from −0.184 to +6.123 with RMSE of 0.884 and WI of 0.907 (Table 3).

The PD under furrows and ridges, for all cropping systems with FYM application ranged from −3.73 to +2.57 while the RMSE was 0.682 and a WI of 0.995. When FYM+RP was applied, the PD ranged from −3.73 to +2.57 with RMSE of 0.872 and WI of 0.993. With application of RP, the PD ranged from −1.51 to +4.994 with a RMSE of 0.685 and WI of 0.957. In the control, the PD ranged from −2.96 to +8.67 with a RMSE of 0.895 and WI of 0.987 (Table 3).

Under tied ridges, for all cropping systems with application of FYM, the PD between observed and simulated values ranged from −1.39 to 3.58 while the RMSE was 0.8286 and WI of 0.955. For the FYM+RP treatment, the PD ranged from −1.633 to +3.078 with RMSE of 0.885 and WI of 0.952. For RP, across all cropping systems and tillage practices, the PD ranged from −1.66 to +0.244 with a RMSE of 0.624 and WI of 0.925. In the control, the PD ranged from −1.05 to +1.55 with a RMSE of 0.687 and WI of 0.972 (Table 3).

The PD between observed and simulated values were less than 9% implying closeness between observed and simulated values. Stockle et al. (2003) noted that simulation models can over-or under-estimate observed values by ±27 percent, without necessarily undermining reasonability of estimates obtained. All the simulated yields were therefore within what can be termed as reasonable estimate of the actual soil moisture.

The low values of RMSE indicate that CropSyst model reasonably simulated soil moisture for different cropping systems, tillage practices and organic inputs. Higher WI values for soil moisture indicate that the model simulated soil moisture reasonably well. CropSyst model has also been reported to simulate soil moisture to a reasonable range as stated by Baroudy et al. (2012) who found an RMSE of 2.5mm and 2.23 mm and a WI of 0.98 and 0.96 while determining soil water for two growing seasons. Similarly Benli et al. (2007) obtained a high WI of agreement with a value of 0.98 and attributed this to agreement between observed and simulated soil moisture values.

Table 3. Statistical comparisons of observed and simulated soil moisture (mm) under different tillage practices, cropping systems and organic fertilizers during sorghum growing season

Treatments	FYM			FYM+RP			RP			Control		
	Observed	Simulated	PD(%)	Observed	Simulated	PD(%)	Observed	Simulated	PD(%)	Observed	Simulated	PD(%)
Oxen plough												
SOR-MONO	21.913	20.82	+0.049	19.5	20.11	-3.128	18.1	17.92	+0.994	14.203	16.816	-0.184
SOR/DOL	49.35	49.00	+0.717	56.9	57.38	-0.844	60.9	59.32	+ 2.59	46.886	44.887	+ 4.24
SOR/CP	46.97	46.41	+7.04	57.13	56.48	+1.138	60.88	61.22	-0.558	37.558	43.41	+4.947
SOR-DOL	40.64	40.88	+3.83	17.41	16.33	+6.203	25.31	24.31	+4.661	32.515	33.367	+1.88
SOR-CP	33.845	35.68	-3.43	44.43	45.75	-2.566	38.97	39.57	-2.002	28.317	25.583	+6.123
RMSE	0.582			0.512			0.487			0.884		
WI	0.989			0.974			0.999			0.907		
Furrows and ridges												
SOR-MONO	36.53	35.59	+2.57	43.05	40.9	+4.994	38.75	36.84	+4.929	29.224	26.69	+8.67
SOR/DOL	79.89	80.84	-1.19	94.16	94.87	-0.754	84.47	86.02	-1.510	63.914	63.34	+ 0.9
SOR/CP	60.126	62.37	-3.73	70.68	69.62	+1.749	63.87	64.83	-0.941	48.101	47.25	+1.77
SOR-DOL	84.755	84.51	+0.29	78.47	77.52	+1.549	89.09	88.80	+1.224	67.804	69.19	-2.96
SOR-CP	66.817	66.51	+0.524	93.04	94.85	-1.9454	70.82	71.31	-0.607	53.454	51.83	+3.03
RMSE	0.682			0.872			0.685			0.895		
WI	0.995			0.993			0.957			0.987		
Tied Ridges												
SOR-MONO	76.94	78.00	-1.39	87.17	85.01	+3.078	84.31	85.71	-1.66	72.8	70.72	+1.03
SOR/DOL	93.77	93.62	+0.16	90.21	89.27	+1.072	93.02	94.12	-1.279	89.90	88.51	+1.55
SOR/CP	75.83	73.11	+3.56	86.32	87.03	-0.787	89.12	90.21	-1.369	72.75	73.00	-0.35
SOR-DOL	83.72	83.10	+0.749	85.01	86.42	-1.633	90.05	89.83	+0.244	80.55	81.39	-1.05
SOR-CP	88.38	87.21	+1.319	93.7	92.56	+1.235	96.08	97.05	-0.109	87.45	87.21	+0.48
RMSE	0.8286			0.885			0.624			0.687		
WI	0.955			0.952			0.925			0.972		

SOR-MONO: Sorghum monocropping; SOR/DOL: Sorghum dolichos intercrop; SOR/CP: Sorghum chickpea intercrop; SOR-DOL: Sorghum dolichos rotation; SOR-CP: Sorghum chickpea rotation; RP: Rock phosphate; FYM: farm yard manure; PD: Percentage differences; RMSE: root mean square error; WI: willmott index.

3.2 Sweet Potato Cropping Systems

The PD ranged from −7.2 to +12.09, −5.003 to +7.539, 4.538 to +8.1 and −6.7 to +6.3, RMSE (1.323, 1.012, 0.973 and 0.753) and WI (0.906, 0.966, 0.953 and 0.946) with application of FYM, RP, FYM+RP and control, respectively for all cropping systems under oxen plough (Table 4) showed good agreement between observed and simulated values of soil moisture. For all cropping systems, under furrows and ridges, when FYM was applied, PD ranged from −4.3 to +2.8 with RMSE of 0.687 and WI of 0.996, for RP, the PD ranged from −3.548 to +4.217 with RMSE of 1.155 and WI of 0.986 while in control PD ranged from −5.8 to +2.6 with RMSE of 0.699 and WI of 0.997 (Table 4).

In all cropping systems under tied ridges the PD ranged from −3.4 to +3.6 with RMSE of 1.249 and WI of 0.739 with FYM application, For the FYM+RP treatments, the PD ranged from −1.902 to +1.788 with RMSE of 0.878 and WI of 0.832, in RP treatment, the PD ranged from −0.815 to +1.888 with RMSE of 0.693 and WI of 0.831 while in control the PD ranged from −3.7 to +3.9 with RMSE of 1.083 and WI of 0.889 (Table 4).

The PD between observed and simulated values for soil moisture in different tillage practices, cropping system and organic inputs were less than ±13% indicating closeness between measured and simulated values. Low PD between observed and simulated values shows good agreement. According to Brassard and Singh (2007), a difference between observed and simulated values of up to ±15% was judged acceptable since there is closeness between the two values.

Tingem et al. (2008) also found a percentage difference between observed and simulated values ranging from 0.6 to −4.5 which are in close agreement with current results. Singh et al. (2008) found CropSyst to predict soil moisture well with low RMSE values.

Table 4. Statistical comparisons of observed and simulated soil moisture (mm) under different tillage practices, cropping systems and organic fertilizers during sweet potato growing season

Treatments	FYM			FYM+RP			RP			Control		
	Observed	Simulated	PC (%)	Observed	Simulated	PC (%)	Observed	Simulated	PC (%)	Observed	Simulated	PC (%)
Oxen plough												
SP-MONO	31.13	27.37	+12.1	28.68	27.41	+4.428	26.44	25.31	+4.274	20.75	21.72	-5.6
SP/DOL	30.98	30.39	+2.0	37.90	35.98	+5.066	34.11	32.88	+3.606	24.79	26.24	-6.0
SP/CP	25.01	26.46	-6.0	30.58	32.11	-5.003	27.53	25.30	+8.100	20.00	18.73	+6.3
SP-DOL	42.43	45.48	-7.2	49.46	50.76	-2.628	44.51	46.53	-4.538	32.68	34.32	-5.0
SP-CP	20.17	20.76	-2.9	24.67	22.81	+7.539	22.20	20.66	+6.937	16.13	17.21	-6.7
RMSE	1.323			1.012			0.973			0.753		
WI	0.906			0.966			0.953			0.946		
Furrows and ridges												
SP-MONO	41.16	42.93	-4.3	48.86	50.13	-2.599	43.97	45.53	-3.548	32.93	34.85	-5.8
SP/DOL	58.64	57.03	+2.8	69.62	67.87	+2.514	62.66	64.64	-3.548	46.91	46.02	+1.9
SP/CP	43.49	43.81	-0.7	51.63	50.20	+3.769	46.47	44.51	+4.217	34.79	33.88	+2.6
SP-DOL	94.42	75.54	-1.3	82.53	80.35	+2.642	81.22	83.79	-3.164	95.66	74.57	+1.3
SP-CP	84.97	84.98	-2.4	87.55	89.52	-2.250	85.79	84.03	+2.052	67.97	68.88	-1.3
RMSE	0.687			1.011			1.153			0.699		
WI	0.996			0.987			0.986			0.997		
Tied Ridges												
SP-MONO	84.19	87.18	-1.1	90.10	91.76	-1.842	88.38	86.89	+1.685	82.07	80.98	-3.7
SP/DOL	87.51	87.97	-3.4	91.17	89.91	+1.382	84.90	83.76	+1.343	83.95	85.71	-3.3
SP/CP	82.93	79.92	+3.6	90.04	88.43	+1.788	87.38	85.73	+1.888	71.68	74.47	+3.9
SP-DOL	86.85	85.71	+1.6	94.11	95.90	-1.902	84.70	85.39	-0.815	69.48	70.33	-1.2
SP-CP	95.10	93.12	+2.1	96.41	95.23	+1.223	92.75	93.45	-0.755	76.08	78.27	-2.9
RMSE	1.249			0.878			0.693			1.083		
WI	0.739			0.832			0.831			0.889		

SP-MONO: Sweet potato monocropping; SP/DOL: Sweet potato dolichos intercrop; SP/CP: Sweet potato chickpea intercrop; SP-DOL: Sweet potato dolichos rotation; SOR-CP: Sweet potato chickpea rotation; RP: Rock phosphate; FYM: Farm yard manure; PD: Percentage differences; RMSE: root mean square error; WI: willmott index of agreement.

3.3 Simulated Soil Moisture in Sorghum and Sweet Potato Cropping Systems

Sorghum cropping system: There were significant ($P < 0.05$) differences in tillage practices, cropping systems and organic inputs across seasons. There were also significant interaction ($P < 0.05$) effects between tillage practices with cropping systems, tillage practices with organic input and tillage practice with cropping systems and organic inputs.

In the first season, simulated soil moisture (101.91 mm) was significantly ($P < 0.05$) high in interactions

interactions involving tied ridges and sorghum/dolichos intercrop when RP+FYM were applied (Figure 1). Simulated soil moisture (13.52 mm) was lowest in sorghum mono crop with no organic input applied under oxen plough (Figure 1).

In the second season, simulated soil moisture was significantly high (108.3 mm) in sorghum/dolichos intercropping with application of FYM+RP under tied ridges (Figure 2). Lowest simulated soil moisture (15.4 mm) was observed in the interaction between oxen plough and sorghum monocropping with no organic input applied (Figure 2).

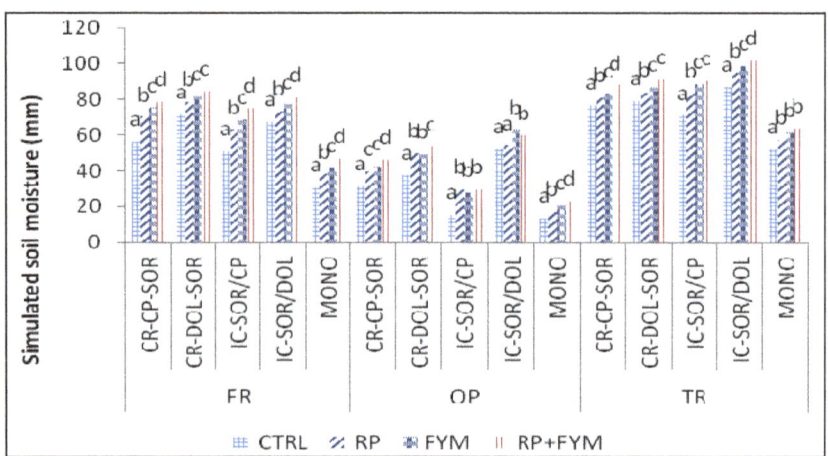

Figure 1. Simulated soil moisture (mm) across tillage practices, cropping systems and organic fertilizers in season 1

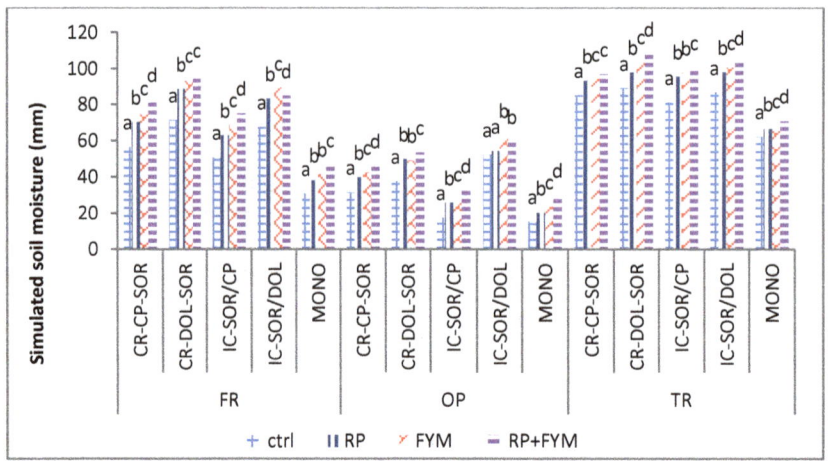

Figure 2. Simulated soil moisture (mm) across tillage practices, cropping systems and organic fertilizers in season 2

The interaction effects between tied ridges, sorghum in rotation with dolichos and FYM+ RP on soil moisture could be attributed to reduced run-off and increased infiltration due to micro- catchment formed by the tied ridges. Sorghum intercropped with dolichos had significantly high soil moisture which could be attributed to reduced evaporation due to dense soil cover provided by the two crops. Higher soil moisture in the FYM+RP could be attributed to improved water retention by the two organic fertilizers.

According to Guzha (2004), tillage practices that increase soil roughness such as tied ridging and ripping can increase soil water storage and availability to crop because they are able to capture rainfall and increase the time for infiltration to take place. Rockstrom (2013) stated that intercropping increases canopy cover and thus reducing evaporation from soil surface. Palm et al. (1997) similarly reported improved soil physical properties such as infiltration and soil moisture retention with application of FYM+RP.

Sweet potato cropping systems: There were significant (P < 0.05) difference in soil moisture in the different tillage practices, cropping systems and organic inputs. Interactions between tillage practice and cropping systems, tillage practice and organic inputs also had significant (P < 0.05) effect.

Tillage practices: In the first season, simulated soil moisture (95 mm) was significantly higher in tied ridges followed by furrows and ridges (68.44 mm) and least (32.49 mm) in oxen plough. In the second season, simulated soil moisture (100.24 mm) was significantly highr in tied ridges followed by furrows and ridges (72.4 mm) and least (34.36 mm) in oxen plough (Figure 3).

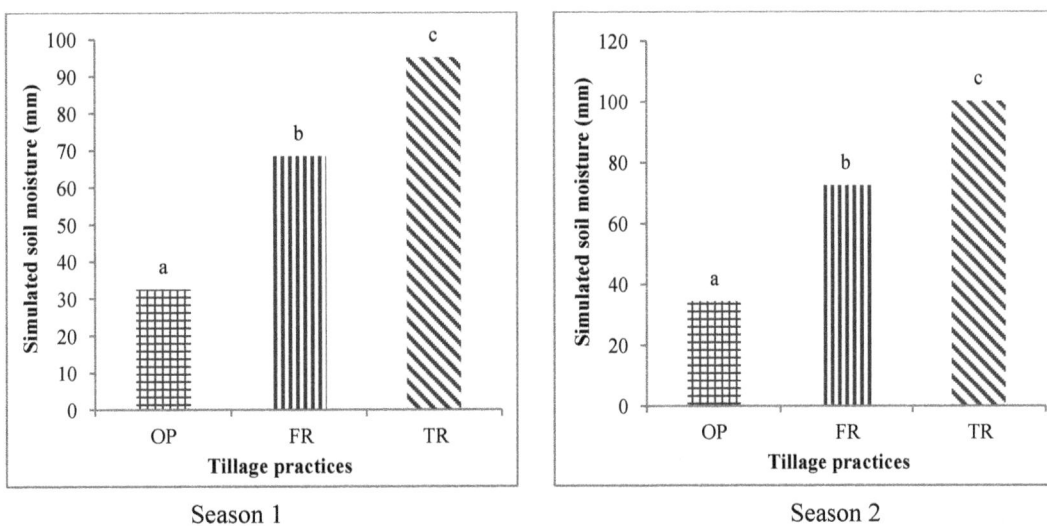

Season 1   Season 2

Figure 3. Simulated soil moisture in the different tillage practices

OP: oxen plough; FR: furrows and ridges; TR: tied ridges.

Tied ridges are able to capture more water compared to oxen ploughed plots and furrows and ridges. The more water collected in tied ridges could be attributed to reduced runoff. According to Taye and Abera (2010), in tied ridges, furrows are blocked with earth ties creating basins that catch and hold rainwater, minimizing surface runoff and improving downward infiltration of water. Tillage can improve the physical and hydro-physical properties of the soil and consequently increase rain water harvesting and crop yields (Gachene & Kimaru, 2003; Strudley et al., 2008).

Cropping systems: In the first season, simulated soil moisture (75.32 mm) was significantly (P < 0.05) high when sweet potato was rotated with dolichos and least (53.46) in the sweet potato monocrop. Simulated soil moisture (79.63 mm) in the second season was highest (55.26 mm) in sweet potato-dolichos rotation and least on sweet potato mono crop (Figure 4).

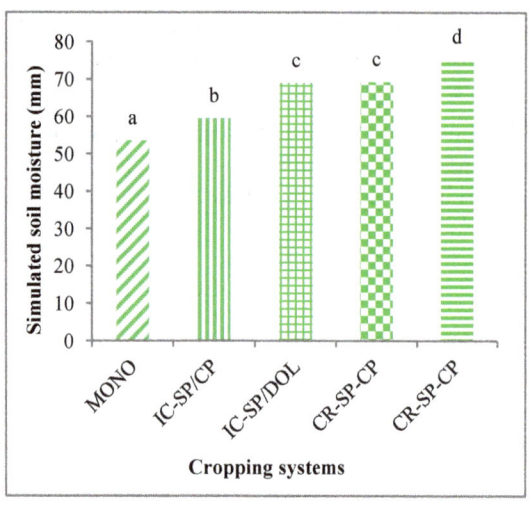 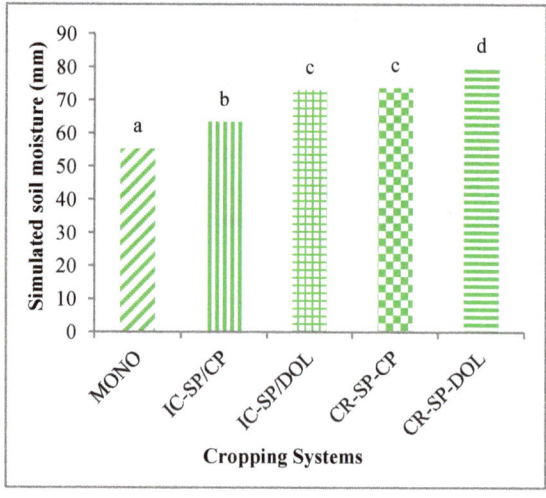

Season 1   Season 2

Figure 4. Simulated soil moisture in the different cropping systems

IC: intercropping; CR: crop rotation; SP: sweet potato; CP: chickpea; DOL: dolichos.

Higher simulated soil moisture in sweet potato- dolichos rotation could be attributed to increased water availability since sweet potato and dolichos had different rooting systems which increased water availability in soil through expanded soil depth. According to Roder (1989) rotation of legumes and cereals, with their different root systems optimizes the network of root channels in soil to deeper soil depths. This leads to increased water penetration, water- holding capacity and available water for crop use.

Organic inputs: Simulated soil moisture (75.03 mm) in the first season was significantly high when RP+FYM was applied and least (52.52 mm) in the control (Figure 5). In the scond season, simulated soil moisture (79.39 mm) was highest in RP+FYM and least (55.39 mm) in the control (Figure 5).

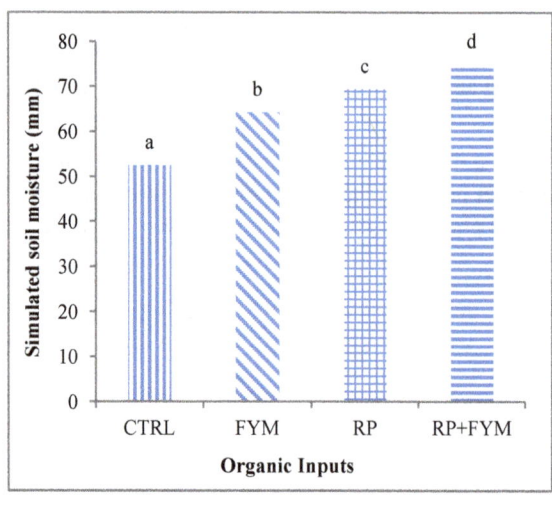 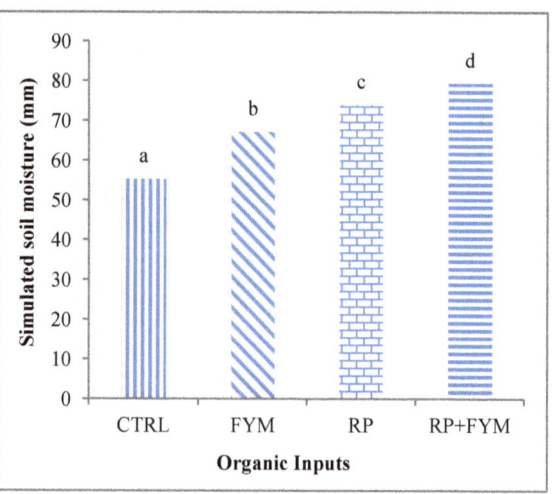

Season 1   Season 2

Figure 5. Simulated soil moisture in the different organic inputs

CTRL: control; FYM: farm yard manure; RP: rock phosphate.

The FYM+RP had high soil moisture and this could be due to improvement of soil structure and hence increased soil water holding capacity. FYM+RP could have improved the soil physical properties particularly water infiltration rate. Lal (1997) and Sharif et al. (2013) also reported that FYM+RP application improves water infiltration rate, water holding capacity, soil aeration and soil moisture.

Tillage practices and cropping systems interaction: In the first season simulated soil moisture (108.08 mm) was significantly (P < 0.05) higher in the interaction between tied ridges and sweet potato intercropped with dolichos and least (23.16 mm) in the interaction between oxen plough and sweet potato mono crop (Figure 6). In the second season, simulated soil moisture (114.48 mm) was significantly high in the interaction between tied ridges and sweet potato intercropped with dolichos and least (25.87 mm) in the interaction between oxen plough and sweet potato monocrop (Figure 6).

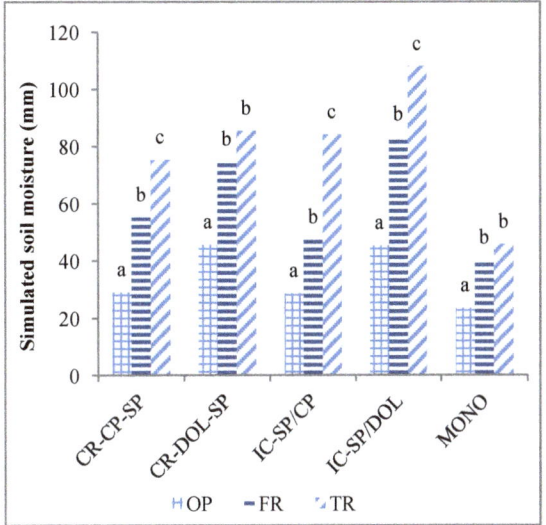 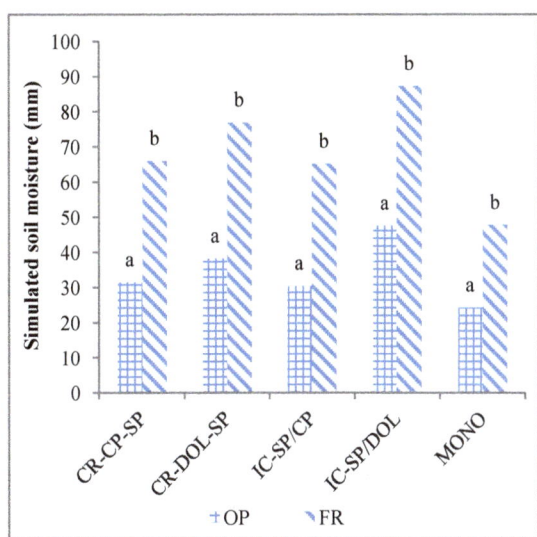

Figure 6. Simulated soil moisture in the interaction between tillage practices and cropping systems

CR; Crop Rotation, IC; Intercropping, CP; chickpea, SP; sweet potato, DOL; dolichos, OP; oxen plough, FR; Furrows and Ridges, TR; Tied Ridges

High simulated soil moisture in tied ridges and sweet potato intercropping could be attributed to reduced run off and reduced evaporation rate as a result of dense canopy created by the two crops. High plant densities in intercropping together with litter-fall block water flow while increased volume of roots further opens up the soil hence improved infiltration. According to Zougmore et al. (2000), intercropping allows for formation of a thick canopy due to higher planting densities. The dense canopy formed helps prevent soil erosion by rain water action. Fewer rain drops reach the soil surface with great impact because the dense canopy intercepts and break-up heavy rain drops. The FYM+RP application had high soil moisture due to improvement of soil structure and this may have led to increased water holding capacity.

Tillage practices and organic input interaction: In the first season, simulated soil moisture (108.57 mm) was significantly higher in interaction between tied ridges and RP+FYM and least (25.77 mm) in the interaction between oxen plough and control. In the second season, simulated soil moisture (112.69 mm) was highest in the interaction between tied ridges and RP+FYM and least (27.43 mm) in the interaction between oxen plough and control (Figure 7).

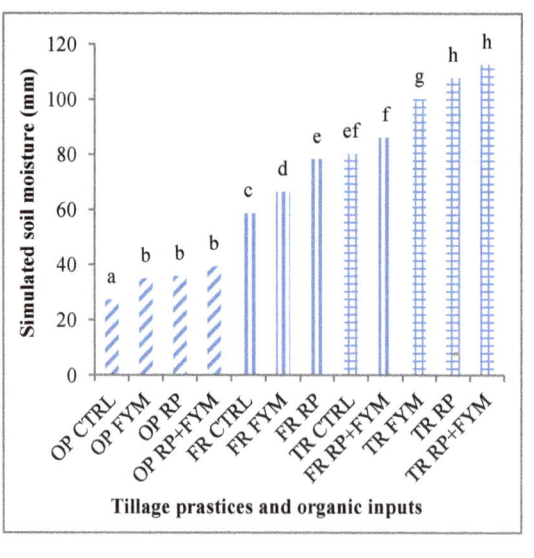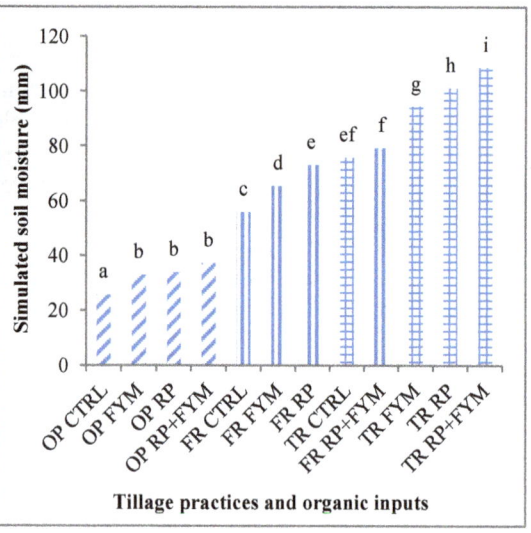

Season 1   Season 2

Figure 7. Simulated soil moisture in the interaction between tillage practices with organic input

OP: oxen plough; FR: furrows and ridges; TR: tied ridges; CTRL: control; FYM: farm yard manure; RP: rock phosphate.

High soil moisture in tied ridges and FYM+RP could be attributed to the fact that tied ridges allow rainwater to be retained on open furrows for longer duration as water infiltrates into soil. This soil management techniques favour prolonged rainwater infiltration and retention thus raising overall soil moisture retention and soil water holding capacity. According to Itabari et al. (2003) tied ridges increase rainwater retention thus increased soil moisture. Combined FYM and RP could have increased water retention in the soil. Manure and rock phosphate have similarly been reported to increase water retention and availability in soil (Silva et al., 2006).

4. Conclusion

In sorghum and sweet potato based cropping systems, simulated soil moisture was highest in tied ridges, intercropping and rotation systems when FYM+RP was applied and least in oxen plough, monocropping when no organic fertilizer was applied. In sorghum based cropping system, soil moisture was high in the interactions involving tied ridges and sorghum intercropped with dolichos when FYM+RP were applied. Whereas in the sweet potato based cropping system, highest soil moisture was observed in interactions involving tied ridges with sweet potato intercropped with dolichos when FYM+RP were applied.

Information on the effects of tillage practices, cropping systems and organic fertilizers could therefore be very useful for water soil conservation purposes. Thus, using simulation models to attain that could be ultimate solution. A good agreement between observed and simulated soil moisture implied that the CropSyst model was capable of investigating sustainable alternatives of increasing soil moisture.

Acknowledgements

The authors appreciate the McKnight Foundation for providing funds to first author to carry out the study which forms part of her master's thesis. We would also like to appreciate the farmers of Matuu division for availing their farms for field research and their willingness to work with us.

References

Anderson, J. M., & Ingram, J. S. (1993). *Tropical Soil Biology and Fertility: A Handbook of Methods*. CAB International, Wallingford, UK.

Benli, B., Pala, M., Stockle, C., & Oweis, T. (2007). Assessment of winter wheat production under early sowing with supplemental irrigation in a cold highland environment using CropSyst simulation model. *Agric. Water Mang., 3*, 45-54. http://dx.doi.org/10.1016/j.agwat.2007.06.014

Biamah, E. K. (2005). Coping with drought: Options for soil and water management in semi arid Kenya. *Tropical resource management papers No. 58*. Wageningen University and research centre publication.

Biamah, E. K., Stroosnijder, L., Sharma, T. C., & Cherogony, R. K. (1998). *Effects of conservation tillage on watershed hydrology in semi-arid Kenya: An application of AGNPS, SCS-CN and Rational Formula runoff models* (Vol. 3, pp. 335-357).

Blake, G. R., & Hartge, K. H. (1986a). Bulk Density. In A. Klute (Ed.), *Methods of Soil Analysis. Part 1 - Physical and Mineralogical Methods* (2nd ed.). American Society of Agronomy, Madison WI.

Blake, G. R., & Hartge, K. H. (1986b). Particle Density. In A. Klute (Ed.), *Methods of Soil Analysis. Part 1 - Physical and Mineralogical Methods* (2nd ed.). American Society of Agronomy, Madison WI.

Brassard, J. P., & Singh, B. (2007). Assessing the impacts of and climate change and CO_2 increase on potential crop yields in Southern Quebec, Canada). *Climate Research, 34*(2), 106-117.

Dillon, J. L. (1992). The Farm as a Purposeful System. *Miscellaneous Publication No. 10*. England, Armidale.

Donatelli, M., Stöckle, C. O., Nelson, R. L., Gardi, C., Bittelli, M., & Campbell, G. S. (1999). Using the software CropSyst and arcview in evaluating the effect of management in cropping systems in two areas of the low Po valley, Italy. *Rev. de Cien. Agric., 22*, 87-108.

El-Baroudy, A. A., Ouda, S. A., & Taha, A. M. (2013). Calibration of CropSyst model for wheat grown in sandy soil under fertigation treatments. *Egypt. J. Soil Sci.* (In Press).

Gachene, C. K. K., & Kimaru, G. (2003). Soil fertility and land productivity. *Technical Handbook* (p. 30). RELMA/Sida Publication, Nairobi Kenya.

Gicheru, P. T., Gachimbi, L. N., Nyangw'ara, M. K., Lekasi, J., & Sijali, I. V. (2004). *Stakeholders consultative meeting on sustainable land management project workshop held at wida highway motel from 1st-2nd November 2004*. Work shop report, Kenya Soil Survey Nairobi.

Grabisch, M. (2003). *Temporal scenario modeling and recognition based on possibilistic logic* (pp. 261-289). Elsevier Science Publishers Ltd. Essex, UK.

Guzha, A. C. (2004). Effects of tillage on soil micro relief, surface depression storage and soil water storage. *Soil and Tillage Research, 76*, 105-114. http://dx.doi.org/10.1016/j.still.2003.09.002

Itabari, J. K., & Wamuongo, J. W. (2003). Water-harvesting technologies in Kenya. *KARI Technical Note Series* (No. 16).

Jaetzold, R., Schmidt, H., Hornetz, B., & Shisanya, C. (2006). Ministry of Agricuture Farm Management. *Handbook of Kenya* (Vol. II, 2nd ed., Part C, Subpart C1). Nairobi, Kenya: Ministry of Agriculture.

Kibunja, C. N., Mwaura, F. B., & Mugendi, D. N. (2010). Long-term land management effects on soil properties 293 and microbial populations in a maize-bean rotation at Kabete, Kenya. *Afr. J. Agric. Res., 5*, 108-113.

Klute, A., & Dirksen, C. (1982). Hydraulic conductivity and diffusivity. In A. Klute (Ed.), *Methods of Soil Analysis*. American Society of Agronomy. Agron. Monogr. 9. ASA, SSSA, Madison, WI. USA.

Lal, R. (1997). Soil quality and sustainability. In R. Lal, W. H. Blum, C. Valentin & B. A. Stewart (Eds.), *Methods for Assessment of Soil Degradation* (pp. 17-31). CRC Press, Boca Raton.

Macharia, P. (2004). *Gateway to Land and Water Information: Kenya National Report*. FAO (Food and Agriculture Organization). Retrieved from http://www.fao.org/ag/agL/ swlwpnr/reports/y_sf/z_ke/ke.htm

Okalebo, J. R, Gathua, K. W., & Woomer, P. L. (2002). *Laboratory methods for soil and plant analysis: A working manual*. TSBF: Nairobi, Kenya.

Palm, C. A., Myers, R. J. K., & Nandwa, S. M. (1997). Combined use of organic and inorganic nutrient sources for soil fertility maintenance and replenishment. In R. J. Buresh et al. (Eds), *Replenishing Soil Fertility in Africa* (SSA Special Publ. 51, pp. 193-217). SSA, Madison, USA.

Purcell, L. C., Edwards, J. T., & Brye, K. R. (2007). Soybean yield and biomass responses to cumulative transpiration: Questioning widely held beliefs. *Field Crops Research, 101*, 10-18. http://dx.doi.org/10.1016/j.fcr.2006.09.002

Roder, W., Mason, S. C., Clegg, M. D., & Kniep, K. R. (1989). Yield-soil water relationships in sorghum-soybean cropping systems with different fertilizer regimes. *Agronomy Journal, 81*, 470-475. http://dx.doi.org/10.2134/agronj1989.00021962008100030015x

Sharif, M., Burni, T., Wahid, F., Khan, F., Khan, S., Khan, A., & Shah, A. (2013). Effect of RP composted with organic materials on yield and phosphorus uptake of wheat and mung bean crops. *Pak.. J. Bot.., 45*(4),

1349-1356.

Silva, P. S. L., Silva, J., Oliveira, F. H. T., Sousa, A. K. F., & Dud, G. P. (2006). Residual effect of cattle manure application on green ear yield and corn grain yield. *Horticultura Brasileira, 24,* 166-169. http://dx.doi.org/10.1590/S0102-05362006000200008

Sinclair, T. R., & Seligman, N. G. (1996). Crop modelling: From infancy to maturity. *Agron. J., 88,* 698-703. http://dx.doi.org/10.2134/agronj1996.00021962008800050004x

Singh, A. K., Tripathy, R., & Chopra, U. K. (2008). Evaluation of CERES-wheat and CropSyst models for water-nitrogen interactions in wheat crop. *Agric. Wat. Mang., 95*(7), 776-786. http://dx.doi.org/10.1016/j.agwat.2008.02.006

Staggenborg, A. S., & Vanderlip, L. R. (2005). Crop Simulation Models Can be Used as Dryland Cropping Systems Research Tools. *Agronomy Journal, 97,* 378-384. http://dx.doi.org/10.2134/agronj2005.0378

Stockel, C. O., Donatelli, M., & Nelson, R. (2003). CropSyst, a cropping simulation model. *European Journal of Agronomy, 18*(2008), 289-307.

Stöckle, C. O., Donatelli, M., & Nelson, R. (2003). CropSyst, a cropping systems simulation model. *Eur. J. Agron., 18,* 289-307. http://dx.doi.org/10.1016/S1161-0301(02)00109-0

Stöckle, S. M., & Campbell, G. S. (1994). CropSyst, a cropping systems model: water/nitrogen budgets and crop yield. *Agric. Syst., 46,* 335-359. http://dx.doi.org/10.1016/0308-521X(94)90006-2

Strudley, W. M., Green, T. R., & Ascough II, J. C. (2008). Tillage effects on soil hydraulic properties in space and time: State of the science. *Soil and Tillage Research, 99,* 4-48. http://dx.doi.org/10.1016/j.still.2008.01.007

Taye, B., & Abera, Y. (2010). Response of Maize (*Zea mays* L.) to tied ridges and planting methods at Goro, Southeastern Ethiopia. *American-Eurasian Journal of Agronomy, 3*(1), 21-24.

Tingem, M., Rivington, M., Bellocchi, G., Azam-Ali, S., & Colls, J. (2008). Comparative assessment of crop cultivar and sowing dates as adaptation choice for crop production in response to climate change in Cameroon. *African Journal of Plant Science and Biotechnology, 1,* 10-17.

USDA. (1993). Soil Survey Manual. *USDA Agricultural Handbook 18.* Superintendent of Documents. Washington, DC.

Willmott, C. J. (1982). Some comments on the evaluation of model performance. *Bull. Amer. Meteor. Soc., 63,* 1309-1313. http://dx.doi.org/10.1175/1520-0477(1982)063%3C1309:SCOTEO%3E2.0.CO;2

World Reference Base for Soil Resources (WRB). (2006). *A framework for international classification, correlation and communication.*

Zougmore, R., Kambou, F. N., Ouattara, K., & Guillobez, S. (2000). Sorghum–cowpea intercropping: An effective technique against runoff and soil erosion in the Sahel (Saria, Burkina Faso). *Arid Soil Research and Rehabilitation, 14,* 329-342. http://dx.doi.org/10.1080/08903060050136441

Notes

Note 1. For Oxen plough only. Similar treatments and cropping sequence was used for TR, and FR.

Note 2. Sweet potatoes were similarly handled as sorghum using same field layout.

5

Winter Cereals as Double Crops in Corn Rotations on New York Dairy Farms

Quirine M. Ketterings[1], Shona Ort[1], Sheryl N. Swink[1], Greg Godwin[1], Thomas Kilcer[1], Jeff Miller[2] & William Verbeten[3]

[1] Department of Animal Science, Cornell University, Ithaca, NY, USA

[2] Cornell Cooperative Extension, Oneida County, Oriskany, NY, USA

[3] Cornell Cooperative Extension, Northwest New York Dairy, Livestock and Field Crops Team, Lockport, NY, USA

Correspondence: Quirine M. Ketterings, Nutrient Management Spear Program, Department of Animal Science, 323 Morrison Hall, Cornell University, Ithaca NY 14853, USA.

Abstract

Weather extremes in 2012 and 2013 impacted corn silage and hay yields for many dairies in the northeastern United States and prompted a growing interest in double cropping of winter cereals for harvest as high quality forage in the spring. Here we report on (1) forage yield ranges of cereal rye and triticale in corn-cereal rotations in New York in 2012-2014, and (2) survey results of 30 New York farm managers who grew winter cereals as double crops with corn silage in 2013. Yields averaged 3.62 and 4.88 Mg ha^{-1} for cereal rye and triticale, respectively. On average, the surveyed farmers planted 8% of their tillable acres to winter cereal with the intent to harvest as forage. Triticale was the most frequently seeded double crop (70%). Most stands were established with a drill (57%). Manure was applied to 37% of the fields. Fertilizer nitrogen (N) was applied at dormancy break by 79% of the farmers with a median application rate of 67 kg N ha^{-1}. The biggest challenge with the double-crop rotation, identified by the farmers, was timely seeding of the double crop in the fall given late corn silage harvest and early onset of frost in the Northeast. Despite challenges encountered and questions about the impact of harvest of the winter cereal on the main crop, 83% of the surveyed farmers planned to continue to grow double crops.

Keywords: corn rotations; cover crops; dairy; double crops; winter cereals

1. Introduction

Interest in seeding winter cereals after corn (*Zea mays* L.) silage harvest as cover crops has been growing over time, as farmers and farm advisors recognize the importance of fall and spring ground coverage for erosion control after corn silage harvest, the potential of overwintering cereals to retain end-of-season nitrogen (N), the need for a growing crop to improve nutrient use efficiency of fall-applied manure, and the addition of carbon (C) to soils through roots and crop residue (Long et al., 2013a, 2013b). However, the yield shortage due to the 2012 drought and the extremely wet growing conditions of 2013 prompted a growing number of farmers to evaluate the potential of overwintering winter cereals as double crops in corn rotations, to be harvested as a forage in May, prior to planting of the next corn silage crop. The two species particularly suited for this use in the northeastern United States (US) are cereal rye (*Secale cereal* L.) and triticale (x *Triticosecale* Wittm.).

Faced with uncertainties in annual weather patterns and greater occurrence of weather extremes, many farmers and advisors ask questions about yield potentials of winter cereals for forage production in corn silage rotations, and about agronomic practices. Given a short growing season in the northeastern US, inclusion of winter cereals for forage production can cause a delay in corn planting and the need for a shorter season corn variety. However, Jemison et al. (2012) showed boot-stage and soft-dough stage harvests of double crop combinations to yield 20 and 33% more total biomass than full season corn in Maine and Vermont, while studies on a western New York (NY) farm indicated a 27% yield increase (Long et al., 2013b). These study results are consistent with findings in Iowa in work by Heggenstaller et al. (2009) that showed a 25% increase in dry matter (DM) yield over full season corn for a double crop corn-triticale rotation.

Information on attainable yields and agronomic management of winter cereals as forage double crops in the northeastern US is limited to a small number of trials. Here we report on (1) the DM yield of nineteen cereal rye fields and 44 triticale fields on commercial farms distributed throughout NY and harvested in May 2012-2014, and (2) farmer motivation for, practices, and experiences with double cropping of winter cereals for forage.

2. Method

2.1 Yield Assessments

Winter cereal yield determinations were done over a period of three years. In May of 2012, one cereal rye and thirteen triticale fields were sampled on commercial farms. At each location, four 96×20 cm frames, spaced at least 50 m apart, were placed 10 cm off the ground and the biomass above the 10 cm mark was harvested. The DM content was determined using a forced-air oven set at 54 degrees C. Fields were located in western and northern NY, two regions where double cropping was practiced in 2011. Fields were identified by the farmer and his/her crop consultant. Fall seeding took place on 21 Sept. for the cereal rye field and between 12 Sept. and 4 Oct. for the triticale fields. Harvest was done in 2012 on 18 May (cereal rye field) and either 4 May (seven triticale fields in western NY) or 18 May (six triticale fields in northern NY). Fields were established and managed by the farmers; the two week difference in harvest window is consistent with regional climate differences within NY. New York is characterized predominantly by USDA Plant Hardiness Zones 4 and 5 but ranges from Zones 3 to 6 (USDA, 2012).

In spring of 2013 and 2014, 49 field harvest assessments were added to the database (35 in 2013 and 14 in 2014). These fields were part of a statewide 5-rate N response project with N rate trials conducted in four replications in each field. Fields were located in northern, eastern, central, western and southern NY. In fall of 2012, seeding took place from 15 Sept. to 10 Oct. (seven cereal rye fields) and from 9 to 20 Oct. (28 triticale fields). The N rate trials were established at dormancy break in the spring. Harvest of the trials took place on 15 or 20 May 2013 for the cereal rye trials, and from 6 to 24 May for the triticale trials. In fall 2013, eleven cereal rye fields (seeded between 30 Aug. and 20 Oct.) and three triticale fields (seeded between 26 and 30 Sept.) were added. Trials were harvested in 2014 between 12 and 21 May for cereal rye, and on 19 or 22 May for triticale. In both years of the N rate study, decisions related to seeding rates, seeding method, and manure management, pest management, and seeding and harvest dates were made by the farmers who hosted the trials. At each trial location, the N rates were applied to 3 by 3 m plots that were established using a randomized complete block design. Nitrogen was applied at dormancy break at rates of 0, 34, 67, 101, or 134 kg N ha^{-1}. Yield in the plots was determined by harvesting the area within three 96×20 cm frames per plot at a 10 cm cutting height. Harvest took place at flag leaf to early boot stage. Results of the N response trials will be documented elsewhere. Here we report on yields at the most economic rate of return to N as determined in the N rate trials.

2.2 Farmer Surveys

The 3-page farmer survey included six components: (1) farm size, acreage in double crops, and number of years of experience of the farmer with double cropping of over-wintering cereals in corn rotations; (2) motivation for adding winter cereals to the corn rotation; (3) agronomic practices; (4) challenges encountered with double cropping; (5) double crop plans for the future; and (6) need for further information. The survey was completed by 30 of 31 NY farmers who participated in the on-farm assessment of yield and crop response to N in the spring of 2013, representing northern (seven farms), eastern (three farms), central (one farm), western (twelve farms), and southern (seven farms) NY. The survey was deemed exempt from Institutional Review Board for Human Participants (IRB) review by the Cornell Institutional Review Board.

3. Results

3.1 Yield Averages and Ranges

Cereal rye fields averaged 3.65 Mg DM ha^{-1} across all three years (nineteen fields), with an average minimum yield of 2.22 Mg DM ha^{-1} and maximum of 5.38 Mg DM ha^{-1} (Table 1; Figure 1). Triticale yields averaged 4.88 Mg DM ha^{-1} (44 fields) with an average minimum yield of 2.37 Mg DM ha^{-1} and maximum of 10.44 Mg DM ha^{-1} (Table 1; Figure 1). The highest producing field was a triticale field that was harvested 18 May 2012 following three weeks with exceptionally good growing conditions.

Table 1. Average yields of cereal rye and triticale seeded after corn silage harvest and harvested for forage in May in New York in 2012-2014

Species		Number of fields	Yield			
			Average	Standard Deviation	Min	Max
			----- Mg DM ha^{-1} -----			
Cereal rye	2012	1	5.38	-	5.38	5.38
	2013	7	3.65	1.23	2.24	5.35
	2014	11	3.47	1.16	1.64	5.40
	All	19	3.63	1.19	2.22	5.38
Triticale	2012	13	5.13	2.51	1.95	10.44
	2013	28	4.82	1.23	2.46	6.76
	2014	3	4.35	1.01	3.34	5.35
	All	44	4.88	1.68	2.37	7.75

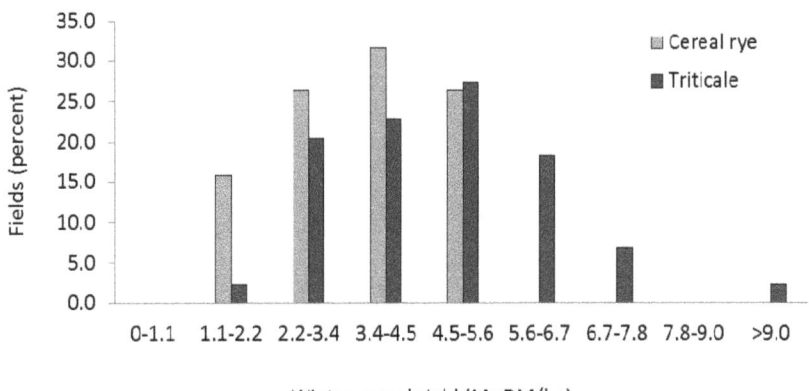

Figure 1. Distribution of yields of 19 cereal rye and 44 triticale fields harvested as forage in May of 2012-2014 in New York

No side-by-side comparisons were done between the two species so we cannot conclude if one species yielded higher than the other in any of the years. The results of the assessment show, however, that yields exceeding 2.24 Mg DM ha^{-1} are common for both species (only two of nineteen cereal rye and one of 44 triticale fields yielded less than 1 ton DM/acre; Figure 1). Replicated side-by-side comparisons are needed to draw conclusions about species selection for optimal yield. Determining the factors that enable yields that exceed 6.5 Mg DM ha^{-1} on some fields (Figure 1) will be critical to increase farmer adoption of double cropping with small grain cereal crops.

3.2 Farmer Survey

3.2.1 Area in Double Crops

The 30 farmers who completed the survey managed from 55 to 2,430 ha of tillable cropland per farm (19,855 ha in total cropland), with 20% managing 203 ha or less versus 23%, 27%, 17% and 13% managing 203-405, 405-810, 810-1,215, and more than 1,215 ha, respectively. For three farms, acres in double crops planted for forage harvest exceeded 203 ha (up to 284 ha). One farmer had terminated the stand as a cover crop while another harvested the winter cereal for grain. Forty percent of all farms seeded 20 ha or less versus 20% who seeded between 20 and 41 ha, 23% with 41-81 ha, and 10% with 81-203 ha in double crops.

Of all farm tillable acres among the 30 farms, 1,526 ha (8%) were double cropped with a winter cereal harvested as forage in May. The survey did not include questions related to percentage of tillable acres in corn silage but in 2012, 468,000 ha of cropland were planted to corn statewide of which 41% was harvested as silage and 59% as grain while hay was harvested on 631,800 ha (USDA-NASS, 2014). If, on average, the farms operated by the 30 surveyed farmers have a similar crop acreage distribution (43% of the total acreage in corn of which 41% is harvested as silage), the 8% of all tillable acres in double crops represents approximately 45% of the acreage in corn silage on the cooperating farms in the study.

3.2.2 Farmer Years of Experience

For 14 (47%) of the 30 farmers in the survey, 2013 was the first year of growing double crops on the farm. Nine farmers (30%) had 2-4 years of experience. Three farmers (10%) had 5-7 years of experience while four farmers (13%) had implemented double cropping for more than 10 years. These results are consistent with the relatively recent introduction of the corn-winter cereal for forage double crop rotation in NY. Seed sales of winter triticale seed have reflected this trend increasing from a total of 810-1,215 ha in 2010 to over 12,150 ha in 2013, and regional seed suppliers continue to increase seed supplies to keep up with this rapidly increasing demand from farmers for this double crop (Bill Verbeten, unpublished).

3.2.3 Species Selection and Rotation

Of all farmers in the survey, 25 (83%) had tried triticale as a double crop versus fourteen farmers (47%) who had experience with cereal rye. Winter wheat had been seeded by 6 farmers (20%) (Table 2). Five farmers (17%) indicated they had tried oats. Because oats winterkill and the fall growing season after corn silage harvest is short in the northeast, this crop is more commonly utilized as a winter-killed cover crop or as a forage double crop after winter wheat with fall harvest of the forage. Oats are also not considered suitable for N management of soils with a high leaching potential in NY (Ketterings et al., 2003) because when planted after corn silage oats can only capture a small amount of the manure N (Graham et al., 2012). It is therefore not surprising that only 17 % of the farmers had tried oats versus 83% and 46% for triticale and cereal rye, respectively. For most farms (70%) triticale was the most frequently seeded double crop. Cereal rye was considered most frequently as well by 37% of the farmers (Table 2). Sixteen farmers (53%) had never tried winter wheat. These results are consistent with the distribution of fields that were sampled to determine achievable forage yields as part of the N rate study in 2012-2013, where 70% of the fields had been seeded to triticale versus 30% to cereal rye (Table 1).

Table 2. Species of winter cereal double crops grown for forage (A) and main crop planted after harvest of double crops for forage (B), ranked in order of frequency by 30 farmers who participated in double crop trials in 2012-2013. Several of the 30 farmers surveyed gave more than one reason for growing double crops. Oats, a non-overwintering cereal, was the only crop listed for the "Other" category. Some farms gave equal ranking to two forage species

(A)			Frequency of double crops selected to grown for forage							
			----Most frequent---		----------2nd----------		----------3rd----------		---------Never-------	
Winter forage	Farms	%	Farms	%	Farms	%	Farms	%	Farms	%
Cereal rye	14	47	11	37	2	7	1	3	11	37
Triticale	25	83	21	70	3	10	1	3	2	7
Winter wheat	6	20	1	3	5	17	0	0	16	53
Other (oats)	5	17	0	0	5	17	0	0	9	30
(B)			Frequency of main crop planted after harvest of double crops for forage							
			--------Most frequent--------		---------------2nd---------------		-------------Never-------------			
Main crop	Farms	%	Farms	%	Farms	%	Farms	%		
Corn silage	24	80	23	77	1	3	1	3		
Small grains	4	13	1	3	2	7	12	40		
Vegetables	4	13	3	10	1	3	14	47		
Alfalfa/grass	3	10	1	3	2	7	14	47		
Soybeans	2	7	2	7	-	-	-	-		

Most (80%) of the fields, that were double cropped and harvested for forage in May on the 30 farms represented in the survey, were subsequently planted to corn. Thirteen percent seeded fields to small grains or vegetables (Table 2). These results reflect our focus on use of double crops in corn silage rotations for dairy farms.

3.2.4 Agronomic Practices

The 35 on-farm N rate trials in 2013 included seven cereal rye and 28 triticale trials. Seeding rates ranged from 67 to 207 kg seed ha^{-1} for triticale and from 67 to 168 kg seed ha^{-1} for cereal rye. In total, 20 fields (57%) were drilled versus fifteen (43%) that were broadcast-seeded. Thirteen fields (37%) received liquid dairy manure, either at the time of double crop establishment (12 fields; 34%) or in Feb (1 field; 3%) with application rates ranging from 23.3 kL to 112.2 kL ha^{-1}. All fields were seeded in either the last three weeks of Sept. (22 fields) or the first three weeks of Oct. (13 fields). Harvest took place between 6 and 29 May, 2013.

Of the 29 farmers who responded to the question about fertilizer use, 23 farmers (79%) applied N fertilizer at dormancy break. Most common fertilizers were urea or urea mixed with ammonium sulfate (48% of the farmers). Ten farmers (34%) used liquid urea ammonium nitrate with or without ammonium thiosulfate while the remaining farmers did not identify the source of N they used. Nitrogen application rates varied from zero (21% of all farmers) to 45-56 kg N ha^{-1} (21%), 56-78 kg N ha^{-1} (29%), 78-90 kg N ha^{-1} (18%), and 90-118 kg N ha^{-1} (11%). The average application rate for those farmers who applied N was 74 kg N ha^{-1} with a median of 67 kg N ha^{-1}. The wide range in N application rates might reflect, among other things, the lack of knowledge about and guidance for N management for these winter cereals grown as forage crop in corn rotations.

Herbicide was applied to the double crops grown as forage in 2013 by only three of the 29 farmers (10%) who responded to this question. None of the farmers indicated use of fungicides or insecticides for the winter cereals. This is not surprising as harvest takes place prior to the onset of common diseases and pests for winter cereals in the Northeast.

3.2.5 Farmer Motivation for Double Crop

Sixteen farmers (53%) listed the desire to increase the forage production on a limited crop area as the main reason for seeding winter cereals (Table 3). Ten (33%) indicated they had seeded double crops primarily to address a feed shortage (emergency feed). Increased farm profits and higher quality feed were listed as reasons for including double crops by five (17%) and four (13%) of the farmers, respectively. Following the experiences of the 2013 growing season, a larger number of farmers identified the desire to increase forage production on a limited acreage (a shift from 53% in fall of 2012 to 63% after the 2013 growing season) while emergency forage needs declined as a prime reason for double cropping from 33% in 2012 to 10% after the 2013 growing season. Of all farmers, 25 (83%) planned to continue to grow winter cereals as a forage crop in the future with an additional five farmers (17%) who said they might consider it. In total, sixteen farmers (53%) planned to increase the acreage planted to double crops in the coming year while another seven (23%) said they may do so but were not sure yet.

Table 3. Main reason for growing double crops as forage in 2013 and primary reason for continuing to grow double crops as forage in the future. Several of 30 farmers surveyed gave more than one reason for growing double crops

Main reason for growing double crop for harvest as forage:		------In 2013------		------ Future ------	
		Farms	%	Farms	%
Increase forage productivity on limited acreage		16	53	19	63
Emergency forage/feed if the need presented itself		10	33	3	10
Increase farm profits (small grain or vegetable rotation)		5	17	5	17
Higher quality forage to feed a certain group(s) of animals		4	13	5	17
Other reasons (in 2013):	(in the future):				
"Profitable cover crop"	"Profitable cover crop"	3	10	2	7
"Soil quality"/"Soil health"	"Build organic matter"/"Soil health"	2	7	3	10
"Need straw for mulch"[x]	"Add cows without increasing land"	1	3	1	3
	"BMP, CAFO requirements"[y]			2	7

Note. [x]This farm planted the double crop for forage, but harvested it for grain and straw; [y]BMP = best management practice; CAFO = concentrated animal feeding operation.

3.2.6 Challenges and Information Needs

The biggest challenge with the double crop rotation identified by the farmers was getting a double crop seeded in time in the fall (Table 4), consistent with the short period between corn silage harvest and onset of cold weather. In addition, nine farmers (32%) pointed to the potential for delay in corn planting following double crop harvest. Five farmers (18%) identified labor and time involved as a constraint while four farmers (14%) pointed to weather challenges during harvest time of the double crops (too wet in spring to get equipment on fields).

Table 4. Greatest challenges with growing winter cereals as double crops for forage according to 30 survey participants with experience in growing double crops. Several respondents described more than one challenge

Greatest challenge	Farms	%
Fall planting timing, getting corn off in time to plant double crop	14	50
Harvest timing of double crop forage to allow for planting of next crop	9	32
Available labor and time	5	18
Weather issues at harvest (too wet)	4	14
Low tonnage, possible need for N stabilizer	1	3
Production costs per ton	1	3
Loss of seed to geese feeding in fall	1	3

Many farmers identified the impact of the double crop on the following crop as the most important aspect of double cropping that they needed to learn more about (Figure 2). Respondents to the survey wanted to know more about the impact of nutrient uptake and removal by the double crop harvest on fertilizer needs of the crop seeded after double crop harvest. This was followed by questions about economics and forage quality (milk production potential of the winter cereals), and harvest methods.

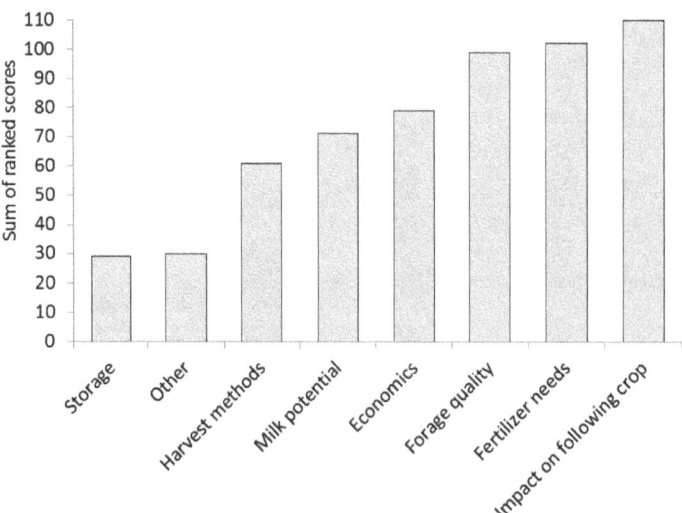

Figure 2. Collective summary of ranking by survey participants of aspects of double crops for forage use that would benefit from more information to aid in decision-making regarding its role in the farm's cropping system. Farmers ranked eight aspects in order of importance: most important (1) to least important (8). To calculate a score, values were assigned in reverse order to the ranking (8 for most important, 1 for least important, 0 if not ranked) and multiplied by the number of responses for each ranking of the particular aspect. The scores for each ranking of an aspect were then added together to obtain the comparative value presented in the bar graph

Other questions raised by farmers related to seeding methods and rates, manure use to supply N needs, fertilizer

N management for high leaching soils, and tradeoffs with shorter season corn varieties. In addition, farmers asked questions about variety selection, integrated pest management and diseases, and on-farm seed collection. Some farmers wanted more information on seeding rate as related to planting date and benefits of using a drill versus broadcasting of the double crop seed. Others raised questions about the costs per acre to plant and harvest the double crop and potential yield trade-offs with a delay in corn planting due to inclusion of the winter cereal, wondering about minimum yields needed to cover the cost of production of the double crop.

4. Discussion and Implications

Yield assessment data showed that 71% of all fields in the study exceeded 3.36 Mg ha^{-1} with an average of 3.62 Mg ha^{-1} for cereal rye and 4.88 Mg ha^{-1}for triticale (Figure 1). In comparison, the average corn silage yield in 2012 in NY amounted to 13.33 Mg ha^{-1} (USDA-NASS, 2014). If we assume no corn yield loss upon shifting to a shorter growing season (later planting), these yield data suggest the potential for a per acre yield increase of 27% for corn and cereal rye rotations versus a 37% yield increase for corn and triticale rotations, consistent with findings of Heggenstaller et al. (2009), Jemison et al. (2012), and Long et al. (2013b). However, a shorter season corn variety might need to be selected as most double crop harvests took place between May 5 and 25, delaying corn planting by one or two weeks. Variety trials conducted in 2012 in NY (Cox et al., 2012) showed 0.56 Mg ha^{-1}lower yields for every 5 day decrease in corn relative maturity. Cox et al. (2013) showed a yield difference between 95-100 d and 106-110 d corn varieties at two NY locations, averaged across all hybrids in the NY assessment, of 2.47 Mg DM ha^{-1}, while 84-90 d corn varieties averaged 0.86 Mg DM ha^{-1} less than 101-106 d corn varieties in the colder regions of the state. However, in both years a large number of short-day corn varieties had equal or higher yields than many of their longer-day counterparts (Cox et al., 2012, 2013). These data suggest an overall (season) yield increase can be obtained with the inclusion of the double crop in corn silage rotations in the northeastern US, even if a shorter day variety is selected. However, economic analyses need to be conducted to evaluate what yield level is needed for a positive economic return on investment.

5. Conclusions

Our study suggests that, despite the often short fall season, NY farmers can successfully implement corn-winter cereal double cropping practices and have done so on an estimated 45% of their corn silage acres, averaging yields of 3.62 Mg ha^{-1} for cereal rye and 4.88 Mg ha^{-1} for triticale. Double cropping with winter cereals can benefit agriculture environmental management and increase per ha forage production.

Acknowledgements

This work was funded through grants from the Northern New York Agricultural Development Program (NNYADP), Federal Formula Funds, USDA-NRCS (Conservation Innovation Grant for the Upper Susquehanna Watershed), and Northeast Region Sustainable Agriculture Research and Education (NESARE). We thank the many farmers who participated in the project as well as our collaborators Paul Cerosaletti, Janice Degni, Dale Dewing, Kevin Ganoe, Mike Hunter, Kitty O'Neil, Ashley Pierce (Cornell Cooperative Extension); Jonathan Barter, Steve Loraine, and Aaron Ristow (Soil and Water Conservation Districts); Shawnna Clark, Martin van der Grinten, and Paul Salon (USDA-NRCS Big Flats Plant Materials Center); Joe Lawrence (Lowville Farmers Co-op); Shawn Bossard (Morrisville State College), and agricultural consultants Peter Barney (Barney Agricultural Consulting), Eric Beaver and Mike Contessa (Champlain Valley Agronomics), Jeff Willard and Jeremy Langer (Agricultural Consulting Service).

References

Cox, W. J., Cherney, J., Atkins, P., & Paddock, K. (2012). New York corn silage hybrid tests – 2012. *Department of Crop and Soil Sciences Extension Series E12-1*. Cornell University, Ithaca NY. Retrieved November 30, 2014, from http://www.fieldcrops.org/VarietyTrials/Documents/2012%20Corn%20Silage%20Report.pdf

Cox, W. J., Cherney, J., Atkins, P., & Paddock, K. (2013). New York corn silage hybrid tests – 2013. *Department of Crop and Soil Sciences Extension Series No. E13-1*. Cornell University, Ithaca NY. Retrieved December 17, 2014 from http://fieldcrops.org/VarietyTrials/Documents/2013%20Corn%20Silage%20Report.pdf

Graham, C., Van Es, H., & Schindelbeck, B. (2012). Rye vs. oat: Environmental benefits vary greatly. *What's Cropping Up? 22*(4), 3-4.

Heggenstaller, A. H., Liebmann, M., & Anex, R. P. (2009). Growth analysis of biomass production in sole-crop and double-crop corn systems. *Crop Science, 49*, 2215-2224. http://dx.doi.org/10.2135/cropsci2008/12/0709

Jemison, J. M., Darby, H. M., & Reberg-Horton, S. C. (2012). Winter grain-short season corn double crop forage production for New England. *Agronomy Journal, 104*, 256-264. http://dx.doi.org/10.2134/agronj2011.0275

Ketterings, Q. M., Klausner, S. D., & Czymmek, K. J. (2003). Nitrogen guidelines for field crops in New York. Second Release. *Department of Crop and Soil Sciences Extension Series E03-16*. Cornell University, Ithaca, NY. Retrieved November 30, 2014, from http://nmsp.cals.cornell.edu/publications/extension/Ndoc2003.pdf

Long, E., Ketterings, Q. M., & Czymmek, K. J. (2013a). Survey of cover crop use on New York dairy farms. *Crop Management*. http://dx.doi.org/10.1094/CM-2013-0019-RS.

Long, E., Van Slyke, K., Ketterings, Q. M., Godwin, G., & Czymmek, K. (2013b). Triticale as a cover and double crop on a New York dairy. *What's Cropping Up? 23*(1), 3-5. Retrieved November 30, 2014, from http://css.cals.cornell.edu/sites/css.cals.cornell.edu/files/shared/documents/wcu/WCUvol23no1.pdf

USDA. (2012). USDA Plant Hardiness Zone Map, 2012. *Agricultural Research Service, United States Department of Agriculture*. Retrieved November 30, 2014, from http://planthardiness.ars.usda.gov

USDA-NASS. (2014). New York State Agricultural Statistics Service; County Estimates. Retrieved November 30, 2014, from http://www.nass.usda.gov/Statistics_by_State/New_York/Publications/County_Estimates/index.asp

6

Genetic Variability, Coefficient of Variance, Heritability and Genetic Advance of Some *Gossypium hirsutum* L. Accessions

Muhammad Zahir Ahsan[1], Muhammad Saffar Majidano[1], Hidayatullah Bhutto[1], Abdul Wahab Soomro[1], Faiz Hussain Panhwar[1], Abdul Razzaque Channa[1] & Karim Buksh Sial[2]

[1] Plant Breeding Section, Central Cotton research Institute Sakrand, Sindh, Pakistan

[2] Agronomy Section, Central Cotton research Institute Sakrand, Sindh, Pakistan

Correspondence: Muhammad Zahir Ahsan, Plant Breeding Section, Central Cotton research Institute Sakrand, Sindh, Pakistan. E-mail: ahsanzahir@gmail.com

Abstract

The present study was conducted in central cotton research Institute Sakrand to analyze the genetic variability, phenotypic, genotypic and environmental coefficient of variation, heritability and genetic advance during summer 2014. In this experiment analysis of variance indicated that significant variation present among the accessions of the upland cotton for all the traits under study. The highest genotypic (GCV) and phenotypic coefficient of variation (PCV) were exhibited by the number of bolls per plant, lint index and seed cotton yield per plant. GCV had similar trend as PCV. High heritability and high genetic advance was observed in the lint index, number of bolls per plant and seed cotton yield per plant. The combination of the high heritability and high genetic advance provide the clear image of the trait in the selection process.

Keywords: genetic advance, heritability, ECV, GCV, PCV, coefficient of variability

1. Introduction:

Cotton is grown in more than 60 counties worldwide and is an important source of natural fiber globally Cotton. Cotton is also famous for it's named as "queen of fiber plants" and white gold. It is very much necessary to understand the gene action and pattern of inheritance of the traits to choose a suitable breeding methodology for crop improvement (Vineela et al., 2013). Effective breeding program depends upon the variation present in the gene pool for the yield enhancing traits. Selection is effective when magnitude of variability in the breeding population is enough.

Cotton was grown in Sindh Pakistan before 2500 B.C. as Excavations of Mohen Jo Daro showed (Khan, 2003). Cotton is of great importance and has a pivotal role in the economy of the Pakistan. Most of our foreign exchange reserves come directly or indirectly through textile channel from cotton. So sustainable cotton production not only become imperative but we must find out the way to enhance the per acre yield to uplift cotton based economy.

The identification and use of genotype with higher genetic potential is a continuous requirement for the production of better cotton. Efforts are going on to enhance the yield as well as the fiber quality. To achieve these objectives comprehensive studies to understand the genetic mechanism to control the plant characters under different environmental conditions is also necessary.

The present research program was imitated to understand the genetic variation of different upland cotton accessions for yield and related traits. Heritability, Genetic Advance, Genotypic, Phenotypic, Environmental variance and response to the Selection also calculated.

2. Materials and Methods

The present study was conducted at Central Cotton Research Institute Sakrand, Sindh, PAKISTAN. The germplasm was comprises of CRIS-664, CRIS-665, CRIS-666, CRIS-667, CRIS-668, CRIS-669, CRIS-670 and CRIS-342. All accessions were sown with P×P distance 30 cm and R×R distance 75 cm with three replications. All agronomic practices were kept same for all the replications and treatments. At maturity ten plants were selected from each replication of each accession and data were recorded for plant height, number of bolls per plant, boll weight, seed index, lint index, ginning out turn and seed cotton yield per plant.

The data was subjected for analysis of variance (Steel et al., 1997). The genotypic and phenotypic correlations were calculated by Kwon and Torrie (1964) technique. The genetic advance in percentage of mean was calculated by using Falconer (1989) formula.

$$Genetic\ Variance\ (Vg) = \frac{Genotype\ Mean\ Square\ (GMS) - Error\ Mean\ Suare\ (EMS)}{Number\ of\ Replications\ (r)} \tag{1}$$

$$Environmental\ Variance = Error\ MeanSuare\ (EMS) \tag{2}$$

$$Phenotypic\ Variance\ Vp = Vg + Ve/r \tag{3}$$

Genotypic Phenotypic and Environmental coefficient of Variation was calculated as

$$GCV\% = \sqrt{\frac{Vg}{\bar{X}}} \times 100;\ \ PCV\% = \sqrt{\frac{Vp}{\bar{X}}} \times 100;\ \ ECV\% = \sqrt{\frac{Ve}{\bar{X}}} \times 100 \tag{4}$$

Where, GCV% = Genotypic Coefficient of variation; Vg = Genotypic Variance; PCV % = Phenotypic Coefficient of variation; Vp = Genotypic Variance; ECV % = Environmental Coefficient of variation; Ve = Environmental Variance.

Heritability (H2) on Entry Mean Basis was calculated as

$$H^2 = \frac{Vp}{Vg} \tag{5}$$

The expected Genetic Advance for each trait was calculated as

$$GA = K\sqrt{Vp}H^2 \tag{6}$$

Where, K = 1.40 at 20% selection intensity for trait; Vp = Phenotypic variance for trait; H_2 = Broad Sense Heritability of the trait; Genetic Advance as percentage of mean is calculated as,

$$GA\% = \frac{GA}{\bar{X}} \times 100 \tag{7}$$

3. Results and Discussions

It is clear from the Table 1that significant variation present between the accessions for all the recorded traits. This variation is very important for the plant breeders and selection is effective when magnitude of variability in the breeding population is too enough. Table 2 depicted that the observation between the accessions were highly significant for all the recorded traits. Genotypic variance, genotypic coefficient of variance, phenotypic variance, phenotypic coefficient of variance, Environmental variance, and environmental coefficient of variance, broad sense heritability and response to the selection for eight recorded traits were shown in Table 3. The knowledge of nature and magnitude of the variability among the accessions for the traits is very important prerequisite for making simultaneous selection on more number of traits to make significant improvement in cotton. The analysis of variance indicted that the significant differences present among the accessions for all the traits. It is difficult to separate the heritable and non heritable variation making difficulty in selection for the breeders. Hence it is very necessary for the breeder to separate the heritable portion from the non heritable part to plan for proper breeding program.

Table 1. Mean and Range performance of the traits among the *G. hirsutum* accessions

Traits	Minimum	Maximum	Average performance
Plant Height (cm)	111	131	121.43
No. of Bolls per plant	25.5	39	30.69
Boll Weight (g)	2.96	3.774	3.51
Seed Index (g)	6.45	8.63	7.54
Lint Index (%age)	3.47	7.06	5.10
Ginning Out Turn (%age)	30.67	46.01	39.59
Seed cotton Yield per Plant (Kg)	83	125.42	107.03

Table 2. Analysis of variance of the different characters among the accessions of the *G. hirsutum*

Source of Variation	Degree of Freedom	Plant height	No. of Bolls per Plant	Boll Weight	Seed Index	Lint Index	G.O.T.%	Seed Cotton Yield per Plant
MSS	7	79.63**	79.63**	0.304**	1.77**	3.98**	61.61**	753.34**
Error	14	1.19	1.38	0.0026	0.00022	0.03	0.65	2.83

** Highly Significant at 0.01% level* Significant at 0.05% level.

Table 3. Mean Sum of squares, genotypic, phenotypic and environmental variance and coefficient of variation, broad sense heritability and response to selection

Trait	G. Variance	GCV %	P. Variance	PCV %	E Variance	ECV%	(H^2)	GA	GA as %age of Mean
Plant Height	130.60	9.4	131	9.4	1.19	0.89	0.9969	15.97	13.11
Bolls/ Plant	79.17	28.99	79.63	29.07	1.38	3.8	0.9942	12.42	40.46
Boll Weight	0.303	15.42	0.304	15.67	0.0026	1.4	0.9971	0.769	21.93
Seed Index	1.76	9.4	1.77	9.4	0.00022	0.89	0.9999	1.86	24.67
Lint Index	3.96	25.6	3.98	25.6	0.03	3.4	0.9974	2.78	54.51
GOT%	61.39	17.6	61.61	17.6	0.65	0.19	0.9964	10.95	27.65
yield/ Plant	752.39	19.7	753.34	19.8	2.83	2.01	0.9987	38.37	35.84

Table 4. Potential donor accessions for yield and other recorded traits

No.	Characters	Potential Donors
1	Plant Height	CRIS-666, CRIS-669, CRIS-670
2	Number of Bolls/Plant	CRIS-667, CRIS-670, CRIS-342
3	Boll Weight	CRIS-667, CRIS-668, CRIS-670
4	Seed Index	CRIS-666, CRIS-669, CRIS-670
5	Lint Index	CRIS-665, CRIS-666, CRIS-670
6	G.O.T.%	CRIS-665, CRIS-666, CRIS-670
7	Seed cotton Yield/ Plant	CRIS-667, CRIS-670, CRIS-342

The trait seed cotton yield per plant exhibit the highest genotypic and phenotypic variance i.e 752.39 and 753.34 respectively and followed by the plant height that have genotypic variance 130.60 and phenotypic variance 131. Lowest genotypic and phenotypic variance was recorded for the traits of boll weight and seed index i.e. 0.303 and 0.304 for boll weight and 1.76 and 1.77 for the seed index respectively. The coefficient of phenotypic, genotypic and environmental variance was also calculated for all the traits under study. The genotypic coefficient of variance was ranged from 9.4% (Plant Height and Seed Index) to 28.99 (Number of Bolls per Plant). Maximum genotypic coefficient of the variation was observed for the number of bolls per plant (28.99%) followed by the lint index (25.6). Phenotypic coefficient of variation also had similar trend as genotypic coefficient of variation. In the present study there was a close correspondence between genotypic and phenotypic coefficient of variation for all the recorded traits it showed that these characters less influenced by the environment. Since the variation depends upon the magnitude of the measuring units of the traits, coefficient of variation is independent of the measuring units so it is more useful in comparing the population. The highest genotypic and phenotypic coefficient of variation observed for the trait number of bolls per plant, lint index and seed cotton yield per plant. It indicates that selection can be applied on the traits to isolate more promising line. Similar type of observations in upland cotton was also reported by the different scientists (Dheva & Potdukhe, 2002; Preetha & Raveendran, 2007; Amir et al., 2012; Abbas et al., 2013). Moderate PCV and GCV were observed for the boll weight and ginning out turn. Girase and Mehatne (2002) and Harshal (2010) also noticed the moderate phenotypic and genotypic coefficient of variation for some traits and suggested that these characters can be improved by the vigorous selection. The traits such as plant height and seed index exhibited low PCV and GCV which indicated that the breeders should go for source of high variability for these traits to make improvement. Similar type of suggestion was given by Kowsalya and Raveendran (1996), Do Thi Ha An et al. (2006) in their experiments. In a population observed variation is due to both factors i.e. genetics and environmental where as genetic variability is the only heritable from generation to the next generation so the heritability alone does not give an idea about the expected gain in the next generation but it has to be considered in conjunction with the genetic advance. The characters those exhibit maximum heritability and high genetic advance as percentage of mean could be used as powerful tool in selection process such characters are controlled by the additive genes and less influenced by the environment (Panes & Sukhatme, 1995). The broad sense heritability was highest for all the recorded traits. For efficient selection we cannot solely believe on heritability the combination of high heritability with high genetic advance will provide a clear base on the reliability of that particular trait in the selection of variable entries. The genetic advance as percentage of means for seven traits ranged from 13.11% to 54. 57%. The higher genetic advance as percentage of mean was recorded by lint index (54.57) followed by the number of bolls per plant (40.46), seed cotton yield per plant (35.84), ginning out turn (27.65), seed index (24.67), boll weight (21.97) and plant height (13.11). High heritability and high genetic advance was observed in traits viz. seed cotton yield per plant, lint index and number of bolls per plant. These traits highly reliable during selection process of the accessions. High heritability and moderate genetic advance was found in ginning out turn, seed index and boll weight these results confirmed by the experiment of Muhammad et al. (2004). In this experiment some accessions were identified as potential donors for the crop improvement of different traits (Table 4). From the results of the present study, it can be concluded that direct selection can be done for most of the yield attributing traits since it exhibited high genetic variability and high range of variation. A high PCV and GCV for the characters studied indicated that environment influences on the expression of these traits were minor.

References

Abbas, H. G., Mahmood, A., & Ali, Q. (2013) Genetic variability, heritability, genetic advance and correlation studies in cotton (*Gossypium hirsutum* L.). *Int. Res. J. Microbiol., 4*(6), 156-161.

Amir, S, Farooq, J., Bibi, A., Khan, S. H., & Saleem, M. F. (2012). Genetic studies of earliness in *Gossypium hirsutum* L. *IJAVMS, 6*(3), 189-207.

Burton, G. W. (1951). Quantitative inheritance in pearl millet (*Pennisetum glaucum*). *Agron. J., 43*, 409-417. http://dx.doi.org/10.2134/agronj1951.00021962004300090001x

Dheva, N. G., & Potdukhe, N. R. (2002). Studies on variability and correlations in upland cotton for yield and its components. *J. Indian Soc. Cotton Improv.,* 148-152.

Do Thi Ha An, & Ravikesavan, R. (2006). Genetic diversity in cotton (*Gossypium sp*) (pp. 11-13). *National conference on plant sciences research and development.* APSI scientist meet, PSG CAS, Coimbatore, India.

Falconer, D. S. (1989). *Introduction to Quantitative Genetics* (3rd ed.). Logman Scientific and Technical, Logman House, Burnt Mill, Harlow, Essex, England.

Khan, N. U. (2003). *Genetic analysis, combining ability and heterotic studies for yield, its components, fibre and oil quality traits in upland cotton (G. hirsutum)* (Ph.D Dissertation). Sindh Agric. Univ. Tandojam, Pakistan.

Kowsalya, R., & Raveendran, T. S. (1996). Genetic variability and D2 analysis in upland cotton. *Crop Res., 12*(1), 36-42.

Kwon, S. H., & Torrie, J. H. (1964). Heritability and interrelationship of two soybean (*Glycine max* L.) populations. *Crop Sci., 4*, 196-198. http://dx.doi.org/10.2135/cropsci1964.0011183X000400020023x

Muhammad, I., Muhammad, A. C., Abdul, J., Muhammad, Z. I., Muhammad-ul-Hassan, & Noor-ul-Islam. (2004). Inheritance of Earliness and other Characters in Upland Cotton. *J. Biol. Sci., 3*(6), 585-590.

Panes, V. G., & Sukhatme, P. V. (1995). *Statistical methods for agricultural workers* (3rd ed., p. 58). ICAR, New Delhi.

Preetha, S., & Raveendran, T. S. (2007). Genetic variability and association studies in three different morphological groups of cotton (*Gossypium hirsutum* L.). *Asian J. Plant Sci., 6*(1), 122-128. http://dx.doi.org/10.3923/ajps.2007.122.128

Steel, R. G. D., Torrie, J. H., & Dicky, D. A. (1997). Principles and procedures of Statistics. *A Biometrical Approach* (3rd ed., pp. 400-428). New York: McGraw Hill Book Co. Inc.

Vineela, N., Samba Murthy, J. S. V., Ramakumar, P. V., & Ratna, K. S. (2013). Variability Studies for Physio

Morphological and Yield Components Traits in American Cotton (*Gossypium hirsutum* L.). *J. Agric. Vet. Sci.,* *4*(3), 7-10.

7

Crossing Borders:
Academe and Cultural Agency in Agricultural Research

Robert W. Blake[1,2], Elvira E. Sanchez-Blake[3] & Debra A. Castillo[4]

[1] Department of Animal Science and director of the Center for Latin American and Caribbean Studies (2009-2014), Michigan State University, East Lansing, Michigan, USA

[2] Professor Emeritus, Department of Animal Science, Cornell University, Ithaca, New York, USA

[3] Department of Romance and Classical Studies, Michigan State University, East Lansing, Michigan, USA

[4] Department of Comparative Literature, Stephen H. Weiss Presidential Fellow, and Emerson Hinchcliff Professor of Hispanic Studies, Cornell University, Ithaca, New York, USA

Correspondence: Robert W. Blake, Department of Animal Science, Michigan State University, East Lansing, Michigan 48824, USA. E-mail: rwblake@msu.edu

Abstract

This article explores social educator actions by academe as cultural agency's natural partner in ways that echo, connect and create plural discourse among the many dimensions and disciplines of society. Based on collaborations with Mexican partners, we argue this goal is achieved with multiplicative effects when students and faculty, key agents themselves and trainers of intercultural agents, learn first-hand by crossing borders to frame issues and work together to articulate collaborative research problems. In so doing a more inclusive worldview becomes integral context in needs assessments. This has been our long-standing pedagogical approach in leading students–undergraduates and graduates–and faculty from around the world on a multidisciplinary, intergenerational examination of rural and urban development in tropical Latin America. Greater academic agency through more alliances of this kind is needed to better achieve equity goals supported by greater investments targeting community engagement and applied problem-solving. We illustrate this learning framework and provide specific livestock research cases in southern Mexico that reveal potentials realized by bringing academe to the field and the field to academe, as part of a reinforcing educational process that promotes understanding and social transformation.

Keywords: academic alliances, agriculture, community engagement, cultural agency, southern Mexico, trading places

1. Crossing Borders

Ours is a story of encounters and problems needing solution. Different worlds reach out to one another on a transformative life stage obtained by crossing borders. For more than four decades Cornell University has led annual groups of 25 to 35 participants totaling about 2000 students–undergraduates and graduates–and faculty from around the world on a two-course multidisciplinary, intergenerational examination of rural and urban development in tropical Latin America. The field component of these courses has been conducted in Puerto Rico, Dominican Republic, Costa Rica, Honduras, Ecuador and Mexico. With library and lecture traded in the second course for mountaintops and mangrove swamps, aspiring intercultural scholars learn first-hand from real-world knowledge providers, some who are highly marginalized.

This year-long experience, which is training ground also for participating faculty, includes a preparatory course (Note 1), followed by an intense research experience (Note 2) that includes at least two weeks in the field and a subsequent on-campus agenda of distillation, analysis, and reporting. Preparation for informed and respectful dialogue in the field by our constituency, including students and faculty from collaborating institutions through videoconferencing, starts with the on-campus course in the preceding semester. This initial course is designed to introduce participants to basic cultural, historical, socio-political, literary, anthropological, linguistic, health, agricultural and food system, and social and family welfare issues that they are likely to observe first-hand. Inequality, possibly the key global social issue of our time, especially in the Americas, constitutes a dominant underlying theme. Besides a gap in earned income, social inequality stems also from unequal access to food, land, education at all levels, health services, markets, credit, capital and justice.

A complementary Spanish section for students with basic language skill was devoted to further discussion of topics from lectures and supported by additional readings in Spanish. In the field, recognizing the authority of "other" voices is crucial. Especially valuable are personal interactions "*in culturally authentic and acceptable ways*" (Meredith, 2010). Some of our hosts are indigenous peoples whose mother tongue is not Spanish, themselves learners of a second language and culture. We see this step towards intercultural practice, made more democratic by plural discourse, as obligatory in a world perceived by some to be increasingly dominated by a "*globalization that flattens everything in its path*" (Godenzzi, 2006). This platform, where academe learns from cultural agents–those whose actions affect collective change (Note 3)–evolved into a shared, live, video-streamed seminar during 2008-11 involving our universities as well as Mexican collaborators. Sandwiched between the two campus-based courses was the field experience itself, where we strengthened our ties to individuals and host institutions, including El Colegio de la Frontera Sur, el Centro de Investigaciones y Estudios Superiores en Antropología Social, and the Instituto Nacional de Investigaciones Forestales, Agrícolas y Pecuarias (INIFAP) (Note 4).

This article demonstrates how encounters during our field explorations helped students and faculty members alike to articulate problems, establish contacts, and develop subsequent research investigations based on these contextualized settings and inputs by our hosts. We illustrate how this consolidated pedagogical and problem-framing approach contributes to the intellectual growth of students and faculty alike with a sample of agricultural research outcomes built upon issues initiated by boots-on-the-ground field experiences. William B. Lacy (Note 5) summarized the achievement of this living laboratory undertaking (Note 6):

> *Experience Latin America is one of the richest learning experiences I have seen in higher education. The dynamic international learning environment is greatly enhanced by bringing together undergraduates and graduates with diverse backgrounds and international experiences with a multidisciplinary, intergenerational group of faculty, administrators, and extension educators. Each of the participants becomes an active learner and teacher...* (Blake, 2001).

Correspondingly, our learning forum is designed to share and to build knowledge, responding to what Godenzzi calls the great challenge of the twenty-first century, "*the construction of an ethic of respect and solidarity*" (2006). It strives to demonstrate tolerance, inclusion and value enhancement through a celebration of cultural difference. Students and faculty typically represent Africa, Asia and Europe as well as half a dozen countries from the Americas (including other regions in the host country). Accordingly, the program objectives are to explore equity-gap challenges, acknowledging rural-urban disparities and aspects of cultural heritage, improving intercultural dialogue, and fostering greater contact and communication among the players, e.g., in-country hosts, students from afar as well as the host country, faculty and other professionals.

Clearly it is not possible in a mere two weeks to thoroughly examine the many facets of the many problems faced by families in many settings. Nevertheless, the field laboratory provides a valuable opportunity to see first-hand how they live, to see their crops and animals, to speak with and learn from them, to visit projects of various institutions designed to serve them, and to listen to professionals who have devoted careers to these challenges. As Professor H. David Thurston, a leader in this enterprise summarized (Thurston, 2001), "*I feel the essence of the field trip... is that the students... have had the chance to 'touch it, feel it and smell it.' The course brings a vast array of experiences into focus in settings that cannot be equaled in a classroom.*" Correspondingly, we follow a beacon similar to the one provided to painter Georgia O'Keefe by her mentor, Professor Arthur W. Dow (Note 7)–structure one's work in better comprehending nature, landscapes, people and their livelihood systems "*by not (only) mastering particular facts, but by seeing, experiencing, and creating (your) own systems or structures.*"

1.1 Course Organization

A hallmark among Cornell's study abroad options, "Agriculture in Developing Nations" evolved with our understanding of agriculture and development from primarily graduate students in the agricultural sciences in the early years, to a gender, disciplinary and culturally-equitable mix of both undergraduate and graduate students from across the university and the globe, now branded *Experience Latin America*. Enrichments to this learning forum included participation also by extension educators, field activities designed to better grasp the major issues, and thematic teams to assess complex issues based on observation, consultation with hosts, and relevant literature.

Students are challenged to reconcile multiple facets with potentially competing goals in order to grasp pragmatically, and in ways witnessed to have touched our hosts, the complex issues of rural development. Jason Ingram summed up the experience of many others (Blake, 2001),

> "*I cannot stress the impact that this class had on me. It was one of the rare times when I felt that, having seen a well-rounded example of an issue, I could form an opinion based on my own observations, not just on what I had been shown in a classroom.*"

The relationships among the sometimes competing goals of poverty alleviation, economic growth and the sustainable use of the environment and its natural resources help define a basic learning framework. Participants grapple for balance among them–and corresponding impacts on human welfare–within the context of the host nation's complex food, environmental, economic and social systems. Culturally and experientially diverse participants also learn to debate more effectively and to better negotiate disparate or conflicting viewpoints into collaboration, aided by hands-on opportunities and guidance in choosing effective criteria to better figure things out for themselves.

From the outset of the field component, participants are organized into multidisciplinary theme groups with faculty facilitators. Each group comprises undergraduate and graduate students in several disciplines, international and US students, and Spanish-language competency. In addition to class discussions to daily process our observations, theme groups continue to deliberate throughout the field trip and subsequently in preparing written projects and their oral team presentations. Findings from rapid appraisals during the field study are also reported using this mechanism. Rapid appraisal exercises include mapping farm resources, mapping community transects, constructing a multi-layer annual calendar of agricultural, household and community activities, and outlining a community and family history of agriculture and resource use. The expectation is that farm visits provide a powerful thread for connecting one's own experiential learning with the learning that comes from reading and discussions also involving our hosts, which may immediately help students to formulate their own projects, including potential return for thesis work.

Themes groups vary with the expertise held by the participating faculty as well as the interests of students. For example, themes in one *Experience Latin America* edition were: Rural Realities: Livelihood systems in Chiapas; Politics, Identity and Society; and Indigenous Cultural Expression and Performance. Another edition focused on complementary livelihoods-related issues: Livelihood Systems in Mexico's Gulf Region: Which are the priority information needs, policies and programs? and Livelihood Systems in Mexico's Gulf Region: How to make research and extension relevant? Associated considerations were the potential "action plans" to increase the impact of research and extension, better understanding how information needs, interests and knowledge systems of resource-limited farmers differ from those with greater endowments, and the roles of farmer organizations and alliances with universities and government and nongovernmental organizations.

2. Research Aim: Agrarian Vulnerability

These living laboratory experiences provided opportunities for our participants to gain awareness and to learn about some of the rural realities that encompass many issues separately addressed by the academy. Farm visits initiated learning and insight about real-world issues concerning food production, land and water, climate change, biodiversity, family and community welfare, and the economic challenges faced by agrarian society. Conversations with farmers, indigenous folks, and other hosts helped articulate needs and contemporary challenges. Recognizing that valuable intellectual work and analysis take place in all disciplines, cultural agency and performance studies were also integrated into our portfolio of pedagogical interactions and the process of articulating researchable problems. Consequently, graduate and undergraduate student researchers developed projects and publications with the dual purposes of pursuing real-world issues and giving something back to our hosts and others with similar needs and interests.

We now illustrate our approach by describing an encounter with the Génesis farmers' organization, which subsequently led to the multi-institutional research collaboration by Absalon et al. (2012a, 2012b). Our group had been invited to the Génesis annual meeting and *barbacoa* (barbecue) at Rancho El Yualito, on the central coastal plain near Cotaxtla, Veracruz. Dozens of these cooperative members, owners of farms with dual-purpose cattle systems to produce milk and beef gathered with their families for business meetings, reporting and festivities. Reports included on-farm technology testing (and viewing of livestock), hearing from INIFAP advisors on technical and financial matters, and reviewing collaborative work plans for the coming year.

The Génesis encounter was graciously arranged and co-hosted by Dr. Francisco Juárez Lagunes, a Cornell alumnus and research scientist for INIFAP and professor at Universidad Veracruzana, along with Génesis farmers. Our El Yualito arrival was like finding a bustling, sunny-day county fair–a parking area for vehicles; streams of people, and unloading cattle from trucks; multi-colored banners; lively music booming from loudspeakers; Génesis men all in red shirts like twins *requete* multiplied (galore); smiling wives, mothers and daughters assuring order over chaos, organizing tables, chairs, projectors and screens; cauldrons of *carnitas* (savory pork) on open fires; and easels displaying posters reporting on-farm research with figures and photographs. Clearly, this was a celebration of successful cooperative action. Upon arrival our international delegation was warmly received by José Ausencio Muñíz Morales, owner of El Yualito, and his nephew, José Miguel Ruíz Espinoza. José Miguel, a former high school exchange student in the US, was urged to address our group on equal ground as Génesis' bi-lingual spokesperson. Greetings and introductory remarks led to a farm tour, presentations, questions and many individual conversations, and ample opportunity to observe Génesis men and women in action with INIFAP advisors.

During the visit our attention seized on a key organizational principle tied to their work-ethic: Génesis insists on a membership comprising only committed, active participants. Subsequent discussions in our small groups highlighted radical differences in agency and voice between the active chorus of Génesis *socios* (members) and the comparatively muted encounters in a previous visit with other farmers just a few hours away. This issue emerged again when visiting the Veracruz highland community of Micoxtla, above the city of Coatepec, where families struggle with seasonal food insecurity and economic instability. Micoxtla residents expressed the desire to be helped to organize cooperatively in order to enter higher-value local and regional markets. This expression and subsequent interaction eventually led to McRoberts' thesis project (2009) and research collaboration with INIFAP (McRoberts et al., 2013).

These and other field encounters helped bring to life the need to help communities and families to better secure their futures, a priority for the Government of Mexico that was emphasized by the United Nations Special Rapporteur (De Schutter, 2011). Despite gains in reducing the percentage of children under five years of age who are underweight (i.e., Millennium Development Goal #1), the Special Rapporteur emphasized the unacceptably uneven progress between rural and urban populaces. About 80% of the 18 million Mexicans living in municipalities characterized by high marginalization are in rural areas. Consequently, the Government of Mexico was urged to improve self-determination in rural communities through greater community participation and capacity building. Recommended actions included implementing mutually reinforcing environmental, agricultural and social protections, among the priorities to improve public policy, education, diets, health care, and family incomes. Calling for a "Third Agrarian Reform" the rapporteur cited a growing income inequality fostered partly by insufficient public support of agriculture. This reform was charged to provide greater public good expenditures, among them greater access to credit and financial services, agricultural technical assistance, and support to producer organizations and cooperatives.

In harmony with De Schutter's recommendations, many students subsequently formulated thesis research projects aimed at helping rural communities diminish their own vulnerability. These projects typically consider multiple goals aligned with the resources and opportunity horizons already dealt to families and communities. In the following sections we summarize outcomes from research endeavors that were initially based on exchanges with families owning livestock. We present results from a set of projects led by five graduate student *Experience Latin America* alumni–Australian, Japanese, US and two Mexicans–who examined some of the nutrient constraints and farming system dynamics at the root of agrarian vulnerability in Mexico's Gulf region. In every case, coalitions with farmers, farmer organizations, communities, local researchers and students enabled problem definition and project field work. Key institutional partners were INIFAP, Universidad Autónoma de Yucatán (UADY), Unión Ganadera Regional de Yucatán, Unión Ganadera Regional del Oriente de Yucatán, Universidad Veracruzana, and Génesis and Tepetzintla farmer organizations in Veracruz (Note 8). Thus, these collaborations (Note 9) constituted a kind of international consortium of its own making.

2.1 Cattle Systems

Animal agriculture is fundamental to the economy of Mexico's Gulf region. Cattle herds, like those owned by Génesis farmers constitute an important livelihood in rural Veracruz, a major supplier of Mexico's beef and milk. However, information for improving the productivity, profitability and sustainability of dual-purpose cattle systems is scarce in tropical Latin America, including Mexico, and likely in tropical agro-ecosystems around the world, especially regarding the benefits and costs of alternative management strategies (Blake & Nicholson, 2004; Blake, 2008). Assisted by INIFAP Génesis members sought to improve farming performance by substituting traditional forages with more nutritious species to increase milk sales from their herds.

Therefore, working together with INIFAP and Génesis herd owners in what are probably the first published tropical case studies to systematically examine complex energetic interactions (Note 10), Absalon-Medina et al. (2012a, 2012b) evaluated the limitations and potential improvements in milk production and profitability from alternative nutritional management. Other students similarly evaluated approaches to overcome productivity bottlenecks in Yucatan beef cattle herds (Baba, 2007) and juvenile female replacements in Tepetzintla herds of the low Huasteca region of Veracruz (Cristóbal-Carballo, 2009). These projects revealed a consistent pattern of key biological (energy) and management constraints on animal performance, which portends broader potential application for improving cattle system performance.

Heretofore unrecognized vulnerabilities were revealed through a study designed to evaluate herd performance limitations parsed by age of cow, physiological status and forage season of the year. The most susceptible management groups were non-lactating cows of all ages and forage seasons, and young cows and herd replacements (heifers) suffering growth retardations. Energy deficits signify repeated opportunity losses across an animal's lifetime, which are manifested in delayed puberty of heifers, fewer offspring born, and less total milk per cow over expected lifetimes. Like past efforts by Génesis farmers, these impediments could be ameliorated by investing in nutritional management and improved forage quality. As a result, diets formulated with better quality grass and legume forages were predicted to increase milk sales by up to 74% with large economic incentives, about $600 to $1100 greater predicted net margin per cow. This increase in net margin is large, equaling or exceeding in value the total milk from an additional full lactation per cow lifetime. A similar dietary strategy to assure normal growth also based on low cost, locally-produced feeds, especially available forages (e.g., grass hay, sugarcane, legumes), resulted in heifers that were 20% younger at first parturition (signifying earlier commencement of milk sales) with lower rearing costs, heavier body weights and greater adipose tissue reserves (Cristóbal-Carballo, 2009).

Large marginal rates of return, the change in net margin per unit increase in variable costs, indicated clear economic incentives to alleviate inherent energy deficits and impaired growth. However, alternative management options may be difficult to implement if they are little practiced, thus generating little knowledge among farmers about potential profitability, and if options are perceived riskier than *status quo* practices. *Ex ante* economic assessment of strategies requiring greater nutrient inputs is important because higher production per animal is not always more profitable (Absalon et al., 2012b).

2.2 Crop-Livestock Systems

Another set of studies examined the nutrient dynamics underlying smallholder systems and the potential of small ruminants to their sustainability. For more than three millennia the shifting cultivation *milpa* system in the Yucatán Peninsula has involved the cutting of forest after a fallow period, burning, and planting of maize mixed with squash and beans. *Milpa* (maize, often multi-cropped with beans) cultivation has been purported to be the only food production method available to farmers in forested areas without draft animals. Slashing and burning clears rocky soils for planting, releases nutrients from slashed vegetation for crop growth, and controls the population of weed seed. A major limitation to the productivity of *milpa* systems, indeed to food production in the developing world, is soil nutrient depletion. Nutrients and organic matter from animal manure—the world's oldest fertilizer—is a vital input for growing food. Agricultural systems of Yucatán have long comprised multiple species of livestock; and the incorporation of hair sheep, a recent practice, is likely driven by market demand for lamb and mutton in the central region around Mexico City. While all adopters of this practice let manure accumulate by corralling animals, only one third of them fertilize with it. Most of these smallholder producers also cultivate a *milpa*, but cannot bear the expense of commercial fertilizer. Parsons et al. (2011c) summarized, "*Farmers have only recently added sheep to their systems to increase household income, and opportunities may exist to develop greater complementarities between these two farming system components, particularly through manure use.*" Thus, a prime research objective was to evaluate the effectiveness of sheep manure fertilization rates combined with weed control in sustaining the productivity of *milpa* cultivation. A study of nutrient fluxes in the *milpa* system of Yucatán with continual maize cultivation and stover removal to feed animals showed that manuring with four metric tons of dry matter per hectare would sustain the soil stock of phosphorus, but not nitrogen or potassium, indicating threats to sustainability from lost fertility (Parsons et al., 2011c).

A companion study suggested that fertility losses and higher weed pressure were important causes of falling maize yields in *milpas*. Chemical control required much less labor than hand weeding, and fewer weeds mean greater maize production. Manure fertilization also increased grain and biomass yields. By third and fourth years of cultivation, high maize yields could be achieved only through a combination of manuring and weed control. "*Small sheep flocks could theoretically provide a sufficient quantity of manure to fertilize a milpa, potentially allowing fertility to be maintained beyond the normal two years. Technologies that increase yield and maintain plots for a longer period have the potential to change elements of the current milpa system. The success of such practices ultimately depends on livelihood needs and aspirations of the households and the communities in which they live*" (Parsons et al., 2009).

Mixed farming systems are enterprises where animal husbandry and crop cultivation are integrated components of one farming system: livestock are fed crop byproducts or residues (e.g., stubble) and significant income is earned by cropping. These systems provide many benefits to low-resource families and although smallholder households produce a large proportion of the food in the tropics, our understanding of the functioning of their farming systems is limited. To address the gap that he previously identified, Parsons et al. (2011a) developed an integrated crop-livestock model to assess biophysical and economic consequences of farming practices incorporated into sheep systems in Yucatán. The resulting dynamic model comprising stocks, flows and feedbacks integrates scientific and practical knowledge of management, flock dynamics, sheep production, nutrient partitioning, labor and economic components. It also accesses information about sheep performance (productivity and manure quantity) and cropping (weather, crop and soil dynamics) obtained from other simulation models.

Thus this simulation model embodies some of the complex interactions occurring between smallholder farmers, crops and livestock; it is a tool for examining selected suites of integrated crop-livestock practices compared to specialized cropping. Studies using this tool revealed that mixed farming scenarios with sheep provide more family income than specialized enterprises. This outcome capitalized on a lower on-farm price of maize grain, efficient utilization of surplus labor, and exploiting the availability of common land. However, more was not always better. It was most profitable to sell excess grain and maize stover, and instead of stover to use common land to feed livestock, thus warning that more integration may not always improve economic outcomes (Parsons et al., 2011b). This systems-oriented approach drew upon local knowledge, synthesizing it in a manner that added value. Humans often have a limited ability to predict outcomes in dynamically complex systems, such as agriculture, where short-term and long-term behaviors may differ.

2.3 Collective Action: Value-added Agricultural Products

Another project embodied a response to rural community interest to organize cooperatively to increase family incomes by accessing higher-value local markets. Communities like Micoxtla in the Veracruz highlands, where most inhabitants work in agriculture, confront multiple livelihood challenges. These include food insecurity, unemployment, and low and variable family incomes, which may be surmounted by the creation of income-generating opportunities. Value-added agricultural products are a potential strategy for earning higher incomes. However, biological and economic uncertainties often must be reduced, especially for this strategy to benefit smallholders. Households may be unable to enter or to compete in high-value agricultural product markets because of low access to market information, seasonal production shortfalls, inconsistent product quality, costly market access and poor infrastructure, all of which increase transaction costs (Holloway et al., 2000). Collective action may help overcome these barriers. Value-added products manufactured and marketed by farmer groups or cooperatives might reduce uncertainty by improving rural livelihoods through collective bargaining, smaller transaction costs, and higher average net incomes.

Most Micoxtla families struggle with seasonal food and economic insecurity. After meeting household needs, the principal sources of cash family income are sales of goat's milk, young goats for meat (*cabrito*), and eggs. Community members identified growing demand for specialty products for the tourist trade in the nearby city of Xico, including aged cheeses made from goat's milk, as one potential component of a rural development project assisted by INIFAP. The community wanted to explore this option to increase incomes, which would require initial funds beyond the capacity of individual families. Further risks from producing and marketing premium cheeses stem from dynamic biological, economic and social processes like weather patterns, market access, and available land to produce forages. Founding an agrarian cooperative supported with startup technical services and training by INIFAP could help reduce these risks.

Consequently, McRoberts (2009) (McRoberts et al., 2013) worked with the community and INIFAP advisors to assess the *ex ante* potential of cheese production and marketing through a dairy cooperative comprising 25 families. This assessment was enabled by participatory group action to develop a dynamic mathematical simulation tool. With caveats acknowledged the resulting analysis indicated that a cooperative has substantial potential to improve community incomes while controlling risk under a broad range of environmental and market conditions (McRoberts et al., 2013). Furthermore, this Micoxtla case supports De Schutter's (2011) admonishment to help foster community self-determination using participatory approaches, in this case through both identification and *ex ante* assessment of potential development interventions. Undertaken with a leading Mexican research and development institution, this case importantly demonstrates a methodological contribution to research and development programming. This approach could be applied more broadly to understand the potential behaviors over time resulting from proposed interventions, to determine their benefits and pitfalls, and to better inform decisions about potential investments by governments, donors, communities and families.

3. Intellectual Gains from Cultural Context

Although these projects cover a limited disciplinary footprint compared to the many needs that were identified, they clearly exemplify learning from cultural agents with efforts to return the favor. Collectively they respond to De Schutter's criteria by providing technical assistance, better understanding of food system function and with methodology and action plans to support local communities' escape from poverty and growth of social capital through collective action. In addition to research publications to inform global audiences, project results were shared with local communities and farmer organizations through our collaborators. On the occasion of an invited presentation (Note 11), Yucatán farmers generously expressed gratitude for the thought-provoking results about cattle system opportunities across the Gulf region. Thus, these agrarian research cases illustrate academe's role in transferring information from the community context to a broader public audience, creating discourse and analysis, and abetting social change along the way.

Our integrative curricular approach, incorporating both formal and communal knowledge producers, has co-evolved in ways that parallel the challenge laid down by Godenzzi (2006),

> "to communicate research results, discourse analysis, and critical reflections with the agents of that education so that these may enrich curricula and pedagogical interactions. In this way, these disciplines will be contributing to the formation of intercultural agents capable of reinventing our life in society."

Taking up this challenge, we submit that academe is cultural agency's natural partner. To better fulfill its social educator role academe must provide curricula and pedagogy in ways that echo, connect and create plural discourse among multiple dimensions and disciplines of society. Multiplicative effects may be gotten when students and faculty, key agents themselves and trainers of intercultural agents, learn first-hand by crossing borders. In so doing a more inclusive worldview about life in society is found through another lens, thus providing critical context for needs assessments.

4. Concluding Remarks

We have illustrated ways in which academe has employed intercultural agency to embrace a more inclusive worldview that helps to frame and effectively address the challenges of agrarian vulnerability and rural life. By allying with hosts from other walks of life who became key professors, students and faculty become co-learners and collaborators charged with social responsibility in delivering voice, knowledge and understanding to extended audiences. We contend that greater academic agency through more alliances like those demonstrated here, and the necessary education investment to foster them, is needed to achieve equity goals through effective community engagement and applied problem-solving. It also helps ensure that all can participate in public policy decisions, which is part of "reinventing our life in society", Godenzzi's challenge to education. Exposed to an enabling cultural landscape, one carrying messages about the substantive contexts surrounding technical intervention and implementation, students winnow and amplify them through their own engagements, lenses and reflections, finally delivering them through an egalitarian process to academe and society writ large.

Acknowledgements

We thank Emily Holley and Charles Nicholson for their helpful feedbacks on drafts of this article.

References

Absalon-Medina, V. A., Blake, R. W., Fox, D. G., Juárez-Lagunes, F. I., Nicholson, C. F., Canudas-Lara, E. G., & Rueda-Maldonado, B. L. (2012a). Limitations and potentials of dual-purpose cow herds in central coastal Veracruz, Mexico. *Tropical Animal Health and Production, 44*(6), 1131-1142. http://dx.doi.org/10.1007/s11250-011-0049-1

Absalon-Medina, V. A., Nicholson, C. F., Blake, R. W., Fox, D. G., Juárez-Lagunes, F. I., Canudas-Lara, E. G., & Rueda-Maldonado, B. L. (2012b). Economic analysis of alternative nutritional management of dual-purpose cow herds in central coastal Veracruz, Mexico. *Tropical Animal Health and Production, 44*(6), 1143-1150. http://dx.doi.org/ 10.1007/s11250-011-0050-8

Baba, K. (2007). *Analysis of productivity, nutritional constraints and management options in beef cattle systems of eastern Yucatan, Mexico: A case study of cow-calf productivity in the herds of Tizimin, Yucatan* (Unpublished master's thesis). Cornell University, Ithaca, NY, USA.

Blake, R. W. (2001). Tradition and transition: INTAG 602 and the graduate field of international agriculture and rural development. *International Agriculture 602 Millennium Conference on Agricultural Development in the 21st Century* (pp. 36-40). International Programs, College of Agriculture and Life Sciences: Cornell University.

Blake, R. W. (2008). Perspectivas de la investigación pecuaria en el mundo tropical: Utilización de recursos genéticos de ganado bovino. In C. V. Durán & R. Campos (Eds.), *Perspectivas de Conservación, Mejoramiento y Utilización de Recursos Genéticos Criollos y Colombianos en los Nuevos Escenarios del Mejoramiento Animal* (pp. 1-17). Valle, Colombia: Universidad Nacional de Colombia.

Blake, R. W., & Nicholson, C. F. (2004). Livestock, land use change, and environmental outcomes in the developing world. In Owen et al. (Eds.), *Responding to the Livestock Revolution—the role of globalization and implications for poverty alleviation* (pp. 133-153). United Kingdom: Nottingham University Press.

Cristóbal-Carballo, O. (2009). *Management of heifer growth in dual-purpose cattle systems in the low Huasteca region of Veracruz, Mexico* (Unpublished master's thesis). Cornell University, Ithaca, NY, USA.

De Schutter, O. (2011). *End of Mission to Mexico: Mexico requires a new strategy to overcome the twin challenges of "food poverty" and obesity, says UN food expert.* Office of the High Commissioner for Human Rights. Retrieved from http://www.ohchr.org/en/NewsEvents/pages/displaynews.aspx?NewsID=11173

Godenzzi, J. C. (2006). The discourses of diversity: Language, ethnicity and interculturality in Latin America. In D. Sommer (Ed.), *Cultural Agency in the Americas* (pp. 146-166). Durham, NC: Duke University Press.

Holloway, G., Nicholson, C. F., Delgado, C., Stall, S., & Ehui, S. (2000). Agroindustrialization through institutional innovation, transaction costs, cooperatives and milk-market development in the east-African highlands. *Agricultural Economics, 23*, 279-288. http://dx.doi.org/10.1111/j.1574-0862.2000.tb00279.x

McRoberts, K. (2009). *Rural development challenges: System dynamics ex ante decision support for agricultural initiatives in southern Mexico* (Unpublished master's thesis). Cornell University, Ithaca, NY, USA.

McRoberts, K. C., Nicholson, C. F., Blake, R. W., Tucker, T. W., & Díaz-Padilla, G. (2013). Group model building to assess rural dairy cooperative feasibility in south-central Mexico. *International Food and Agribusiness Management Review, 16*(3), 55-98.

Meredith, R. A. (2010). Acquiring cultural perceptions during study abroad: The influence of youthful associates. *Hispania, 93*(4), 686-702.

Parsons, D., Ketterings, Q. M., Cherney, J. H., Blake, R. W., Ramírez-Aviles, L., & Nicholson, C. F. (2011c). Effects of weed control and manure application on nutrient fluxes in the shifting cultivation milpa system of Yucatán. *Archives of Agronomy and Soil Science, 57*(3), 273-292. http://dx.doi.org/10.1080/03650340903307236

Parsons, D., Nicholson, C. F., Blake, R. W., Ketterings, Q. M., Ramírez-Aviles, L., Fox, D. G., … Cherney, J. H. (2011a). Development and evaluation of an integrated simulation model for assessing smallholder crop-livestock production in Yucatán, Mexico. *Agricultural Systems, 104*, 1-12. http://dx.doi.org/10.1016/j.agsy.2010.07.006

Parsons, D., Nicholson, C. F., Blake, R. W., Ketterings, Q. M., Ramírez-Aviles, L., Cherney, J. H., & Fox, D. G. (2011b). Application of a simulation model for assessing integration of smallholder shifting cultivation and sheep production in Yucatán, Mexico. *Agricultural Systems, 104*, 13-19. http://dx.doi.org/10.1016/j.agsy.2010.07.006

Parsons, D., Ramírez-Aviles, L., Cherney, J. H., Ketterings, Q. M., Blake, R. W., & Nicholson, C. F. (2009). Managing maize production in shifting cultivation milpa systems in Yucatán through weed control and manure application. *Agriculture Ecosystems and Environment, 133*(1-2), 123-134. http://dx.doi.org/10.1016/j.agee.2009.05.011

Sommer, D. (2006). Introduction: Wiggle room. In D. Sommer (Ed.), *Cultural Agency in the Americas* (pp. 1-28). Durham: Duke University Press.

Thurston, H. D. (2001). A living laboratory: Field study in agriculture and agricultural development. *International Agriculture 602 Millennium Conference on Agricultural Development in the 21st Century* (pp. 33-35). International Programs, College of Agriculture and Life Sciences, Cornell University.

Notes

Note 1. This cross-college course was entitled *Bridging Worlds: Rural and Urban Realities*. Retrieved from http://ip.cals.cornell.edu/courses/iard4010/

Note 2. *Experience Latin America.* Retrieved from http://ip.cals.cornell.edu/courses/iard6010/

Note 3. Practitioners of cultural agency, as defined by Sommer (2006), exploit "*a vehicle for agency through creative actions and reflections that influence collective change*". Other variants affecting collective action and change may include political agency and community agency.

Note 4. Chiapas field coordinators were Dr. Carlos Riqué Flores and Blanca Concepción (Conchita) Guzmán de Riqué. Host institutions in 2005-08 were the Universidad Autónoma de Yucatán, Universidad Veracruzana and INIFAP, also with student and faculty participation partially supported through a Training, Internships, Exchanges and Scholarships project funded by USAID-Mexico through Higher Education for Development (http://tiesmexico.cals.cornell.edu/).

Note 5. Vice Provost, University Outreach and International Programs, University of California, Davis. Dr. Lacy was the Director of Cornell Cooperative Extension and Associate Dean of the Colleges of Agriculture and Life Sciences, and Human Ecology at Cornell University, 1994-1998.

Note 6. Personal communication appearing in Blake (2001).

Note 7. Professor Dow's advice was part of the 1999 exhibition of Georgia O'Keefe's work at the Phillips Museum in Washington, D.C. Thereafter it was added to our course strategy materials.

Note 8. These organizations are widely known as a Grupo Ganadero de Validación y Transferencia de Tecnología (GGAVATT), or GGAVATT Génesis and GGAVATT Tepetzintla.

Note 9. Decision Support of Ruminant Livestock Systems. Retrieved from http://tiesmexico.cals.cornell.edu

Note 10. These interactions involve dietary energy balance, milk production and expected growth, and indirectly, their potential effects on herd reproduction.

Note 11. "*Limitaciones y Manejo Alternativo para la Producción de Carne en los Trópicos*" (Limitations and Alternative Management for Beef Production in the Tropics), invited presentation by R. W. Blake (2010) at *Día del Ganadero 2010* (Cattleman's Field Day 2010), INIFAP Sitio Experimental Tizimín, Tizimín, Yucatán, México.

8

Competitiveness in Michoacán: A Proposal for an International Positions in Agroindustrial Sector

Odette V. Delfin-Ortega[1] & Joel Bonales Valencia[1]

[1] Institute of Economic Research and Enterprise from Universidad Michoacana de San Nicolás de Hidalgo Morelia, Michoacán, México

Correspondence: Odette V. Delfin-Ortega, Institute of Economic Research and Enterprise, Universidad Michoacana de San Nicolas de Hidalgo, Morelia. Mich., México.

Abstract

The present research has as aim, to determine the ways in which are affected the quality, the price, the technological innovation, the environmental management, the market and the public agro industrial policies in the international competitiveness of the agro industrial sector of Michoacan. It was located 51 agroindustrial companies that are exporting. It was used as instrument of compilation of information a questionnaire composed of 80 items. Once the information was processed, it was determined the correlational analysis, linear regression and attempts at hypothesis. We can see with the results that the state is competitive in this sector and that the variables explain 97% of the competitiveness. The variable that determined the competitiveness with greater proportion was the technological innovation, so it requires public policies that strengthen this indicator and the sector can be even more competitive. In the other hand public polices in agro business got the lowest score and there is a gap in the implementation of programs and modernization in the sector that causes both nationally and internationally doesn´t be strong in the area of agribusiness Michoacan, so it must strengthen these policies from the national development plan.

Keywords: competitiviness, agroindustrial, Michoacan

1. Introduction

Agriculture is a vulnerable sector in Mexico and therefore in Michoacan state, so its development represents an economic and social balance. To the extent that quality, price, technological innovation, environmental management, market and public policy impact on the competitive development of the state; agribusiness firms will be more profitable and the industry is constantly growing. International markets every day become more demanding and regulations coupled with quality are a challenge for local supply (Padilla et al., 2010; Carvalho & Santos, 2006). Competitiveness is essential for economic growth. Productivity growth is a major element of sustained competitiveness and is largely linked to the adoption of new technologies or other innovations (Toming, 2010; OECD, 2011).

Market forces directly influence in export supply and the agricultural sector is one such example through technological development and has been able to improve their different phases of production and industrialization (Ball et al., 2006; Latruffe, 2010). Food and Agriculture Organization (FAO) says that "the industrialization of agriculture and agribusiness development is a common process that is creating an entirely new type of industrial sector" (FAO, 2014).

Michoacan State is the largest producers of fruits in the country with a production of 2 million tons per year (SAGARPA, 2009). But agribusiness has not developed in the same direction and we can see that Michoacan ranks 18 place with respect all the country, so it has to direct the attention to improve the development and encourage greater competitiveness (Agrointernet, 2009); aimed not only at national but also international markets. Therefore, this research has as main objective to determine how affect the quality, price, technological innovation, environmental management, market and public policies on the competitiveness of the agribusiness sector in Michoacan State. The results obtained can be used for decision making in the private and government sector (Pallares, 1998); therefore government programs will be aimed in that direction. This leads to increased business in the industrialized agricultural products in order to strengthen the sector (López & Castrillón, 2007).

The hypothesis is that "The quality, price, technological innovation, environmental management, market and agro-industrial policies are the main causes of the agribusiness exports of Michoacan state have increased international competitiveness".

2. Methods

The study of foreign trade has been initiated on the theoretical framework of the economy explained by Adam Smith, David Ricardo and John Stuart Mill. Even as the economic, political and social environment in which the current classical studies arise are very different from the reality of our days, the analysis of the classical approach is the foundation for understanding the theory and logic of post developments in international trade theory. This theory emerges as the liberal response to mercantilist restrictions against free trade, since its inception Smith shows that a small difference in cost may be enough to benefit from the exchange between countries; Ricardo Smith reinforces the idea of considering the absolute advantage as a special case of a less stringent notion, such as comparative advantage, to reaffirm the benefits of trade (Ricardo, 1971).

Ricardo and Mill idea is that international trade will result in a complete specialization in the production of goods in which they have comparative advantage from the others with some products. For agribusiness Michoacan, an agricultural state, has a specialization in the fruit and vegetable and it has a comparative advantage over other states of the republic; therefore at the time of trade between countries will get more profit if the exported products are more competitive than other countries that send the same or similar products (Dini, 2010).

Michael Porter is an author in which this research is based; considering as a starting point his "Competitive Advantage" (Porter, 2007) and assigning the value that a company is able to create for its applicants; as lower than those of its competitors or the equivalent provision of differentiated products, where income benefits outweigh the costs prices. In its brief states that the value is the amount that buyers are willing to pay for what the company provides. Then a company is profitable if the value that it gets from its sales is greater than its cost of production. The goal of every business is to create products where the value is greater than its costs. To analyze this process, he used what he called "value chain" or a series of activities from which arises a value.

Furthermore competitive strategy Michael Porter is the way that a company has to compete; through which the company objectives are developed and it is trying to achieve in a market; taking into account the policies needed to achieve them. So the competition is a benchmark which can lead us to success or failure, so for strategy, it is important that the company is related to its environment (Fernández, 2005).

2.1 Methodology

This research seeks to measure competitiveness for that, in a first step, a questionnaire is used as a tool to obtain the primary information sources. We obtained the variables from literature review, obtaining as a result the following variables:

Dependent Variable: competitiveness;

Independent Variables: quality, technological innovation, environmental management, price, market and public policy;

Universe: Currently there are 18,119 companies in Michoacán established in the Registration of Business Information System of Mexico; of which only 345 firms export;

Population: The population was obtained from the following sources: Business Information System of Mexico through the Ministry of Economy; Cexporta and Customs Broker Barrenechea. We obtained 51 export agribusiness companies in the perishable area in Michoacan state.

The Likert Scale was used for processing data with SPSS; this scale is a very useful tool as it is designed to measure attitudes. From the arithmetical point of view it is a summation scale and the score or measure of each person in the attitude in question is given by the sum of their responses to questions that was implemented.

For this study we work with 80 items, which were reviewed in detail and were developed according to each study variable.

It can get different kind of score for each variable between the maximum and minimum values as shown in the Table 1:

Table 1. Response values

Variable	Minimum value	Maximum value
Quality	10	50
Technological innovation	14	70
Environmental management	15	75
Price	10	50
Market	16	80
Agribusiness Public Policy	15	75
Competitiveness	80	400

Source: Own calculations based on the results of likert scale.

The next step was to verify the reliability of the instrument with Cronbach Alpha technic; where the following formula was used:

$$\alpha = \frac{\kappa}{\kappa-1}\left(1 - \left|\frac{\sum S_i^2}{S_x^2}\right|\right) \tag{1}$$

Where:

K: is the number of the items; $\sum S_i^2$: Is the sum of the variance of the items; S_x^2: It is the variance of the total score.

The survey results were processed using SPSS; giving as result:

Table 2. Cronbach's alpha result

Reliability	
Cronbach´s Alpha	No. of Items
0.951	80

Source: Results obtained on base SPSS.

This means that the degree of reliability of the test is high.

2.2 Quantitative Analysis Techniques

Once registered the information, it continues with the process of analysis and interpretation of data. The importance of them is that they are useful for organizing, describing and analyzing the data collected by the instruments of investigation tools. To test the hypothesis the ordinary least squares (OLS) method is used.

The basics of traditional econometrics and time series are composed of mathematical and statistical methods to estimate the dependence of an endogenous variable of other exogenous variable and quantification or association that may exist between them. Therefore application allows express and economic measures, and enables an economic theory expressed in mathematical form (single equation or multi-equation models), and verify statistical indicators (hypothesis testing) (Loria, 2007).

Within the mathematical methods applied to economics and business highlights the least squares under it with him to do regression analysis and correlation, both simple and multiple.

With the regression analysis, it can determine the dependence of a variable (Y) of another variable (X) when it is a single regression or several variables (X, Z, Q) when it is a multiple regression. In the first, Y, is called the dependent variable (also called endogenous, explained or returned); the second, (X), is known as an independent variable, explanatory, exogenous or regressor. With the correlation analysis, the relation, the association or degree in which X explains Y.

The dependence of X of Y is expressed mathematically as follows: $Y = f(X)$ and is called functional. The mathematical development and especially its application in computer currently available to the researcher many functional forms for regression analysis and correlation. Among the best known is that of the straight, the parabola, exponential, hyperbola, reciprocal, etc. (Loria, 2007).

The equation describing this model is:

$$Y_t = \beta_0 + \beta_1 X_{1t} + \cdots + \beta_i X_{it} + \beta_k X_{kt} + \varepsilon_t \qquad (T = 1, \ldots T) \qquad (2)$$

The Y_t variable is the dependent or endogenous variable, $X_{it,i}$, i = 1, ... K are the explanatory variables and ε_t is the random disturbance, c (K + 1) are the parameters associated with each of explanatory variables called the regression coefficient and measures the impact of each variable on the behavior of the endogenous variable. β_0 is the independent term and may be said to be the parameter associated with an explanatory variable that takes the value 1 for all observations and therefore does not appear explicitly in the equation.

The statistic test used for this hypothesis is the statistical "t" is calculated as the ratio estimator and its standard error and allows the hypothesis that the coefficient is equal to zero ($H_{0:}$ $\beta_i = 0$ and $H_1 : \beta_i \neq 0$) and therefore, determine whether the variable is individually significant to the dependent variable (Carrascal, González, & Rodriguez, 2001).

Now the method of least squares (Carrascal et al., 2001) is based on the fulfillment of the following assumptions or classical assumptions:

1. Implicit in the specification of the model equation is the linearity of the relationship and the constancy of the parameters.

2. There are no exact linear relationships among the explanatory variables or regressors, besides that these are not random variables.

3. There exact linearity between variables only when the independent variable is raised to the pwer of 1 (Gujarati, 1990) and x2 terms are excluded, among others. Of the two interpretations of the parameters linearity is the most important in the theory and regression means that the parameters are increased to the first power.

4. The random disturbances are variable (random or stochastic) independent and equally distributed normal with zero mean and some variance.

5. There is no autocorrelation (they are independent) to each random perturbations Ui.

6. All random perturbations have equal variance, in other words, there is homoscedasticity, the Ui.

7. There zero covariance (COV) between the explanatory variable (Xi) and the variable or random disturbance (Ui), in other words, uncorrelated so that their significance in the dependent variable (Yi) is separated and additive. When X and Ui are correlated (positively or negatively) is difficult to isolate the individual influence of Xi and Yi Ui.

On the significance observed in these cases where the correlation involved, it must be said that the quantification of the relationship between Y and X, regardless of whether they are qualitative or quantitative variables is done with statistics like the correlation coefficient R, the determination R2 and adjusted determination.

It is said that when R tends to one, there is a strong relationship or correlation between X and Y, such that X is a good explanatory variable Y. That is, we can explain the behavior of Y based on the behavior of X. In this case if X increases, Y also does.

3. Results

This research is based on the assumption that the dependent variable competitiveness is impacted by independent variables: quality, technological innovation, environmental management, pricing, market and public policy. The questionnaire consists of 80 items that include variables, indicators and dimensions, and together account for international competitiveness. The responses were weighted according to the Likert Scale (Fernandez, 2005), representing a very low competitiveness with value 1 and highly competitive with the value 5. It was begun with the dependent variable "International competitiveness", in Table 3 the results are observed

Table 3. Competitiviness descriptive statistics

Index	Value
Mean	273.01
Median	274
Mode	268
Standard Deviation	18.42
Variance	339.49
Asymmetry	-0.899
Kurtosis	1.161
Range	86
Minimum	217
Maximum	303
Sum	13924

Source: Based on data from field study.

Table 3 shows the values obtained from measurements of central trend and dispersion; the average is 273.01 and very close to this value is the median that had a value of 274 which is also located in the competitive range, however, the mode is located between two values: one is 268 (which is located in the box Regular competitiveness) and the other is 275 (which is in the competitive range). Standard deviation indicates how the values are far from the mean; in this case is 18.42. The distance is as follows: from the middle to both sides: (273.01 + 18.42) and (273.01 − 18.42) being the results from: 291.43 and 254.59 and is where the highest percentage of response was observed with a 80.39%; addition to the value obtained is 18.42 less than the value 64, which is the distance between each of the Likert response scale processed.

Furthermore the asymmetry coefficient indicates that if (g1 > 0) the curve is asymmetrically with negative values tend to gather more on the right side of the mean that is what happens in this case because it gave a result the value of -0899. Kurtosis determine the degree of concentration values presented in the central region of the distribution. In this case the result is 1.161 coefficient value is greater than "zero" (g2 > 0) so that in this case the distribution is leptokurtic.

The frequency distribution of the minimum response value was 217 and 303 having a maximum distance between them of 86 points; the values were below the mean, representing 45.09% of total values. The rest of the results, 54.91%, are the average upward, reaching the competitive picture. Therefore the overall result agribusiness export of Michoacan, according to the Likert scale is considered "competitive" as shown below:

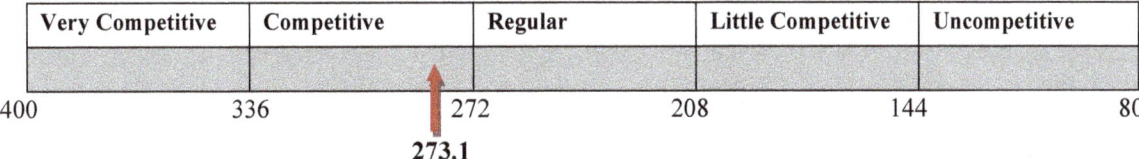

Very Competitive	Competitive	Regular	Little Competitive	Uncompetitive

400   336   272   208   144   80

273.1

3.1 Hypothesis Testing

To test hypotheses presented below so bivariate dependent variable "competitiveness" with each of the independent variables, the method used Least squares with Eviews program as shown in Table 4.

Dependent Variable: Competitiviness;

Method: Least Squares;

Sample: 1 51;

Included observations: 51.

Table 4. Agro-business competitiviness in Michoacan

Variable	Coefficient	Std. Error	t-Statistic	Prob.
C	0.618352	1.642228	0.376532	0.7083
Quality	0.118593	0.031893	3.718445	0.0006
Environmental management	0.176276	0.018131	9.722264	0.0000
Technological innovation	0.188143	0.021742	8.653331	0.0000
Market	0.153997	0.013658	11.27553	0.0000
Agribusiness Public Policy	0.153409	0.018187	8.435290	0.0000
Price	0.173260	0.029797	5.814588	0.0000
R-squared	0.971360	Mean dependentvar		45.05882
Adjusted R-squared	0.967454	S.D. dependentvar		3.081634
S.E. of regression	0.555940	Akaikeinfocriterion		1.790561
Sum squaredresid	13.59905	Schwarzcriterion		2.055714
Log likelihood	-38.65931	Hannan-Quinncriter.		1.891884
F-statistic	248.7168	Durbin-Watson stat		2.369547
Prob(F-statistic)	0.000000			

Source: Own calculations based on the results of the eviews program.

After running the model results are analyzed and the null hypothesis is rejected, because applying the t test, it is found that the values obtained $\neq 0$ ($H_1 : \beta_i \neq 0$) and the probability $p < 0.05$, $p = 0.0000$ being so there is enough statistical evidence to reject the null hypothesis H0, accepting the alternative hypothesis H1. However it is observed that the result is 0.97 square R by indicating that the variables explain 97% of the dependent variable competitiveness. Therefore the hypothesis is adopted based on the data presented.

The coefficient shown indicates the change that the endogenous variable (Y1); to a unit change in the explanatory variable (X1) while all other variables remain constant; this way if the quality increases one unit, competitiveness increases 0.11 units, in the case of environmental management increased by 0.17, with technological innovation increases 0.18; in the case of market and public policy increases 0.15.

4. Discussion

This research was conducted in the agro-export sector in which we sought to measure competitiveness at sector and subsector for each processed product export, as the competitiveness of a product depends on the constituent companies and their capacity to produce goods that meet the requirements of buyers, in addition to its ability to capture markets. This measurement helps to identify the factors that contribute to strengthening or weakening the competitiveness of agri-business industry.

The results of the data processing show a picture of how it is agro export sector in the state. The results of the Likert scale shows that the industry is competitive. Indeed the first instance Michoacan has fresh fruits and vegetables with a strong position in international markets that give direct entry into processed products online; however trends are focused markets increasingly demand products such as pasta, juices and purees, as was observed in this investigation, it is unfortunate there are small number of companies exporting processed products in this case were counted 51; and that results that are not adding value to exports of agricultural products from Michoacan as there are many crops that can be harvested. Also it is found in low numbers industrialization of the sector, since in 2012 the share of the primary sector was 24,607.5 million pesos vs 4171.6 million pesos in agribusiness.

Michael Porter theory was the fundamental part in achieving competitive advantage which is based on the technological and human capacity to produce more. It is observed in this research that technological innovation variable has a high correlation with competitiveness and more companies where implemented this factor in their enterprises, in order to have a higher competitive values. In other hand the profitability of exports of agro products is one of the objectives of the companies, however there are external factors that are not always possible to reach, these results as in the case of high input costs, speculation price target markets, non-tariff barriers that require changes in processes and procedures which require capital investments and available to meet them.

Michael Porter mentioned four factors that may be determinants of competitiveness: the introduction of a new product or an existing differentiation; the introduction of a new production process, opening new markets, changes in industrial organization. However in Michoacan it is not easy to implement because the companies don´t have money to develop these strategies so the government need to implement agro industries policies because the exporters require capital, training, access to research and technology, and the only way they can get them is through government programs.

Several studies have been conducted on measuring competitiveness as presented below:

Wijnands, Bremmers, van der Meulen, and Poppe (2008) made a measurement of competitiveness of food industry of Australia, Brazil, Canada, and the United States in 1996-2004. They used trade data of world market. The results showed that United States less level of competitiviness than Brazil. In this research, we made the analysis of Mexico and one of the variables were market and opposite occurred with Wijnands et al., this one was the most representative variable for the competitiveness in which we obtained the highest score.

Other research was made by Bojnec and Fertö (2009), who realized a measurement of international competitiviness of agribusiness sector in eight countries: Bulgaria, Croatia, Czech Republic, Hungry, Poland, Rumania, Slovakia, Slovenia in 1995-2007. They used trade data using four products: raw commodities, processed intermediates, horticulture and consumer-ready food. The results showed that raw commodities had export competitive advantage for all countries. In this research one of the products that was assessed, was the horticulture and like this research, this subsector was not as representative, specifically in our case was most impressive fruit sector.

Qineti, Rajcaniova, and Matejkova (2009) made a research where computed the competitiveness of agrifrut sector of Slovak, United States, Russia and Ukraine in the period 2002-2006. The main conclusions were that commodity group had comparative advantage in the United States and opposite occurred with Russia and Ukraine which had declined in this group. In this research we got competitiveness index like these authors and we observed that in the state there is a high level of competitiveness; however it was not included in the study a commodity sector just fruit and horticulture sectors.

Finally Van Berkum (2009) realized a competitiveness study of agrobusiness products exported of several european countries. He concluyed the Baltic countries and Poland had increased their export surpluses since the 1990s. In our research, the factor of export of agro products was key to measuring competitiveness and its position in the international markets as well as the study of these authors and it was increasing through the years like this research.

5. Conclusions

Michoacan currently requires a big push to grow as a state and be more competitive, the agricultural sector has been a central axis within the same economic boost, particularly in fruit and vegetable products, which is much currently positioned in international markets. The industrialization of agricultural products is a fundamental part of economic and social developments.

To determine the competitiveness of the agribusiness sector of Michoacan, it was used a questionnaire with 80 items and it was processed through the Likert scale, in order to obtain the degree of competitiveness. In a second stage the OLS technique was applied to test the hypothesis.

Furthermore, the results of the questionnaire using the Likert scale was observed that the agribusiness sector of Michoacan is competitive, since an average of 273 points being in the competitive range was obtained

The hypothesis presented at the beginning, was confirmed because the results obtained from the least squares technique showed that quality, technological innovation, environmental management, price, market and public policy determine international competitiveness of the agribusiness sector. Since the results obtained in the coefficient of determination r^2 was 0.97.

The variable that determined the competitiveness with greater proportion was the variable technological innovation, so they require public policies that strengthen this indicator and the sector can be even more competitive. There is a gap in the implementation of programs and modernization in the sector that causes both nationally and internationally, Michoacán isn't strong in the area of agribusiness only fresh products.

The government through agricultural policies, boosted production chains with the creation of the product system however although the purpose was to strengthen each link in the chain, strengthening occurs only at producing fresh, making reflect on the importance of resource programs that strengthen each of the activities specifically in promoting income new markets with higher levels of requirement on par with higher value-added products are developed.

References

Agrointernet. (2009). *Agronegocios-analisis Económicos.* Retreived from http://www.agrointernet.com/index.php/agronegocios/240-agronegocios-analisis-economico-.html

Ball, E., Butault, J. P., San Juan, M. C., & Mora, R. (2006). *Productivity and International Competitiveness of European Union and United States Agriculture, 1973-2002.* Paper presented at the AIEA2 International Meeting Competitiveness in agriculture and the food industry: United States and EU perspectives, Bologna.

Bojnec, S., & Latruffe, L. (2009). Determinants of technical efficiency of Slovenian farms. *Post-Communist Economies, 21*(1), 117-124. http://dx.doi.org/10.1080/14631370802663737

Carvalho, L., & Santos, R. (2006). Estudios sociedade de agricultura. *Programa de Apoio a núcleos de excelencia.* Brasil: Nucleo de Estudos Agrários e desenvolvimento.

Carrascal, A. U., González, Y., & Rodríguez, B. (2001). *Análisis Econométrico con Eviews.* RA-MA editorial. Madrid

Dini, M. (2010). *Competitividad, redes de empresas y cooperación empresarial* (72nd ed.). Santiago, Chile: Series CEPAL.

Food and Agriculture Organization of the United Nations (FAO). (2014). *Perspectivas de la agricultura y del desarrollo rural en las Américas: una mirada hacia América Latina y el Caribe 2014.* Retreived from http://www.fao.org/publications/card/es/c/209f117b-debf-5170-ac4a-9bf5ac1efabe

Fernández, I. (2005). *Construcción de escala aditiva tipo Likert.* España. Retreived from http://www.mtas.es/insht/ntp/ntp_015.htm

Fernández, G. C. (2005). *Qué significa la competitividdad en negocios internacionales.* Retreived from http://www.alafec.unam.mx/mem/cuba/Negocios_internacionales/negint05.swf

Latruffe, L. (2010). Competitiveness, Productivity and Efficiency in the Agricultural and Agri-Food Sectors. *OECD Food, Agriculture and Fisheries Papers* (No. 30). OECD Publishing. http://dx.doi.org/10.1787/5km91nkdt6d6-en

Loría, E. (2007). *Econometría con Aplicaciones.* Universidad Autónoma de México. Editorial Pearson Educación, México.

López, M. F., & Castrillón, J. (2007). In E. Gratuita (Ed.), *Teoría Económica y algunas experiencias latinoamericanas relativas a la agroindustria.* Retreived from http://www.eumed.net/libros/2007b/304/indice.htm

OECD. (2011). *Fostering Productivity and Competitiveness in Agriculture.* OECD publishing. http://dx.doi.org/10.1787/9789264166820-en

Padilla, B., Luz, R. R., Agustin, P. V. O., & Reyes-Rivas, E. (2010). Competitiveness of Zacatecas (Mexico) Protected Agriculture: The Fresh Tomato Industry. *International Food and Agribusiness Management Review, 13*(1).

Pallares, F. (1988). Las políticas públicas: El sistema político en acción. *Revista de estudios políticos (nueva época), 62,* 141.

Porter, M. (2007). *Competitive Strategy: Techniques for Analyzing Industries and Competitors.* New York: Free Press.

Qineti, A., Rajcaniova, M., & Matejkova, E. (2009), The competitiveness and comparative advantage of the Slovak and the EU agri-food trade with Russia and Ukraine. *Agricultural Economics, 55*(8), 375-383.

Ricardo, D. (1971). *The principles of political economy and taxation (1817).* Bltimore: Penguin.

Secretaría, de A., Ganadería, D. R., & Pesca, A. (2009). *Monitor agroeconómico 2009 del estado de Michoacán.* Retreived from http://www.sagarpa.gob.mx/agronegocios/Estadisticas/Documents/MICHOACAN.pdf?Mobile=1&Source=%2Fagronegocios%2FEstadisticas%2F_layouts%2Fmobile%2Fview.aspx%3FList%3Dea4191c6-15b5-4625-afe9-be7e6cce2216%26View%3Df5c8d175-3fb9-49f2-86e6-c9db05b29bfb%26Curren

Toming, K. (2007). The impact of EU accession on the export competitiveness of Estonian food processing industry. *Post-Communist Economies, 19*(2), 187-207. http://dx.doi.org/10.1080/14631370701312170

Wijnands, J., Bremmers, H., van der Meulen, B., & Poppe, K. (2008), An economic and legal assessment of the EU food industry's competitiveness. *Agribusiness, 24*(4), 417-439. http://dx.doi.org/10.1002/agr.20167

9

Characterization of Oil Yield and Quality in Shatter-Resistant Dwarf Sesame Produced in Virginia, USA

Harbans L. Bhardwaj[1], Anwar A. Hamama[1], Mark E. Kraemer[1] & D. Ray Langham[2]

[1] Agricultural Research Station, Virginia State University, Petersburg, Virginia, USA

[2] Sesaco Corporation, San Antonio, Texas, USA

Correspondence: Harbans L. Bhardwaj, Agricultural Research Station, Virginia State University, Petersburg, Virginia 23806, USA. E-mail: hbhardwj@vsu.edu

Abstract

Sesame has potential as an alternative crop for former tobacco farmers in eastern USA to increase agricultural diversification and enhance farm incomes. Oil yield and quality of five shatter-resistant and dwarf sesame cultivars when grown using rows 37.5 cm or 75 cm apart were evaluated. Sesame was planted on 23 May and 8 Jun. (PT1) during 2011 and on 17 and 9 Jul. (PT2) during 2012. Early planting (Late May to early June) resulted in 716.6 kg oil/ha as compared to 479.6 kg oil/ha from late planting (early June to mid-July). The closer row spacing of 37.5 cm out-yielded the wider row spacing of 75 cm by about 34% for oil yield. Early planting increased the contents of C16:0, C20:0, C18:1, and C20:1 fatty acids whereas late planting increased the contents of C18:2 and C18:3 fatty acids. Contents of total saturated fatty acids and mono-unsaturated fatty acids were also greater after early planting as compared to those after late plantings. Closer row spacing of 37.5 cm resulted in significantly higher contents of C20:1 and saturated fatty acids in the oil as compared to the wider row spacing of 75 cm. Sesame seed produced in Virginia contained 6.8% more oil than that produced in Texas (45. 5 vs. 42.6%, respectively). The results indicated that sesame could be produced in the mid-Atlantic region of USA.

Keywords: *Sesamum indicum* L., fatty acids, oil yield, planting dates, row spacings, new crops

1. Introduction

Sesame (*Sesamum indicum* L. Pedaliaceae) is one of the oldest crops known to humans. There are archeological remnants of sesame dating to 5,500 BP in the Harappa valley in the Indian subcontinent (Bedigian & Harlan, 1986). Sesame was a major oilseed in the ancient world because of its ease of extraction, its great stability, and its drought resistance. Through the ages, sesame seeds have been a source of food and oil, about 65% of the annual sesame crop is processed into oil and the remainder is used in food (AgMRC, 2013). In the United States, sesame seed production has been limited to the Southwest, primarily due to the lack of mechanically harvestable cultivars suited to other climates. Almost all commercial production is in Texas and Oklahoma. In 2007, sesame was grown commercially on 29 farms, 25 of them were in Texas and the remaining four were in Oklahoma. All together, U.S. sesame farms produced over 2.9 million pounds of sesame during 2007, on about 4,978 acre (in contrast, less than 2,500 acres were planted to sesame in 1987). The pounds of sesame have doubled since 2002, with the addition of only seven farms. Most U.S. sesame is grown on contract (USDA, 2009).

Sesame was introduced to the US from Africa and was called beni/benne/benni. Betts (1999) quotes letters from Thomas Jefferson that document his trials with sesame between 1808 and 1824. Jefferson stated that sesame "... *is among the most valuable acquisitions our country has ever made. ... I do not believe before that there existed so perfect a substitute for olive oil*" (Langham & Wiemers, 2002). In the 1940s, research was started in South Carolina (Martin, 1949), Nebraska (Hoffman & Claassen, 1949), and Texas (Kalton, 1949). A major issue with sesame was its' tendency to shatter. In 1978, Sesaco, Inc. was established to bring D.G. Langham's work from Venezuela to the US. However, it became clear that in the US complete mechanization would be necessary. By 1982, the first mechanized cultivars were released (Langham & Wiemers, 2002). In 2010, Sesaco had more than 100,000 acres of sesame under contract production with growers in southwest Kansas, Oklahoma and central Texas. This production has been facilitated by availability of dwarf, shatter-resistant sesame cultivars.

International demand for sesame continues to increase every year. The world's traded sesame seed recently surpassed one million tons per year and was valued at approximately $850 million. In the last 15 years, world trade in sesame has increased by nearly 80 percent. The United States imports more sesame than it grows. In 2010, the United States imported sesame seed valued at $69.9 million which was relatively unchanged from 2009 (AgMRC, 2013). This observation indicates that there is potential for enhancing sesame production in the United States.

Due to loss of tobacco and other crops such as peanut (*Arachis hypogea* L.), the New Crops Program of Virginia State University has been evaluating several summer crops that can be grown in Virginia to increase agricultural diversification and enhance farm incomes. In this regard, sesame (*Sesamum indicum* L. Pedaliaceae) is considered a potentially lucrative crop due to its use in foods including the production of hummus. Goals of the current research effort were to determine if sesame could be successfully produced in Virginia. Specifically, the objectives were evaluation of sesame cultivars for their oil yields and characterization of the effects of planting date and row spacing on the oil and fatty acid content in sesame seed.

2. Materials and Methods

2.1 Plant Material

Five proprietary sesame cultivars (22K, S26, S28, S30, and S32) from Sesaco Corporation (San Antonio, Texas, USA) were used in this study. These cultivars are dwarf and shatter-resistant and are intellectual property of Sesaco, Inc. (San Antonio, Texas, USA). These were the only shatter-resistant cultivars available to us for our studies. We were interested only in shatter-resistant cultivars because these cultivars can be mechanically harvested as compared to traditional cultivars that shatter and need to be harvested manually.

2.2 Plant Production

This field study was conducted at Randolph Farm of Virginia State University located in Ettrick, Virginia. The five cultivars were planted in four-row plots with row spacings of 37.5 and 75 cm in an Abel Sandy Loam soil. Sesame was planted on 23 May and 8 Jun. (PT1) during 2011 and on 17 and 9 Jul. (PT2) during 2012 using a split-plot experimental design with planting dates as main plots, row spacings as sub-plots and cultivars as sub-sub-plots with four replications of a Randomized Complete Block Design. The experimental field received a pre-plant incorporated application of approximately 2 liters of Trifluralin herbicide per hectare. The experimental area received a fertilizer application of 112 kg per hectare of N, P, and K. All plots were harvested manually at maturity, generally in November to December of each year.

2.3 Oil extraction, Preparation and Analysis of FAME

The oil was extracted from 1 g of ground seed at room temperature by homogenization for 2 min in 10 mL hexane/isopropanol (3:2, v/v) with a Biospec Model 985-370 Tissue Homogenizer (Biospec Products, Inc. Racine, WI, USA) and centrifuged at 4000 g for 5 min, as described by Hamama et al. (2005). The oil extraction was repeated for each sample for three times to ensure full oil recovery and the three extractions were combined. The hexane-lipid layer was washed and separated from the combined extract by shaking and centrifugation with 10 mL of 1% $CaCl_2$ and 1% NaCl in 50% methanol. The washing procedure was repeated and the purified lipid layer was removed by aspiration and dried over anhydrous Na_2SO_4. The oil was stored under nitrogen at -10 °C until analysis.

Fatty Acid Methyl Esters (FAME) were prepared by an acid-catalyzed transestrification method as described by Hamama et al. (2005). The oil samples (5 mg) were vortexed with 2 mL sulfuric acid/methanol (1:99, v/v) in 10-mL glass vials containing a Teflon boiling chip. The open vials were placed in a heating block at 90 °C until the sample volume was reduced to 0.5 mL. After cooling to room temperature, 1 mL of hexane, followed by 1 mL of distilled water was added. The mixture was vortexed and the upper hexane layer containing the FAME was taken and dried over anhydrous Na_2SO_4. The hexane phase containing FAME was transferred to a suitable vial and kept under N_2 at 0 °C for gas chromatographic analysis.

Analyses of FAME were carried out as described by Hamama et al. (2005) using a SupelcoWax 10 capillary column (25 m × 0.25 mm i.d. and 0.25 mm film thickness, SupelcoWax, Inc., Bellefonte, PA, USA) in a Varian model Vista 6000 GC equipped with a Flame Ionization Detector (FID) (Varian, Sugar Land, TX, USA). An SP-4290 Integrator (Spectra Physics, San Jose, CA, USA) was used to determine relative concentrations of the detected fatty acids. Peaks were identified by reference to the retention of FAME standards and quantified by the aid of heptadecanoic acid (17:0) as an internal standard. The concentration of each fatty acid was calculated as a percentage (w/w) of the total fatty acids.

2.4 Statistical Analysis

All data were analyzed using the Analysis of Variance procedures in version 9.3 of SAS (SAS, 2013) significance of mean squares was tested at 1 and 5% levels. The means were compared using Fisher's Protected Least Significant Difference at a 5% level of significance.

3. Results and Discussion

3.1 Cultivars, Planting Dates, and Row Spacing Effects on Oil Yield

Oil yields were significantly affected by planting dates and row spacings (Table 1). Early planting (late May to early June) resulted in 716.6 kg oil/ha as compared to 479.6 kg oil/ha from late planting (early June to mid-July), an increase of about 1.5 times (Table 2). Differences among five sesame cultivars for oil yield were not significant (Table 3). The closer row spacing of 37.5 cm out-yielded the wider row spacing of 75 cm by about 34 % for oil yield (Table 4).

Table 1. Partial analysis of variance (Mean squares) for oil yield and contents of fatty acids in five sesame cultivars grown using two planting dates and two row spacings in Virginia during 2011 and 2012

Variable	Planting dates (PD)	Row spacings (RS)	Cultivars (C)	PD x RS	PD x C	RS x C	R^2%	CV%	Mean
Oil yield (kg/ha)	*	**	ns	**	*	*	87.7	31.6	598.1
C16:0[x]	**	ns	*	ns	ns	ns	98.4	0.97	8.8
C18:0[x]	ns	ns	ns	**	ns	*	98.9	0.68	5.2
C20:0[x]	**	*	ns	**	ns	ns	94.3	2.02	0.57
C22:0[x]	ns	*	ns	**	ns	ns	94.9	20.83	0.11
C16:1[x]	ns	ns	ns	**	*	ns	93.2	9.86	0.11
C18:1[x]	*	ns	*	**	ns	ns	97.7	1.70	38.3
C18:2[x]	*	ns	**	ns	ns	ns	99.2	0.91	46.2
C18:3[x]	*	ns	ns	ns	ns	ns	88.7	9.21	0.39
C20:1[x]	*	**	ns	**	ns	ns	98.7	9.15	0.17
SFA[y]	**	**	**	*	ns	ns	98.8	0.60	14.7
UFA[y]	*	ns	*	**	ns	ns	90.1	0.62	85.3
MUFA[y]	*	ns	*	**	ns	ns	97.7	1.69	38.6
PUFA[y]	*	ns	**	ns	ns	ns	99.3	0.90	46.6
RATIO[y]	ns	ns	ns	ns	ns	ns	86.3	9.04	119:1

Note. *, **: Mean squares significantly different from residual mean squares at 5 and 1 percent levels, respectively.

x: Fatty acids as percentage of total fatty acids.

y: SFA = Saturated fatty acids; UFA = Unsaturated fatty acids, MUFA = Mono-unsaturated fatty acids; PUFA = Poly unsaturated fatty acids; Ratio = C18:2 / C18:3.

3.2 Cultivars, Planting Dates, and Row Spacing Effects on Fatty Acids

This study revealed that planting dates significantly effected contents of several fatty acids sesame oil produced in Virginia but did not effect contents of C18:0, C22:0, and C16:1 fatty acids (Table 1). Early planting resulted in increased contents of C16:0, C20:0, C18:1, and C20:1 fatty acids whereas late planting resulted in increased contents of C18:2 and C18:3 fatty acids. Contents of total saturated fatty acids and mono-unsaturated fatty acids were also greater after early planting as compared to those after late plantings (Table 2). Thus, early planting resulted in a 2.2% increase in content of C16:0, 1.7% increase in the content of C20:0, 7.0% increase in the content of C18:1, 26.7% increase in content of C20:1, and 7.0% increase in the content of mono-unsaturated fatty acids. Conversely, late planting resulted in increased contents of C18:2 (+7.6%), C18:3 (+10.8%), and poly-unsaturated fatty acids (+7.8%) over early planting. Content of omega-3 fatty acid, which is considered healthy for human nutrition (Rennie et al., 2003; Freeman, 2000; Holm et al., 2001), in sesame oil was only around 0.4% as compared to other oilseeds such as canola (Approximately 8%). However, increased concentrations of poly-unsaturated fatty acids, such as C18:3) is known to cause oxidative instability in oils Nardini et al., 1995) which necessitates the use of an anti-oxidant such as vitamin E (Gil, 2002).

Table 2. Effects of planting dates on contents of oil yield and fatty acids in five sesame cultivars grown using two row spacings in Virginia during 2011 and 2012

Variable	Early planting (PT1)	Late planting (PT2)
Oil yield (Kg/ha)	716.6 a*	479.6 b
C16:0[x]	8.90 a	8.7 b
C18:0[x]	5.30 a	5.20 a
C20:0[x]	0.57 a	0.56 b
C22:0[x]	0.11 a	0.10 b
C16:1[x]	0.12 a	0.11 a
C18:1[x]	39.6 a	37.0 b
C18:2[x]	44.5 b	47.9 a
C18:3[x]	0.37 b	0.41 a
C20:1[x]	0.19 a	0.15 b
SFA[y]	14.9 a	14.5 b
UFA[y]	84.8 b	85.6 a
MUFA[y]	39.9 a	37.3 b
PUFA[y]	44.8 b	48.3 a
RATIO[y]	121:1 a	117:1 a

Note. *: Means followed by similar letters within rows were not different according to Least Significant Difference test at 5 percent level of significance.

x: Fatty acids as percentage of total fatty acids.

y: SFA = Saturated fatty acids; UFA = Unsaturated fatty acids; MUFA = Mono-unsaturated fatty acids; PUFA = Poly unsaturated fatty acids; Ratio = C18:2 / C18:3.

In general, contents of various fatty acids in the oil from five cultivars didn't differ significantly perhaps due to limited number of cultivars included in these studies. We expect that evaluations conducted with larger number of cultivars would indicate cultivar differences for oil yield and quality. However, oil of 22K and S32 cultivars contained significantly higher contents of C18:2 and poly-unsaturated fatty acids as compared to the oil of S26 and S28 cultivars (Table 3).

Table 3. Cultivar differences for contents of oil yield and fatty acids in sesame produced using two row spacings and two planting dates in Virginia during 2011 and 2012

Variable	22K	S26	S28	S30	S32
Oil yield (Kg/ha)	534.4 a	577.6 a	543.0 a	669.5 a	666.0 a
C16:0^x	8.75 bc	8.87 ab	8.95 a	8.67 c	8.72 bc
C18:0^x	5.25 a	5.29 a	5.29 a	5.23 a	5.19 a
C20:0^x	0.57 a	0.57 a	0.57 a	0.57 a	0.56 a
C22:0^x	0.10 a	0.11 a	0.10 a	0.11 a	0.11 a
C16:1^x	0.11 a	0.12 a	0.12 a	0.11 a	0.12 a
C18:1^x	38.1 b	38.81 a	38.3 b	38.2 b	38.2 b
C18:2^x	46.6 a	45.6 c	46.0 bc	46.4 ab	46.5 a
C20:1^x	0.17 a	0.17 a	0.17 a	0.17 a	0.18 a
SFA^y	14.7 b	14.8 a	14.9 a	14.6 b	14.6 b
UFA^y	85.3 a	85.1 ab	84.9 b	85.2 ab	85.4 a
MUFA^y	38.3 b	39.1 a	38.6 b	38.4 b	38.5 b
PUFA^y	47.0 a	46.0 c	46.3 bc	46.7 ab	46.9 a
RATIO^y	113:1 a	119:1 a	121:1 a	123:1 a	119:1 a

Note. *: Means followed by similar letters within rows were not different according to Least Significant Difference test at 5 percent level of significance.

x: Fatty acids as percentage of total fatty acids.

y: SFA = Saturated fatty acids; UFA = Unsaturated fatty acids; MUFA = Mono-unsaturated fatty acids; PUFA = Poly unsaturated fatty acids; Ratio = C18:2 / C18:3.

It was observed that closer row spacing of 37.5 cm resulted in significantly higher contents of C20:1 and saturated fatty acids in the oil as compared to the wider row spacing of 75 cm. However, the wider row spacing resulted in a significant increase in oil yield (Table 4). In general, effects of closer and wider row spacings on sesame oil quality were not that pronounced. This indicates that sesame can be produced in Virginia by using either of these row spacings. The objectives were to identify optimal row spacing for sesame production in Virginia because farmers, generally, resist purchase of new equipment when initially evaluating production of new, non-traditional crops. Most Virginia farmers own soybean planters either configured for 37.5 cm or 75 cm rows and our results indicate that sesame in Virginia could be planted with existing planters. Given that amount of seed for planting is small, it might be desirable to use 37.5 cm rows because of better weed control due to shading. Moreover, if sesame is to be produced organically using cultivation for weed control then 75 cm rows might be desirable.

Table 4. Row spacing effects on contents of oil yield and fatty acids in five sesame cultivars produced using two planting dates in Virginia during 2011 and 2012

Variable	37.5 cm	75 cm
Oil yield (kg/ha)	684.4 a	511.9 b
C16:0[x]	8.8 a	8.8 a
C18:0[x]	5.25 a	5.25 a
C20:0[x]	0.57 a	0.57 a
C22:0[x]	0.11 a	0.10 a
C16:1[x]	0.11 a	0.12 a
C18:1[x]	38.1 a	38.5 a
C18:2[x]	46.3 a	46.1 a
C18:3[x]	0.39 a	0.39 a
C20:1[x]	0.19 a	0.15 b
SFA[y]	14.7 a	14.7 a
UFA[y]	85.1 a	85.1 a
MUFA[y]	38.4 a	38.8 a
PUFA[y]	46.7 a	46.5 a
RATIO[y]	119:1 a	119:1 a

Note. *: Means followed by similar letters within rows were not different according to Least Significant Difference test at 5 percent level of significance.

x: Fatty acids as percentage of total fatty acids.

y: SFA = Saturated fatty acids; UFA = Unsaturated fatty acids; MUFA = Mono-unsaturated fatty acids; PUFA = Poly unsaturated fatty acids; Ratio = C18:2 / C18:3.

3.3 Cultivars, Planting Dates, and Row Spacing Effects on Ratio of C18:2 and C18:3 Fatty Acids

We also analyzed the oil of sesame produced in Virginia for ratio of C18:2 and C18:3 fatty acids. The ratio of n-6 and n-3 fatty acids is important for human health and should be in the range of 1:1 to 4:1 whereas in the Western diet it is now estimated to be 10:1 to 30:1 (Raper & Cronin, 1992; Watkins, 2004). Flaxseed oil has an omega-6:omega-3 ratio of 0.3:1. As intake of n-6 fatty acids increased in developed countries, consumption of foods rich in n-3 fatty acids steadily declined. Omega-3 fatty acids constitute a minuscule portion (< 1%) of the total fatty acids in U.S. food supply whereas the omega-6 fatty acids (Linoleic acid, 18:2) constitute a significant majority. Watkins (2004) summarized the information relative to n-3 and n-6 fatty acids and concluded, without suggesting that insufficiency of n-3 fatty acids in the diet may be the only cause of increased incidence of many human disorders, that insufficiency n-3 fatty acids in Western diets may be at the root cause of many behavioral, learning, memory, and neurological human disorders.

The ratio of C18:2 to C18:3 fatty acids in oil of sesame seeds produced in Virginia averaged 119:1 which is considerably different than the desired ratio from 1:1 to 4:1. In this study, the ratio of C18:2 to C18:3 fatty acids was not affected by planting dates, row spacings, or cultivars. We do not feel comfortable drawing any substantial conclusions based on the results related to the ratio between C18:2 and C18:3 fatty acids in sesame oil but present the results for use by future investigators.

3.4 Comparison of Seed Produced in Virginia and Texas

A comparison of concentrations of oil and fatty acids indicated significant differences between seed produced in Virginia and Texas (Data not presented). Oil content in sesame produced in Virginia was 6.8% higher than that produced in Texas (45.5 vs. 42.6%, respectively). With regards to individual fatty acids, seed produced in Virginia had significantly higher contents of C18:2 (45.7 vs. 41.5%), C18:3 (0.40 vs. 0.36%), and C22:0 (0.134 vs. 0.128%) over seed produced in Texas whereas seed produced in Texas had significantly higher contents of C16:0 (9.3 vs. 8.8%), C16:1 (0.13 vs. 0.11%), C18:0 (5.5 vs. 5.2%), C18:1 (42.2 vs. 38.9%), and C20:0 (0.60 vs. 0.58%) fatty acids over seed produced in Virginia.

We attribute these differences to soil and agro-climatic conditions in Virginia and Texas. Temperature data for Uvalde, Texas where seeds of cultivars included in this study were produced during 2010 indicates that this location was hotter than Petersburg, Virginia location where the five cultivars were evaluated during 2011 (Mean temperatures (°F) of 78, 85, 84, 87, 80, and 72 at Uvalde, Texas during May, June, July, August, September, and October, respectively during 2010 as compared to mean temperatures of 67, 77, 81, 80, 78, and 61 at Petersburg, Virginia during May, June, July, August, September, and October, respectively during 2011. The soil at Uvalde, Texas location is Uvalde silty clay loam (Fine-silty, mixed, hyperthermic Aridic Calciustoll) with a pH 8.2 whereas the soil at Petersburg, Virginia location is Abel sandy loam (Fine loamy mixed thermic Aquatic Hapludult) with a pH of 6.2. These data indicate that cooler temperature during growing season and sandy soils might be beneficial for better sesame oil yield and quality.

4. Conclusions

Overall, the results indicate that Virginia might be a better location for sesame production than Texas for higher oil content and higher content of linolenic acid (C18:3) which is considered desirable for human nutrition. The crop, in general, was identified to be healthy. We didn't observe any insects-pest or disease problems in sesame plots and suggest that sesame could be developed as an alternative summer cash crop for former tobacco farmers in Virginia and elsewhere. It is recommended that sesame be produced by planting early in the summer using a closer row spacing. Our studies show, for the first time, that sesame could be produced in the mid-Atlantic region of USA.

Acknowledgements

Financial support from Virginia Tobacco Commission (Virginia Tobacco Indemnification and Community Revitalization Commission) is gratefully acknowledged).

References

AgMRC. (2013). *Sesame Profile.* Ag Marketing Resource Center. Iowa State University, Ames, IA. Retrieved June 19, 2014, from http://www.agmrc.org/commodities__products/grains__oilseeds/sesame-profile/

Bedigian, D., & Harlan, J. R. (1986). Evidence for cultivation of sesame in the ancient world. *Econ. Bot., 40,* 137-154. Retrieved June 19, 2014, from http://link.springer.com/article/10.1007/BF02859136#

Betts, E. M. (1999). *Thomas Jefferson's garden book (1766–1824).* Thomas Jefferson Memorial Foundation, Inc., Charlottesville, VA. Retrieved June 19, 2014, from http://www.amazon.com/Thomas-Jeffersons-Garden-Book-1766-1824/dp/1432588893

Freeman, M. P. (2000). Omega-3 fatty acids in psychiatry: A Review. *Annals of Clinical Psychiatry, 12,* 159-165. Retrieved June 19, 2014, from http://www.ncbi.nlm.nih.gov/pubmed/10984006

Gil, A. (2002). Polyunsaturated fatty acids and inflammatory diseases. *Biomedicine and Pharmacotherapy, 56,* 388-396. Retrieved June 19, 2014, from http://www.ncbi.nlm.nih.gov/pubmed/12442911

Hamama, A. A., Bhardwaj, H. L., & Starner, D. E. (2003). Genotype and Growing Location effects on Phytosterols in Canola Oil. *J. Am. Oil Chem. Soc., 80,* 121-1126. http://dx.doi.org/10.1007/s11746-003-0829-3

Hoffman, A., & Claassen, C. E. (1949). *Sesame research in progress at the University of Nebraska* (pp. 36-37). I Proc. First Int. Sesame Conf., Clemson Agricultural College, Clemson, SC. Retrieved from http://www.hort.purdue.edu/newcrop/ncnu02/v5-157.html

Holm, T., Andreassen, A. K., Aukrust, P., Andersen, K., Geiran, O. R., Kjekshus, J., … Gullestad, L. (2001). Omega-3 fatty acids improve blood pressure control and preserve renal function in hypertensive heart transplant recipients. *European Heart Journal, 22,* 428-436. http://dx.doi.org/10.1053/euhj.2000.2369

Kalton, R. (1949). *Sesame, a promising new oilseed crop for Texas* (pp. 62-66). Proc. First Int. Sesame Conference, Clemson Agricultural College, Clemson, SC. Retrieved from http://www.hort.purdue.edu/newcrop/ncnu02/v5-157.html

Langham, D. R., & Wiemers, T. (2002). Progress in mechanizing sesame in the US through breeding. In J. Janick & A. Whipkey (Eds.), *Trends in New Crops and New Uses* (pp. 157-173). ASHS Press, Alexandria, VA. Retrieved June 19, 2014, from http://www.hort.purdue.edu/newcrop/ncnu02/v5-157.html

Martin, J. A. (1949). *Improvement of sesame in South Carolina* (pp. 71-73). Proc. First Int. Sesame Conf., Clemson Agricultural College, Clemson, SC. Retrieved from http://www.hort.purdue.edu/newcrop/ncnu02/v5-157.html

Nardini, M., DAquino, M. D., Gentili, V., DiFelice, M., & Scaccini, C. (1995). Dietary fish oil enhances plasma and LDL oxidative modification in rats. *Journal of Nutritional Biochemistry, 6*, 474-480. http://dx.doi.org/10.1016/0955-2863(95)00081-A

Raper, N. R., Cronin, F. J., & Exler, J. (1992). Omega-3 fatty acid content of the US food supply. *J. Am. Coll. Nutr., 11*, 304-308. http://dx.doi.org/10.1080/07315724.1992.10718231

Rennie, K. L., Hughes, J., Lang, R., & Jebb, S. A. (2003). Nutritional management of rheumatoid arthritis: A review of the evidence. *Journal of Human Nutrition and Dietetics, 16*, 97-109. http://dx.doi.org/10.1046/j.1365-277X.2003.00423.x

SAS. (2013). SAS System for Windows. SAS Institute, Inc., Cary, NC. Retrieved from http://www.sas.com/en_us/software/sas9.html

USDA. (2009). *Census of Agriculture*. National Agricultural Statistical Service (NASS), US Department of Agriculture, Washington, D.C.

Watkins, C. (2004). Fundamental fats. *INFORM, 15*, 638-640. Retrieved June 19, 2014, from http://aocs.files.cms-plus.com/inform/2004/10/fundamental.pdf

10

Influence of Dietary Supplementation of Coated Sodium Butyrate and/or Synbiotic on Growth Performances, Caecal Fermentation, Intestinal Morphometry and Metabolic Profile of Growing Rabbits

A. Hassanin[1], M. A. Tony[2], F. A. R. Sawiress[3], M. A. Abdl-Rahman[3] & Sohair Y. Saleh[3]

[1] Department of Cytology and Histology, Faculty of Veterinary Medicine, University of Sadat City, Meonofya, Egypt

[2] Department of Nutrition and Clinical Nutrition, Faculty of Veterinary Medicine, Cairo University, Giza, Egypt

[3] Department of Physiology, Faculty of Veterinary Medicine, Cairo University, Giza, Egypt

Correspondence: Sohair Y. Saleh, Department of Physiology, Faculty of Veterinary Medicine, Cairo University, Giza-12211, Egypt. E-mail: sohair_saleh@hotmail.com

Abstract

The aim of the present experiment was to study the synergistic effects of dietary supplementation with coated slow released sodium butyrate (CM3000®) and a commercial synbiotic (Poultry-Star®) on the productive performance and intestinal morphometry of the growing rabbits. Thirty two apparently healthy male New Zealand rabbits with average body weight of 544 ± 9 g were divided randomly into four dietary treatments at weaning (28th day of age). The control group (C) was fed on standard basal diet with no supplementation. Rabbits in the second group (T1) received the same basal diet supplemented with CM3000® 500 g/ton feed. Animals in the third group (T2) consumed the basal diet containing Poultry-Star® 500 g/ton feed. Rabbits in the fourth group (T3) were fed on the basal diet enriched with mixture of CM3000® and Poultry-Star®, 250 g/ton feed for each. Feed and water were offered *ad-libitum* during 70 days experimental period. Body weight and feed consumption were recorded biweekly to calculate body weight gain and feed conversion. At the end of the experimental period blood and caecal content samples were collected from all animals. Duodinal tissue samples were collected for histomorphometry. The results revealed that additives used improved significantly live body weight compared to the control group. Rabbits in T3 group showed the highest body weight gain. In addition, supplementation of the basal diet with a mixture of additives revealed significant increase of feed intake. The blood urea level was reduced significantly in bucks of T1. The rabbits in T3 group recorded the highest level of blood glucose. Caecal pH revealed a significant decrease in T1 and T3. The mixture of additives has positive results on the intestinal morphometry. Coated butyrate and synbiotic are capable of improving performance, enhancing intestinal health.

Keywords: rabbits, sodium butyrate, synbiotic, zootechnical performance, intestinal morphology

1. Introduction

Rabbit meat for its dietetic and nutritional characteristics is accepted by the consumer as a product of high quality meat. Rabbit meat is lean and its lipids are highly unsaturated (60% of the total fatty acids), rich in protein (20-21%) with high biological values, very low in cholesterol and sodium and rich in potassium and magnesium (Dalle Zotte, 2000). Under intensive production conditions (high genetic makeup breeds, high density of breeding unit, developing feeding mills, early weaning, etc), multi-factorial stressors (environmental, nutritional, managerial, etc) affects and suppress the health and production of all rabbit breeds. Improvement of the rabbit health and productive performance are very important to maximize the production at low cost. Recently using of growth promoters and feed additives to maintain rabbit health and maximize economics of rabbit meat production has increased.

Short review of European Union (2003) recorded that the rabbit's gut microflora are the key factors to maintain rabbit's health, zootechnical performance and reproduction. This review pay attention to feed additives that favour better rabbit gut microbiota with special emphasize on their effects on growth performance and health status as safe, reliable and efficacious alternatives to antibiotics in rabbit nutrition.

It is reported that butyrate derived from the fermentation of non-starch polysaccharides is considered to be important for normal development of epithelial cells with improved gastrointestinal health and reduced incidence of enteric diseases (Bron et al., 2002). Sakata (1987) reported that infusion of butyrate into fistulated rats increased the proliferation of crypt cells in both the small and large intestines. Sharma et al. (1995) suggested that the effect on crypt cell growth might reflect changes in the gut microflora, which is known to be a major modulator of epithelial cell activity. It has been demonstrated that short chain organic acids produced from fibre fermentation can inhibit the growth of bacteria of the group Enterobacteriaceae. Hume et al. (1993) showed that butyric acid has a higher diffusion coefficient through the pathogenic cell wall than other acids with shorter chains, which allows it to pass through the bacterial membrane more easily. Furthermore, in a recent work by Sunkara et al. (2011), the hypothesis that sodium butyrate is capable of inducing host defence peptides (HDPs) and enhancing disease resistance. Host defence peptides are natural broad spectrum antimicrobials and an important first line of defence in almost all forms of life.

Unfortunately, the natural levels of butyrate from fibre fermentation are quite low in the intestine and caeca (van der Wielen, 2000). Moreover, uncoated butyrate feed additive is immediately absorbed in the first part of the digestive tract before reaching the large intestine. In order to exert the influence in the large intestine, dietary butyrate should slowly be released over the gastrointestinal tract (Hu & Guo, 2007).

Probiotics, prebiotics and synbiotics have proven to be effective in reducing mortality in domestic animals (Collier et al., 2010) by maintaining the microbial balance in the digestive tract and reducing potentially pathogenic bacteria (Corcionivoschi et al., 2010). Several strains of probiotic bacteria and yeast have reduced mortality in various species of animals (Collier et al., 2010; Peret et al., 1998), and beneficial effects are more pronounced in conditions of stress or in herds with high mortality (Ewing, 2008). However, few studies have evaluated the effectiveness of synbiotics in rabbits.

Low information is available regarding the effect of adding coated or slowly released butyrate products with or without combination with synbiotic in rabbit diets. Based on this concept, the goal of the present study was to investigate the synergistic effects of dietary supplementation with coated slow released sodium butyrate (CM3000®) and a commercial synbiotic (Poultry-Star®) on the productive performance and intestinal morphometry of the postweaning growing rabbits.

2. Materials and Methods

2.1 Feed Additives Used

1) CM3000® is a commercial 30% microencapsulated sodium butyrate. CM3000® keeps releasing slowly and continuously in both small and large intestine. CM3000® is manufactured by Hangzhou King Techina Feed Co., Ltd, China.

2) Poultry-Star® is a microencapsulated synbiotic product for feed application. Poultry-Star® contains multi-species probiotic microorganisms (*Enterococcus faecium*, *Pediococcus acidilactici*, *Bifidobacterium animalis*, *Lactobacillus reuteri* and *Lactobacillus salivarius*) combined with prebiotic fructo-oligosaccharide (FOS). Poultry-Star® produced by BIOMIN America, Inc, USA.

2.2 Experimental Animals, Housing and Diets

Thirty two apparent healthy, weaned male New Zealand white bucks 35 day age and average body weight of 544 ± 9 g were used. The rabbits were weighed individually and randomly allocated to four dietary treatments (8 animals/group). The rabbits were housed individually in commercial cages (55 × 60 × 34 cm), equipped with automatic drinkers and j-feeders. Daily lighting regime was 10-12 hour photoperiod/day through both natural and fluorescent lighting. The study was conducted in the experimental rabbitry of Physiology Department, Faculty of Veterinary Medicine Cairo University, Egypt.

Basal diet was formulated and analysed to cover the nutrient requirements of growing rabbits as recommended in NRC (1977) (Table 1).

Feed and water were provided *ad-libitum* for 70 days experimental period. Rabbits in the first group were offered non-supplemented basal diet and served as a control group (C). Animals in the second group were reared on the basal diet supplemented with CM3000® 500 g/ton feed (T1). Rabbits in the third group consumed the basal diet containing Poultry-Star® 500 g/ton feed (T2). While rabbits in the fourth group were fed on the basal diet enriched with both CM3000® and Poultry-Star®, 250 g/ton feed for each additive (T3). Individual body weights for all animals as well as the rest of feeds were recorded biweekly. Body weight gain and feed conversion were calculated.

Table 1. Composition percentage, calculated nutrients profile and chemical analyses of the basal diet

Ingredients (%)	%	Calculated analysis (%)[**]	
		Crude protein	17.5
		Crude fiber	14.0
Berseem hay	30.0	Ether extract	2.7
Barley grain	21.0	Nitrogen free extract	56.4
Yellow corn	5.0	Digestible energy (kcal/kg)	2600
Wheat bran	21.1	Chemical analysis (%)[***]	
Soybean meal	17.5	DM	90.1
Molasses	3.0	Moisture	9.9
CaCl$_2$	1.5	Crude protein	17.7
NaCl	0.4	Crude fiber	13.8
Vit.&Min. Premix[*]	0.3	Total ash	3.5
DL-Methionine	0.2	Ether extract	2.3
		Nitrogen free extract	52.8

Note. [*] The Rabbit's vitamin and mineral premix/kg contained the following IU/g for vitamins or minerals: A-4,000,000, D3-5000,000, E-16,7 g, K-0.67 g, B1-0.67 g, B2-2 g, B6-0.67 g, B12-0.004 g, B5-16.7 g, Pantothinc acid-6.67 g, Biotein-0.07 g, Folic acid-1.67 g, Choline chloride-400 g, Zn-23.3 g, Mn-10 g, Fe-25 g, Cu-1.67 g, I-0.25 g, Se-0.033 g, and Mg-133.4 g (Rabbit premix); [**] Based on NRC (1977); [***] According to AOAC (1999).

2.3 Blood and Caecal Content Samples

Blood samples were collected at the end of the experimental period (at 105 days of age) from the ear vein of all animals at the morning before accesses to feed and water. Citrated plasma (in a ratio of 1 volume sodium citrate (3.8%): 9 volumes blood) Serum was obtained for determination of certain haematological metabolic parameters namely, glucose (Trinder, 1969), triglycerides (Wahlefeld, 1974), total cholesterol (Allian, 1974), total protein (Henry, 1964), albumen and urea (Fawcett & Scott, 1960).

Caecal content samples were collected at the end of the experimental period post slaughtering for determination of caecal fermentation pattern namely, pH, total short chain fatty acids (Eadie et al., 1967), individual volatile fatty acids (Samuel et al., 1997) and ammonia concentration (Chaney & Marbach, 1962; Abdl-Rahman et al., 2010).

2.4 Histological Investigation of the Intestinal Wall

At the end of the experimental period rabbits were slaughtered. Tissue samples were collected from the duodenum, 10 cm from the pyloric junction, flushed with physiological saline and fixed in10% buffered neutral formaldehyde solution for 48 hr, dehydrated in a graded series of ethanol, cleared in xylene and embedded in paraffin. Cross sections 5 μm thickness were cut and mounted on slides and stained with hematoxylin and eosin.

2.5 Histomorphometry

Villus height was measured by averaging the height of 10 intact villi, from the tip of the villus to the end of the crypt depth. The duodenal gland lenght was determined as the distance between the lamina muscularis mucosa and the tunica musuclaris externa, while the villus width was measured as the distance beween the eoithelium at the middle of the villus. Morphological indices were measured using image processing and analysis system (Version 1, Leica Imaging System Ltd., Cambridge, UK).

2.6 Statistical Analysis

All data were statistically analysed using IBM SPSS® version 19 software for personal computer (2010). Means were compared by one way ANOVA ($p < 0.05$) using Post Hoc test according to Snedecor and Cochran (1980).

3. Results

3.1 Zootechnical Performance

The live body weight and body weight gain in the treated groups were significantly improved ($p < 0.05$) in comparison with the control group (Tables 2 and 3).

Table 2. Effects of sodium butyrate and/or synbiotic on live body weight of growing male rabbit (g/rabbit) raised to70 days of age (means±SE)

Age (day)	Control	T1 Sod. butyrate	T2 Synbiotic	T3 Mixture
35	544.37 ± 6.3[a]	544.38 ± 6.3[a]	544.37 ± 6.3[a]	539.38 ± 8.6[a]
49	656.25 ± 22.0[a]	800.00 ± 23.1[bc]	768.75 ± 33.9[b]	850.00 ± 16.3[c]
63	875.00 ± 19.41[a]	1325.00 ± 21.1[c]	1181.25 ± 16.2[b]	1378.12 ± 24.7[c]
77	1387.50 ± 18.2[a]	1643.75 ± 17.5[c]	1568.75 ± 24.8[b]	1708.75 ± 19.4[d]
91	1671.87 ± 17.5[a]	2030.62 ± 17.5[b]	2000.00 ± 17.5[b]	2184.37 ± 17.5[c]
105	2011.87 ±20.6[a]	2350.00 ± 13.3[c]	2300.00 ± 13.4[b]	2468.75 ± 13.1[d]

Note. Different superscripts within a row indicate a significant treatment effect ($p < 0.05$).

Table 3. Effects of sodium butyrate and/or synbiotic on body weight gain of growing male rabbit (g/rabbit) raised to70 days of age (means±SE)

Age (day)	Control	T1 Sod. butyrate	T2 Synbiotic	T3 Mixture
49	111.88 ± 4.5[a]	255.62 ± 4.2[b]	224.38 ± 3.3[b]	310.62 ± 5.2[c]
63	218.75 ± 4.7[a]	525.00 ± 7.5[c]	412.50 ± 9.8[b]	528.12 ± 6.3[c]
77	512.50 ± 12.3[c]	318.75 ± 11.2[a]	387.50 ± 13.4[b]	330.63 ± 16.5[a]
91	284.37 ± 17.3[a]	386.87 ± 19.5[c]	431.25 ± 20.0[b]	475.62 ± 18.2[c]
105	340.00 ± 18.1[c]	319.38 ± 15.6[b]	300.00± 17.8[a]	284.38 ± 18.6[a]
Total	1467.50 ±21.8[a]	1805.62 ± 18.3[c]	1755.63 ± 17.5[b]	1929.37 ± 19.3[d]

Note. Different superscripts within a row indicate a significant treatment effect ($p < 0.05$).

Live body weight showed significant differences ($p < 0.05$) at the end of the experiment in the following order: T3 > T1 > T2 > C (2468.75 ± 13.1, 2350.00 ± 13.3, 2300.00 ± 13.4 and 2011.87 ± 20.6 g/rabbit respectively). Both additives CM3000® (T1) and Poultry-Star® (T2) significantly improved live body weight. The highest body weight was shown in rabbits fed on the mixture of CM3000® and Poultry-Star® (T3) group. In the same way body weight gain showed significant differences at the end of the experiment in the following order: T3 > T1 > T2 > control group.

Tables 4 and 5 show the feed consumed and feed conversion in the different groups. Rabbits reared on the diet containing mixture of CM3000® and Poultry-Star® (T3) consumed more feed than the control and the other treated groups. In addition, feed conversion was improved by the additives used to record 3.30 in T3 and T2 versus 3.53 in the control group.

Table 4. Effects of sodium butyrate and/or synbiotic on feed intake of growing male rabbit (g/rabbit) raised to70 days of age (means±SE)

Age (day)	Control	T1 Sod. butyrate	T2 Synbiotic	T3 Mixture
49	381.06 ± 16.8^a	381.10 ± 15.5^a	381.06 ± 12.3^a	377.57 ± 11.6^a
63	643.13 ± 21.0^a	784.00 ± 25.3^c	753.38 ± 24.9^b	833.00 ± 22.4^d
77	1157.50 ± 19.41^a	1298.50 ± 27.2^b	1157.63 ± 23.6^a	1350.56 ± 21.7^c
91	1359.75 ± 28.9^a	1610.88 ± 35.1^c	1537.38 ± 33.6^b	1674.58 ± 30.2^c
105	1638.43 ± 28.2^a	1990.00 ± 25.3^b	1960.00 ± 21.4^b	2140.68 ± 20.5^c
Total	5179.87 ± 20.6^a	6064.48 ± 13.3^c	5789.45 ± 13.4^b	6376.39 ± 13.1^d

Note. Different superscripts within a row indicate a significant treatment effect ($p < 0.05$).

Table 5. Effects of sodium butyrate and/or synbiotic on feed conversion (FCR) of growing male rabbit raised to70 days of age (means±SE)

Age (day)	Control	T1 Sod. butyrate	T2 Synbiotic	T3 Mixture
49	3.41	1.49	1.70	1.22
63	2.94	1.49	1.83	1.58
77	2.26	4.07	2.99	4.08
91	4.78	4.16	3.56	3.52
105	4.82	6.23	6.54	7.53
Total	3.53	3.36	3.30	3.30

3.2 Caecal Fermentation Parameters

The results clearly demonstrate decrease in the pH in groups T1 and T3 compared to the control group ($p < 0.05$) (Table 6).

Table 6. Effects of sodium butyrate and/or synbiotic on caecal fermentation of growing male rabbit raised to70 days of age (means±SE)

Parameter	Control	T1 Sod. butyrate	T2 Synbiotic	T3 Mixture
pH	6.79 ± 0.12^a	6.32 ± 0.03^b	6.92 ± 0.07^a	6.29 ± 0.03^b
Ammonia	17.35 ± 1.11^c	14.58 ± 0.96^b	17.32 ± 0.33^c	13.87 ± 0.81^a
TVFA	64.0 ± 6.11^a	126.67 ± 4.81^a	46.0 ± 1.15^a	124.67 ± 3.71^a
Acetic acid mol%	55.17 ± 0.93^a	58.56 ± 1.96^c	62.35 ± 0.22^d	56.98 ± 1.82^b
Propionic acid mole%	26.56 ± 1.14^c	14.81 ± 2.10^a	14.34 ± 0.57^a	21.31 ± 0.34^b
butyric mol%	7.78 ± 0.71^a	26.09 ± 0.33^c	22.56 ± 0.54^b	20.70 ± 0.21^b

Note. Different superscripts within a row indicate a significant treatment effect ($p < 0.05$).

Each treatment modified the VFAs profile within the caecum in a particular fashion. As for acetic acid proportion, the highest percent was recorded in T2 group compared to the C group, while all supplements lowered propionic acid molar proportion compared to C group. Regarding caecal NH3-N concentrations, both T1 and T3 groups demonstrated a decrease significantly ($p < 0.05$) with reference to control group.

3.3 Blood Parameters

The results obtained from serum analyses are shown in table 7. An increase in the glucose concentration was reported in all treated groups and the highest was recorded in T3 group. Serum triglycerides showed highest values in the treated group (T1). Ammonia concentration recorded low values in T1 and T3 groups compared with the control one ($p < 0.05$).

Table 7. Effects of sodium butyrate and/or synbiotic on some blood parameters of growing male rabbit raised to70 days of age (means±SE)

Parameter	Control	T1 Sod. butyrate	T2 Synbiotic	T3 Mixture
Albumin	3.72 ± 0.28^a	4.42 ± 0.34^b	3.15 ± 0.23^a	4.80 ± 0.27^c
Total protein	6.04 ± 0.20^a	6.25 ± 0.28^a	5.67 ± 0.40^a	6.53 ± 0.38^a
Urea	35.14 ± 4.27^b	30.61 ± 3.75^a	63.29 ± 5.18^d	37.49 ± 4.60^c
Triglycerides	122.03 ± 19.20^b	288.28 ± 49.52^c	36.68 ± 5.83^a	119.68 ± 17.56^b
Cholesterol	127.03 ± 20.27^a	96.73 ± 13.72^a	109.00 ± 9.70^a	81.38 ± 12.67^a
Glucose	137.64 ± 4.95^a	139.02 ± 5.37^b	139.85 ± 6.48^b	179.74 ± 14.24^c

Note. Different superscripts within a row indicate a significant treatment effect ($p < 0.05$).

3.4 Histological Investigation of the Intestinal Wall

The effect of dietary treatments on the duodenal morphology (villus height, villus width and duodenal gland lenght) is shown in Table 8 and Figure 1. Dietary supplementations influenced the histomorphological measurements of duodenal villi and gland lenght comparing with the control. Feed additives increased significantly ($p < 0.05$) the villus and gland lengths in all treated groups when compared with the control. Furthermore, the mixture of sodium butyrate and synbiotic (T3) supplementation increased the gland lenght and villus width numerically compared with other additives.

Table 8. Effects of feed additive supplementations on the histomorphological parameters of the rabbit intestine (means±SE)

Parameter	Control	T1 Sod. butyrate	T2 Symbiotic	T3 Mixture
Villus height (µm)	349 ± 11^a	395 ± 12^b	378 ± 12^b	389 ± 13^b
Villus width (µm)	43 ± 1.4^a	44 ± 1.9^{ab}	49 ± 2.3^{bc}	51 ± 1.7^c
Gland lenght (µm)	177 ± 10^a	250 ± 20^b	236 ± 14^b	286 ± 19^b

Note. Different superscripts within a row indicate a significant treatment effect ($p < 0.05$).

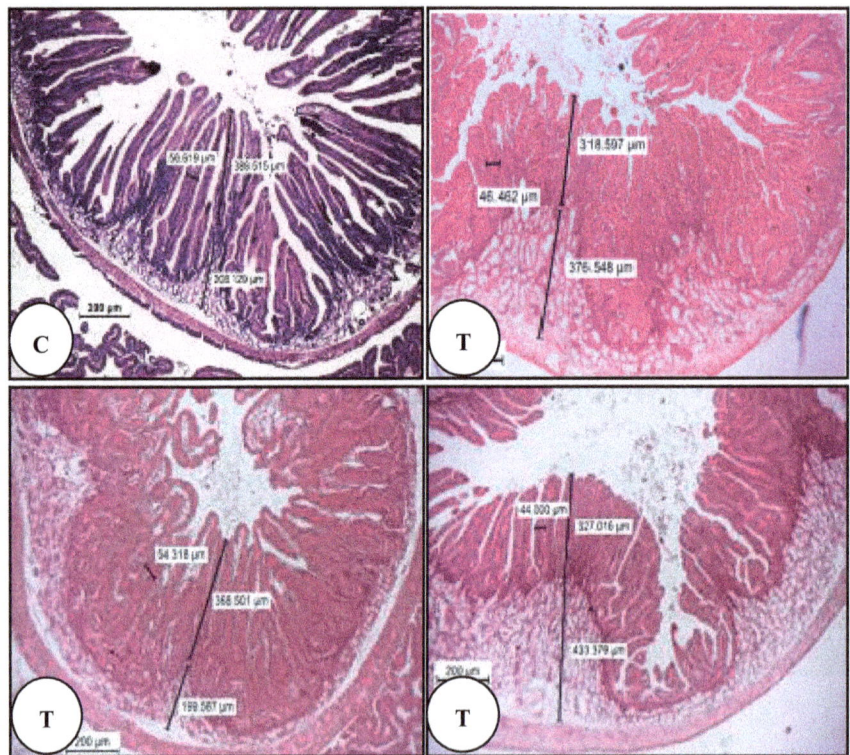

Figure 1. Photomicrographs of the small intestine (duodenum) from male New Zealand white rabbits showing the intestinal villi and duodenal glands. (C) Control group; (T1) Sodium butyrate supplemented group; (T2) Symbiotic supplemented group; (T3) Mixture of butyrate and symbiotic group, (H&E) staining

4. Discussion

From the obtained results it is clear that rabbits performance was enhanced by the additive used. These findings can be explained in the light of that short chain fatty acids (SCFA) produced by microbial fermentation from dietary fibre stimulate epithelial cell proliferation resulting in a larger absorptive surface (Sakata, 1988; Leeson et al., 2005; Hu & Guo, 2007; Panda et al., 2009). Moreover, normal colonic epithelia derive 60 to 70% of their energy supply from SCFA, particularly from butyric acid (Scheppach et al., 1992). Butyric acid induces cell differentiation and regulates the growth and the proliferation of normal colonic mucosa (Treem et al., 1994) while suppressing the growth of cancer cells (Clausen et al., 1991). Synbiotic products contain viable bacterial cultures that establish early in the gut while the prebiotic present in them serve as a source of nutrients for the probiotics in addition to dietary sources (Mohnl et al., 2007; Zhang et al., 2006). Probiotics and/or synbiotics could have positive effects on bacterial population in the gastrointestinal tract (Smirnov et al., 2005), and the addition of probiotic to diets has been found to improve growth performance (Jin et al., 1997; Wenk, 2000). Gut microflora changes actively by adding prebiotics and significantly reduces gut pH which improve rabbit's performance through influencing gut microbial population (Rahmani & Speer, 2005). Hooge (2004) reported that positive effects of mannan oligosaccharides on animal performance could be more visible during stressful, high temperature as that situation in Egypt, high density and weak management conditions. Prebiotics are potential alimentary supplements which reduce harmful effects of putrefactive factors and increases nutrition output (Fooks & Gibson, 2002). Also it has been reported that using prebiotics increases nutrient absorbance area via increasing gut length and thus improves bird performance (Santin et al., 2001). In the present study butyrate and synbiotic mixture used affected the productive performance of rabbits synergistically and revealed positive results on rabbit growth and zootechnical performance.

The obtained results of feed conversion reflect the efficient utilization of feed by the additives used. These results coincide with Tony et al. (2014), who concluded that coated and slowly released sodium butyrate (CM3000®) could significantly improve broiler performance and feed conversion. They mentioned that the oral administration of CM3000® enhances resistance to *Salmonella* Enteritidis challenge and markedly reduce *Salmonella* shedding. In the present study butyrate and synbiotic mixture used affected the productive performance of rabbits synergistically and revealed positive results on rabbit growth and zootechnical performance.

The analyses of caecal contents revealed significant decrease in the pH in T1 and T3 compared with C group. This observation may be due to an increase in TVFA concentration reported in former groups. These short-chain carboxylic acids in the gastrointestinal tract of non-ruminants reflect the amount consumed and the rate of intraluminal production by anaerobic microorganisms from fermentable substrate (Smulikowska et al., 2009). These acids have a number of important regulatory functions related to gastrointestinal functionality, among others mucosal development, proliferation, differentiation, maturation and apoptosis of enterocytes and colonocytes (Mroz et al., 2006).

Propionic acid is a valuable substrate in glucose synthesis in many species (Bergman, 1990) as it contributes in gluconeogenesis and formation of long-chain fatty acids in the liver and its intermediate products   change and participate in regulation of a series of processes, including ketogenesis, gluconeogenesis and ureogenesis (Remesy et al., 1995). Despite lower propionic acid molar proportion, the present study failed to demonstrate a decrease in serum glucose concentration in the experimental groups, but rather an increase in the glucose concentration was reported in T3 group which may be due to better digestibility and absorption of carbohydrates, secondary to the improvement in intestinal morphometry. Concerning butyrate, all experimental groups showed higher concentrations compared to C group with highest % recorded for T1 group. As butyrate is an essential precursor in lipogenesis (Remesy et al., 1995), serum triglycerides showed highest values in the former group (T1). Moreover, butyrate is recognized as the most effective source of energy for epithelial cells proliferation (Mroz et al., 2006), where sodium butyrate has been reported to be helpful in maintenance of intestinal villi structure after coccidial challenge (Leeson et al., 2005).

According to Macfarlane and Gibson (1995) a series of factors could influence NH3-N concentrations within the caecum, including H2 pressure, chyme reaction, and carbohydrates availability. In comparison with ruminants, proteolytic activity in the rabbit caecum is relatively higher (Gidenne, 1997). The lower recorded ammonia concentration in T1 and T3 group could be attributed to either increased nitrogen retention by enterocytes and colonocytes in these groups which may be connected with greater epithelial cell proliferation in gastrointestinal tract as suggested by Smulikowska et al. (2009) or better ammonia utilization in liver for protein production, this seems true since albumen concentrations were increased in T1 and T3 groups. The decreased in caecal pH, the decreased in ammonia-N concentration and the higher VFA concentration suggest high fermentation activity, caecal microbial synthesis, gut health and high nitrogen retention. This observation is consistent with the growth performance result discussed above, and is also in agreement with the results of Garcia et al. (2000).

Concerning serum total proteins all supplements did not alter serum values reflecting that the experimental animals were in a good nutritional status and liver has no pathological lesions (Abdl-Rahman et al., 2010).

The histomorphological changes in the intestine of growing rabbit reported in the present study provide new information regarding the potential for using synbiotics and probiotics in rabbit feed. Increasing the villus height suggests an increased surface area capable of greater absorption of available nutrients (Caspary, 1992). Feeding of probiotics has been shown to induce gut epithelial cell proliferation in rats (Ichikawa et al., 1999). The intestinal epithelial cells originating in the crypt migrate along the villus surface upward to the villus tip and are extruded into the intestinal lumen within 48 to 96 h (Imondi and Bird, 1966; Potten, 1998). A shortening of the villi may lead to poor nutrient absorption, increased secretion in the gastrointestinal tract, and lower performance (Xu et al., 2003). In contrast, increases in the villus height and villus height:crypt depth ratio are directly correlated with increased epithelial cell turnover (Fan et al., 1997). It is understood that greater villus height is an indicator that the function of intestinal villi is activated (Langhout et al., 1999; Shamoto & Yamauchi, 2000). This fact suggests that the villus function is activated after feeding of dietary probiotic and synbiotic.

In the present study, coated sodium butyrate supplementation was associated with increased intestinal morphological parameters compared with the control (Table 8 and Figure 1). Sodium butyrate has been reported to be helpful in maintenance of intestinal villi structure after coccidial challenge (Leeson et al., 2005). Moreover, Mroz et al. (2006) reported that butyrate is recognized as the most effective source of energy for epithelial cells proliferation. Per contrary, in pigs and chickens, the effect of sodium butyrate on small intestinal epithelium is often insignificant (Biagi et al., 2007; Hu & Guo, 2007).

5. Conclusion

Exogenous administration of coated, slow-release butyrate (CM3000®) combined with synbiotic (Poultry-Star®) is capable of improving zootechnical performance, enhancing intestinal health. Butyrate with synbiotic mixture may be used as alternatives for antibiotics in rabbit nutrition to improve growth performance that would be a valuable feeding strategy in developing countries.

Acknowledgements

The authors thank Hangzhou King Techina Feed Co., Ltd, China, Ms. Shuyi Li and Dr. Tony Niu for supplying of CM3000® used in the study and for their support. The authors also thank BIOMIN America, Inc, USA for supplying of Poultry-Star®.

References

Abdl-Rahman, M. A., Sawiress, F. A. R., & Sohair, Y. S. (2010). Effect of Kemzyme–Bentonite Co-supplementation on Cecal Fermentation and Metabolic Pattern in Rabbit. *Journal of Agicultural Science, 2*, 183-188.

Allain, C. C., Poon, L. S., Chan, C. S. G., Richmond, W., & Fu, P. C. (1974). Enzymatic determination of total serum cholesterol. *Clin Chem, 20*, 470.

AOAC. (1999). *Official Methods of Analysis*. Association of Official Analytical Chemists, Washington DC.

Bergman, E. N. (1990). Energy contributions of volatile fatty acids from the gastrointestinal tract in various species. *Physiological Reviews, 2*, 567-590.

Biagi, G., Piva, A., Moschini, M., Vezzali, E., & Roth, F. X. (2007). Performance, intestinal microflora, and wall morphology of weanling pigs fed sodium butyrate. *J. Anim. Sci., 85*, 1184-1191. http://dx.doi.org/10.2527/jas.2006-378

Bron, F., Kettlitz, B., & Arrigoni, E. (2002). Resistant starches and the butyrate revolution. *Trends Food Sci. Technol., 13*, 251-261. http://dx.doi.org/10.1016/S0924-2244(02)00131-0

Caspary, W. F. (1992). Physiology and pathophysiology of intestinal absorption. *Am. J. Clin. Nutr., 55*, 299S-308S.

Chaney, A. L., & Marbach, E. P. (1962). Modified reagents for determination of urea and ammonia. *Clinical Chemistry, 8*, 130-132.

Clausen, M. R., Bonnen, H., & Mortensen, P. B. (1991). Colonic fermentation of dietary fibre to short chain fatty acids in patients with adenomatous polyps and colonic cancer. *Gut, 32*,923-928. http://dx.doi.org/10.1136/gut.32.8.923

Collier, C. T., Carroll, J. A., Ballou, M. A., Starkey, J. D., & Sparks, J. C. (2010). Oral administration of Saccharomyces cerevisiaeboulardii reduces mortality associated with immune and cortisol responses to Escherichia coli endotoxin in pigs. *Journal of Animal Science, 89*, 52-58. http://dx.doi.org/10.2527/jas.2010-2944

Corcionivoschi, N., Drinceanu, D., Mircea, P. I., Stack, D., Ştef, L., Julean, C., & Bourke, B. (2010). The effect of probiotics on animal health. *Animal Science and Biotechnologies, 43*, 35-41.

Dalle Zotte, A. (2000). Main factors influencing the rabbit carcass and meat quality. *Proc. 7th World Rabbit Congress* (Vol. A-B, pp. 507-537). Valencia, Spain.

Eadie, J. M., Hobson, P. N., & Mann, S. O. (1967). A note on some comparisons between the rumen content of barley fed steers and that of young calves also fed on high concentrate rations. *Animal production, 9*, 247-250. http://dx.doi.org/10.1017/S0003356100038514

European Commission. (2003). *Opinion on the use of certain micro-organisms as additives in feedingstuffs*. Report prepared for Health & Consumer Protection Directorate-General.

Ewing, W. N. (2008). *The living gut* (p. 192). Nottingham, UK: Nottingham University Press.

Fan, Y., J. Croom, V., Christensen, B., Black, A., Bird, L., Daniel, B., ... Eisen, E. (1997). Jejunal glucose uptake and oxygen consumption in turkey poults selected for rapid growth. *Poult. Sci., 76*, 1738-1745. http://dx.doi.org/10.1093/ps/76.12.1738

Fawcett, J. K., & Scott, J. E. (1960). A rapid and precise method for the determination of urea. *J. Clin. Pathol., 13*,156-159. http://dx.doi.org/10.1136/jcp.13.2.156

Fooks, L. J., & Gibson, G. R. (2002). Probiotics as modulators of the gut flora. *British Journal Nutrition, 88*, 39-49. http://dx.doi.org/10.1079/BJN2002628

Garcia, J., Carabano, R., Perez-Alba, L., & de Blas, J. C. (2000). Effect of fiber source on cecal fermentation and nitrogen recycled through cecotrophy in rabbits. *J. Anim. Sci., 78*, 638-646.

Gidenne, T. (1997). Caeco-colic digestion in the growing rabbit: impact of nutritional factors and related disturbances. *Livestock Production Science, 51*, 73-88. http://dx.doi.org/10.1016/S0301-6226(97)00111-5

Henry, R. J. (1964). Colorimetric determination of total protein. *Clinical Chemistry* (p. 181). Harper and Row Publ., New York, USA.

Hooge, D. (2004). Meta-analysis of broiler chicken pen trials evaluating dietary mannan oligosaccharide, 1993-2003. *International Journal of Poultry Science, 3*, 163-174. http://dx.doi.org/10.3923/ijps.2004.163.174

Hu, Z., & Guo, Y. (2007). Effects of dietary sodium butyrate supplementation on the intestinal morphological structure, absoptive function and gut flora in chickens. *Aniemal Feed Science and Technology, 132*, 240-249. http://dx.doi.org/10.1016/j.anifeedsci.2006.03.017

Hume, M. E., Corrier, D. E., Ivie, G. W., & Deloach, J. R. (1993). Metabolism of [14C] propionic acid in broiler chicks. *Poult. Sci., 72*, 786-793. http://dx.doi.org/10.3382/ps.0720786

Ichikawa, H., Kuroiwa, T., & Inagaki, A. (1999). Probiotic bacteria stimulate epithelial cell proliferation in rat. *Dig. Dis. Sci., 44*, 2119-2123. http://dx.doi.org/10.1023/A:1026647024077

Imondi, A. R., & Bird, F. H. (1966). The turnover of intestinal epithelium in the chick. *Poult. Sci., 45*, 142-147. http://dx.doi.org/10.3382/ps.0450142

Jin, L. Z., Ho, Y. W., Abdullah, N., & Jalaludin, S. (1997). Probiotic in poultry: Modes of action. *World's Poult. Sci. J., 53*, 351-368. http://dx.doi.org/10.1079/WPS19970028

Langhout, D. J., Schutte, J. B., Van, L. P., Wiebenga, J., & Tamminga, S. (1999). Effect of dietary high and low methylated citrus pectin on the activity of the ileal microflora and morphology of the small intestinal wall of broiler chickens. *Br. Poult. Sci., 40*, 340-347. http://dx.doi.org/10.1080/00071669987421

Leeson, S., Namkung, H., Antongiovanni, M., & Lee, E. H. (2005). Effect of butyric acid on the performance and carcass yield of broiler chickens. *Poult. Sci., 84*, 1418-1422. http://dx.doi.org/10.1093/ps/84.9.1418

Macfarlane, G. T., & Gibson, G. R. (1995). Microbiological aspects of the production of short-chain fatty acids in the large bowel. In J. H. Cummings, J. L. Rombeau & T. Sakata (Eds.), *Physiological and Clinical Aspects of Short-chain fatty acids* (pp. 87-105). *Cambridge Univ Press*, London.

Mohnl, M., Acosta Aragon, Y., Acosta Ojeda, A., Rodriguez Sanchez, B., & Pasteiner, S. (2007). Effect of synbiotic feed additive in comparison to antibiotic growth promoter on performance and health status of broilers. *Poult. Sci., 86*(Suppl. 1), 217.

Mroz, Z., Koopmans, S. J., Bannink, A., Partanen, K., Krasucki, W., Øverland, M., & Radcliffe, S. (2006). Carboxylic acids as bioregulators and gut growth promoters in nonruminants. In R. Mosenthin, J. Zentek, T. Żebrowska (Eds.), *Biology of Nutrition in Growing Animals* (pp. 81-133). Elsevier, Edinburgh, UK. http://dx.doi.org/10.1016/S1877-1823(09)70091-8

NRC. (1977). *Nutrient requirements of rabbits*(2nd ed.). National Academy of Sciences. Washington, DC.

Panda, A. K., Rama Rao, S. V., Raju, M. V. L. N., & Shyam Sunder, G. (2009). Effect of Butyric Acid on Performance, Gastrointestinal Tract Health and Carcass Characteristics in Broiler Chickens. *Asian-Aust. J. Anim. Sci., 22*, 1026-1031.

Peret, F. L. A., Penna, F. J., Bambirra, E. A., & Nicoll, J. R. (1998). Dose effect of oral *Saccharomyces boulardii* treatments on morbidity and mortality in immunosuppressed mice. *Journal of Medical Microbiology, 47*, 111-116. http://dx.doi.org/10.1099/00222615-47-2-111

Potten, C. S. (1998). Stem cells in the gastrointestinal epithelium: Numbers, characteristics and death. *Philos. Trans. R. Soc. Lond. B. Biol. Sci., 353*, 821-830. http://dx.doi.org/10.1098/rstb.1998.0246

Rahmani, H. R., & Speer, W. (2005). Natural additives influence the performance and humoral immunity of broilers. *International Journal of Poultry Science, 4*, 713-717. http://dx.doi.org/10.3923/ijps.2005.713.717

Remesy, C., Demigne, C., & Morand, C. (1995). Metabolism of short-chain fatty acids in the liver. In J. H. Cummings, J. L. Rombeau, & T. Sakata (Eds.), *Physiological and Clinical Aspects of Short-chain fatty acids* (pp. 171-190). Cambridge University Press, London.

Sakata, T. (1987). Stimulatory effect of short-chain fatty acids on epithelial cell proliferation in the rat intestine: a possible explanation for tropic effects of fermentable fibre, gut microbes and luminal tropic factors. *Br. J. Nutr., 58*, 95-103. http://dx.doi.org/10.1079/BJN19870073

Sakata, T. (1988). Chemical and physical trophic effects of dietary fibre on the intestine of monogastrics animals.

In L. Buraczewska, S. Buraczewski, B. Pastuszewska, T. Zebrowska (Eds.), *Digestive Physiology in the pig* (pp. 128-135). Polish Academy of Science, Joblonna, Poland.

Samuel, M., Sagathewan, S., Thomas, J., & Mathen, G. (1997). An HPLC method for estimation of volatile fatty acids of ruminal fluid. *Indian J. Anim. Sci., 67*, 805-807.

Santin, E., Maiorka, A., Macari, M., Grecco, M., Sanchez, J. C., Okada, T. M., & Myasaka, A. M. (2001). Performance and intestinal mucosa development of broiler chickens fed diet containing Sccharomyces cerevisiae cell wall. *Journal of Applied Poultry Research, 10*, 236-244. http://dx.doi.org/10.1093/japr/10.3.236

Scheppach, W., H., Sommer, T., Kirchner, G. M., Paganelli, P., Bartram, S., Christl, F., ... Kasper, H. (1992). Effect of butyrate enemas on the colonic mucosa in distal ulcerative colitis. *Gastroenterology, 103*(1), 51-56.

Shamoto, K., & Yamauchi, K. (2000). Recovery responses of chick intestinal villus morphology to different refeeding procedures. *Poult. Sci., 79*, 718-723. http://dx.doi.org/10.1093/ps/79.5.718

Sharma, R., Schumarcher, U., Ronaasen, V., & Coates, M. (1995). Rat intestinal mucosal responses to a microbial flora and different diets. *Gut., 36*, 206-214. http://dx.doi.org/10.1136/gut.36.2.209

Smirnov, A., Perez, R., Amit-Romach, E., Sklan, D., & Uni, Z. (2005). Mucin dynamics and microbial populations in chicken small intestine are changed by dietary probiotic and antibiotic growth promoter supplementation. *J. Nutr., 135*, 187-192.

Smulikowska, S., Czerwiński, J., Mieczkowska, A., & Jankowiak, J. (2009). The effect of fat-coated organic acid salts and a feed enzyme on growth performance, nutrient utilization, microflora activity, and morphology of the small intestine in broiler chickens. *Journal of Animal and Feed Sciences, 18*, 478-489.

Snedecor, F. W., & Cochran, W. G. (1980). Statistical methods 7th ed. *Lowa State Univ. Press* Ames .I.A.

Sunkara, L. T., Achanta, M., Schreiber, N. B., Bommineni, Y. R., Dai, G., Jiang, W., ... Zhang, G. (2011). Butyrate enhances disease resistance of chickens by inducing antimicrobial host defense peptide gene expression. *PLoS One, 6*, e27225. http://dx.doi.org/10.1371/journal.pone.0027225

Tony, M. A., Hamoud, M. M., Bailey, C. A., & Hafez, H. M. (2014). Effects of coated sodium butyrate (CM3000®) as a feed additive on zootechnical performance, immune status and *Salmonella Enteritidis* shedding after experimental infection of broiler chickens. *Proceeding of Poultry Science Association meeting.* Corpus Christi, Texas, USA.

Treem, W. R., Ahsan, N., Shoup, M., & Hyams, J. S. (1994). Fecal short-chain fatty acids in children with inflammatory bowel disease. *J. Pediatr. Gastroenterol. Nutr., 18*(2), 159-164. http://dx.doi.org/10.1097/00005176-199402000-00007

Trinder, P. (1969). Determination of glucose in blood using glucose oxidase with an alternative oxygen acceptor. *Ann. Clin Biochem., 6*, 24. http://dx.doi.org/10.1177/000456326900600108

Van der wielen, P. (2000). Dietary strategies to influence the gastrointestinal microflora of young animals and its potential to improve intestinal health. *Nutrition and Health of the Gastrointestinal Tract* (pp. 37-60). M.C. Blok, H.A.

Wenk, C. (2000). Recent advances in animal feed additives such as metabolic modifier, antimicrobial agents, probiotics, enzymes and available minerals. *Asian-Aust. J. Anim. Sci., 13*, 86-95. http://dx.doi.org/10.5713/ajas.2000.86

Xu, Z. R., Hu, C. H., Xia, M. S., Zhan, X. A., & Wang, M. Q. (2003). Effects of dietary fructooligosaccharide on digestive enzyme activities, intestinal microflora and morphology of male broilers. *Poult. Sci., 82*, 1030-1036. http://dx.doi.org/10.1093/ps/82.6.1030

Zhang, G., Ma, L., & Doyle, M. P. (2006). *Effect of Probiotics, Prebiotics and Symbiotics on Weight Increase of Chickens (Gallus domesticus).* Retrieved from http//www.ugacfs.org/research/pdf/poultry 2006.pdf

11

Influence of Extra Weight and Tire Pressure on Fuel Consumption at Normal Tractor Slippage

Vidas Damanauskas[1], Algirdas Janulevičius[1] & Gediminas Pupinis[1]

[1] Institute of Power and Transport Machinery Engineering, Aleksandras Stulginskis University, Lithuania

Correspondence: Algirdas Janulevičius, Aleksandras Stulginskis University, Studentų Str. 15, 53361 Kaunas-Akademija, Lithuania. E-mail: algirdas.janulevicius@asu.lt

Abstract

Tire pressure and wheel load are both easily managed parameters which play a significant role in tillage operations for limiting slip which involves energy loss. To a great extent, this aspect affects the fuel consumption and the time required for soil tillage. The study was focused on the tire pressure and extra weight variation effect on fuel consumption and work productivity for soil tillage at normal tractor wheels slippage (7-15%). The experimental research unit composed of an 82.3 kW 4WD tractor and a reversible 4-bodies plough is presented. Tests were carried out on a stubble loam, where slip of tractor driving wheels was < 15%, tractor front ballast mass was varied in the range from 0 to 520 kg and inflation pressure in the tires from 240 kPa to 100 kPa. Dependences of tractor performance indicators on ballast mass and tires inflation pressure are presented. When tractor tire slip varies in the range from 7 to 15 percent (which is normal slip in the soil), reducing the tires inflation pressure decreases the driving wheel slip and fuel consumption, while increases work productivity. Increasing the additional mass of the tractor (adding ballast weights) decreases the driving wheel slip, increases work productivity, but also increases fuel consumption and soil compaction.

Keywords: tractor, slippage, fuel consumption, extra weight, tire pressure, tillage operation

1. Introduction

Energy systems, transport and agriculture are named as the main sectors that need more attention for the appropriate measures in order to reduce fuel consumption and unfriendly impact on the environment (Dagiliūtė & Juknys, 2012; Szendrő & Török, 2014). Agricultural mechanization is required to sustain food production with high productivity, but fuel resource limitation has spurred both tractor manufacturers and users to address their fuel consumptions. Fuel consumption and exhaust emissions, including harmful components, can be reduced only by complex optimization of technological processes and tractor operating modes (Backman, Oksanen, & Visala, 2013; Janulevičius, Juostas, & Pupinis, 2013; Magalhães, Souza, Santana, & Sabbag, 2013; Moitzi, Haas, Wagentristl, Boxberger, & Gronauer, 2013; Khambalkar, Pohare, Katkhede, Bunde, & Dahatonde, 2010; Kheiralla, Azmi, Zohadie, & Ishak, 2004).

Agricultural tractors combined with diverse implements are the basic tools used in field production to conduct different field operations. Agricultural tractors commonly employ a four-wheel drive (4WD) transmission. Four-wheel drive tractors offer a number of advantages over two-wheel drive; the main advantage of the front-wheel assist is that it improves the tractor's ability to cross soft, wet, slippery and/or uneven terrain (Molari, Bellentani, Guarnieri, Walker, & Sedoni, 2012; Patterson, Gray, Bortolin, & Vantsevich, 2013). However, Wong (2010) and Vantsevich (2008) noted that under certain circumstances, a tendency exists for four-wheel drive tractors to suffer a reduction in power delivery efficiency and an increase in fuel consumption as a result of interaction between front and rear wheels being less than optimal.

Pulling ability of any tractor depends on many factors, includings engine power, tractor mass, contact area between tires and the ground and soil strength (Lyasko, 2010; Moitzi et al., 2013; Senatore & Sandu, 2011; Srinivasa Rao, Ramji, & Naidu, 2012; Stoilov & Kostadinov, 2009; Wong, 2010). When the soil is strong (dry), the cohesion is good, which results in greater pulling force, less wheel slip and lower rolling resistance (Moitzi et al., 2013). While the soil is at the plastic state (wet), the cohesion is good, but the wheels slip and the rolling resistance is high, which causes greater power losses and reduces tractor pulling efficiency (Battiato & Diserens, 2013; Wong, 2010).

Driving-wheel slip and rolling resistance are regarded as the main sources of power loss. The research indicates that 20 to 55% of available tractor power is lost in the process of interaction between tires and soil surface (Peca et al., 2010; Šmerda & Čupera, 2010; Taghavifar & Mardani, 2014). The rolling resistance has on approximately constant relation with low velocities. The rolling resistance of the wheel is increasingly influenced by inflation pressure and vertical load than velocities in agricultural works. Also, Taghavifar and Mardani (2013) noted that increase of inflation pressure suggests reverse relation with rolling resistance particularly at higher values of vertical load. A single-wheel test facility (Taghavifar & Mardani, 2013) was utilized to investigate the effect of velocity, tire inflation pressure, and vertical load on rolling resistance of wheel. The results showed that rolling resistance is less affected by velocities of tractors in farmlands but is much influenced by inflation pressure and vertical load (Taghavifar & Mardani, 2013).

Fuel consumption during tractor operation is highly dependent on the engine rotational speed and load characteristics. In most cases, the most productive and cost-effective work is obtained when engine load is less than 80% of its rated power and the engine rotational speed does not exceed 80% of its rated rotational speed (Grisso et al., 2011; Janulevičius et al., 2013; Juostas & Janulevičius, 2014; Lacour, Burgun, Perilhon, Descombes, & Doyen, 2014; Moitzi et al., 2013). In order to reach maximum economic efficiency of works performed by agricultural equipment, tractors with higher pulling power are unavoidable for usage. Draft depends on the pulling power and running speed. Basically, operational speeds of tractors in the farm cannot exceed to high levels. Operational speed of 3-15 km h^{-1} prevails in agricultural work. Deviations from the operational speed deteriorate the quality of work and increase energy consumption. For example, faster tilling greatly increases the dynamic effects on the soil, the earth is thrown strongly and more energy is consumed (Hashemi, Ahmad, Othman, & Sulaiman, 2012; Khambalkar et al., 2010; Moitzi et al., 2013). For agricultural work to be carried out at the operational speeds, especially on soft soil, the traction power is limited by the grip between driving wheels and the soil. In order to effectively use the engine power and not to deviate from the operational speed, working width has to be increased (Lacour et al., 2014; Moitzi et al., 2013). That is, the tractor has to be loaded with higher traction force. As a rule, when a tractor is loaded with high traction force, the slip of driving wheels exceeds the permissible limits. Terramechanics points out two essential ways to reduce the slip. One possibility is to increase tractor's mass by adding ballast. The other possibility is to enlarge the contact area between tires and terrain. With enlargement of the contact area between tires and terrain, tractor tires make less negative effect on the field and the result is less compacted soil under the tracks (Saengprachatanarug, Ueno, Taira, & Okayasu, 2013; Srinivasa Rao et al., 2012; Way & Kishimoto, 2004). Furthermore, due to the larger contact area, rolling resistance is smaller in soft soil (Battiato & Diserens, 2013; Molari et al., 2012; Taghavifar & Mardani, 2013).

Rolling resistance of driving wheels and slip of driving wheels are two factors that influence tractor pulling power, and these factors are interrelated. For the tractor driving on a hard-surface road, rolling resistance of the wheels becomes lower when inflation pressure in the tires is increased. On the soil, the lower the inflation pressure in the tires, the more shallow the track and less rolling resistance. For the tractor driving at low speeds (e.g., for soil tillage operations) pulling force is limited by contact area between tires and the soil. Driving wheels are not able to transfer all available engine power due to the fact that the grip between driving wheels and the soil realizes smaller propulsive force (Lyasko, 2010; Xia, 2011). In order to increase the pulling force, it is necessary to improve the conditions for the grip between driving wheels and the soil.

Totally different results are obtained when the tractor is working at higher speeds. For example, when stubble is being tilled at a speed higher than 15 km h^{-1}, pulling power is limited by engine power. This means that almost all power of the engine can be converted into pulling power (Wong, 2010; Zoz & Grisso, 2003). In such conditions, the ballast mass does not give an additional effect (Janulevičius & Giedra, 2008). Therefore, what positive effect is obtained by using ballast mass depends on the soil characteristics, inflation pressure in the tires, tractor working speed, etc.

Variations in soil structure and surface roughness affect variations in implement resistance and pulling force. In case of tractors with manual transmissions, engine torque reserve helps to overcome the increased resistance to pulling (Battiato & Diserens, 2013; Grisso et al., 2011). In order to not overload the engine and let it operate normally, work is usually done by not utilizing the full load. When pulling force increases, engine works in a range of higher torque, so greater traction is ensured even without switching into lower gear. Power reserve in the conditions of agricultural production varies in the range from 6 to 18 percent of total engine power. Most of the tractors usually are operated at 60-70 percent of maximum load (Grisso et al., 2011; Janulevičius et al., 2013; Lacour at al., 2014). Test results show that when a tractor is used in a step less variable transmission control mode, the engine can be loaded almost 100% (Macor & Rossetti, 2013).

In tillage work, draft can be increased by up to 15 percent depending on the ballast mass and the place where it is mounted (Janulevičius & Giedra, 2008; Pranav & Pandey, 2008). Results of tests carried out by researchers of tractors show that applying ballast mass is not the best solution to reduce tractor slip. This method has a very important drawback – A danger always remains of excessive compacting of the soil and damaging its structure at great depths (much deeper than it is tilled), which can reduce soil productivity (Keller, 2005; Saengprachatanarug et al., 2013; Way & Kishimoto, 2004). Extra mass makes the tires sink deeper into the soil and leave tracks. Depth of the tracks depends on the mass, soil hardness, tire dimensions and tire inflation pressure. Tire inflation pressure also influences how much the wheels slip and is regarded as on important factor affecting tractor field performance indicators, such as draft power (Senatore & Sandu, 2011; Srinivasa Rao et al., 2012; Šmerda & Čupera, 2010; Taghavifar & Mardani, 2013).

Currently, tractor performance researchers recommend reducing the inflation pressure in the tires, thus increasing contact area between the tire and the ground. It means that tractor mass is distributed over a larger contact area and the wheels' pressure on the soil decreases. Driving wheels sink less into the soil, tracks are not so deep and the rolling resistance is reduced. For example, resistance of 8-12 cm depth track corresponds to driving up a slope of 10%. Normally, slip of driving wheels should not exceed 15%, otherwise it causes lower productivity, cost-effectiveness, and intensive destruction of the soil (Keller, 2005; Moitzi et al., 2013; Saengprachatanarug et al., 2013; Šmerda & Čupera, 2010). If slip of driving wheels in the soil is low (less than 6-7%), it is also unacceptable, as traction power is not utilized and energy consumption per unit of performed work increases. Slip is low when driving wheels are loaded with too big weight force. In this case the power is used to carry the excess mass and press the soil, and fuel consumption may increase by 15% (Wong, 2010). Fuel consumption for carrying excess mass increases significantly when working at higher speeds (Battiato & Diserens, 2013; Moitzi et al., 2013; Taghavifar & Mardani, 2013). Analysis of research materials shows that optimal tractor slip in soil should be in the range of 8-12% (Battiato & Diserens, 2013; Keller, 2005; Moitzi et al., 2013).

The wheel slip is a critical parameter for fuel consumption and field performance. Many researchers in their works solve the problem of tractor tire slip normalization by adding ballast masses and reducing the tire pressures. However, the influence of variations in tire pressures and extra mass on tractor fuel consumption when tire slip is in the normal range (7-15%) is considered just moderately.

Purpose of the study was to test the effect of inflation pressure in the tires and additional (ballast) mass on the slip of driving wheels, fuel consumption and work productivity for a 4 WD tractor when slip varies in the range from 7 to 15 percent and the tractor is engaged in tillage operations.

2. Materials and Methods

2.1 Equipment, Site and Layout

To test effect of tractor driving wheels slip, fuel consumption and productivity on inflation pressure in the tires and ballast mass, a tractor FORD 8340 SLE and four bodies reversible plough KONGSKILDE VARIANT VPS (Kongskilde) was composed. Specifications of the tractor used for the test are listed in Table 1, and specifications of the plough are listed in Table 2.

Table 1. Tractor (Ford 8340 SLE) specifications

Engine:	
Engine model	Ford 7.5L 6-cyl diesel
Engine type	6 cylinder, liquid-cooled, in-line, turbocharged
PTO power (rated engine speed):	82.3 kW at 2200 rpm
Transmission & Chassis:	
Drive type	4WD
Transmission type	Electro Shift/Pulse Command
	four-speed power shift, 16 forward and reverse
Clutch	wet disc
Final drives:	inboard planetary
Differential lock:	electro-hydraulic
Front tires	440/65 R 24
Rear tires	18.4 R 38
Tractor weight (operating)	4936 kg
Wheelbase	2.61 m
Hydraulics	closed center pressure flow compensating (pressure: 200 bar)

Table 2. Plough specifications

Plough model	KONGSKILDE VARIANT VP–S
Type	Heavy duty 4 furrow version
Working width of furrow	0.3–0.52 m
Type of beam	Auto-reset
Weight	1850 kg

The tractor was equipped with a data recorder – accumulator SKRT–21 Lite (TECHNOTON) with electronic clock and the software SKRT–MANAGER. Tractor sensors were used to measure engine speed, ground speed and conditional drawbar pull force. The sensors were connected in parallel to the installed instrument system SKRT. Fuel consumption (1 or 1 h⁻¹) was measured by AIC-4004 VERITAS (AIC SYSTEMS AG) fuel flow meter. Calibration of all devices was performed according to standard procedures (Barzdžiukas, Augutis, & Žilinskas, 2012; Bručas, Šiaudinytė, Rybokas, & Grattan, 2014; Naranjo, Sandu, Taheri, & Taheri, 2014). Error did not exceed ±2%. Detailed specifications of measurement devices are presented in Table 3.

Engine speed, fuel consumption, actual speed of the tractor, conditional draft indicator and time were measured during the test. Average values of measured parameters (in 15 s intervals) were recorded as a time function and stored in data storage. Data from data storage were then transferred in a digital form to the computer, into Microsoft Excel program (Figure 1).

Figure 1 shows how the Microsoft Excel program window displays the recorded data: Column "A" shows the recording time, columns "B" and "C" show draft (from sensors on the right and left arms respectively), column "D" – actual speed, column "E" – travelled distance during the period, column "F" – fuel consumption during the period, column "G" – hourly fuel consumption, and column "H" – engine speed readings. Lines 560-573 show tilling process parameters for one direction of travel. Lines 574-578 are for turning at headlands. Line 579 accounts for starting a new test. This test was carried out for the unit traveling the other direction in the field. In such a way data were continuously recorded during the total test period. The results for each test were calculated for 3 minutes of traveled distance, i.e., each test consisted of 12 intervals, each 15 seconds long. In Figure 1 we can see data of one of the test, corresponding to recorded lines 561-572. Transitional test intervals recorded in lines 560 and 573 were ignored.

Table 3. Specifications of measurement devices

Instrumentation	Measurements	Range	Accuracy
PTO dynamometer, AW Type 400, (Perfect Power Control)	Speed	0–1500 rpm	±0.1%
	Torque	0–2850 Nm	±0.1%
	Power	0–298 kW @ 1000 rpm	±0.2%
Axis scales, WPD-2, (ZEMIC EUROPE)	Mass	5–15000 kg	±0.1%/1,0 kg
Fuel flow meter, AIC-4004 VERITAS, (AIC SYSTEMS AG)	Fuel consumption	2.0–80 l h^{-1}	±1%
		1.0–2.0 l h^{-1}	+1/–2%
Data recorder, SKRT–21 Lite, (TECHNOTON)	Software	SKRT–MANAGER	
	Information channels	8	
	Period limits	5–180 s	
	Memory capacity of each channel	750 h	
Penetrometer, PENETROLOGGER, (PENETROLOGGER)	Penetration force	0–100 N	±1%
Moisture sensor ML2x-UM-1.21, (Equitensiometers)	Soil moisture	m^3m^{-3} or%vol.	± 0.5 m^3m^{-3}

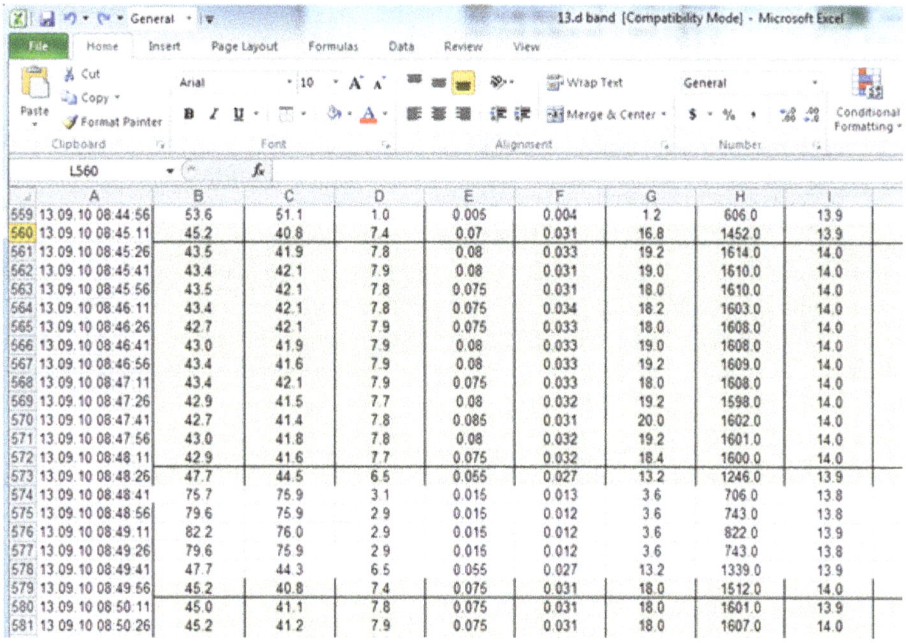

	A	B	C	D	E	F	G	H	I
559	13.09.10 08:44:56	53.6	51.1	1.0	0.005	0.004	1.2	606.0	13.9
560	13.09.10 08:45:11	45.2	40.8	7.4	0.07	0.031	16.8	1452.0	13.9
561	13.09.10 08:45:26	43.5	41.9	7.8	0.08	0.033	19.2	1614.0	14.0
562	13.09.10 08:45:41	43.4	42.1	7.9	0.08	0.031	19.0	1610.0	14.0
563	13.09.10 08:45:56	43.5	42.1	7.8	0.075	0.031	18.0	1610.0	14.0
564	13.09.10 08:46:11	43.4	42.1	7.8	0.075	0.034	18.2	1603.0	14.0
565	13.09.10 08:46:26	42.7	42.1	7.9	0.075	0.033	18.0	1608.0	14.0
566	13.09.10 08:46:41	43.0	41.9	7.9	0.08	0.033	19.0	1608.0	14.0
567	13.09.10 08:46:56	43.4	41.6	7.9	0.08	0.033	19.2	1609.0	14.0
568	13.09.10 08:47:11	43.4	42.1	7.9	0.075	0.033	18.0	1608.0	14.0
569	13.09.10 08:47:26	42.9	41.5	7.7	0.08	0.032	19.2	1598.0	14.0
570	13.09.10 08:47:41	42.7	41.4	7.8	0.085	0.031	20.0	1602.0	14.0
571	13.09.10 08:47:56	43.0	41.8	7.8	0.08	0.032	19.2	1601.0	14.0
572	13.09.10 08:48:11	42.9	41.6	7.7	0.075	0.032	18.4	1600.0	14.0
573	13.09.10 08:48:26	47.7	44.5	6.5	0.055	0.027	13.2	1246.0	13.9
574	13.09.10 08:48:41	75.7	75.9	3.1	0.015	0.013	3.6	706.0	13.8
575	13.09.10 08:48:56	79.6	75.9	2.9	0.015	0.012	3.6	743.0	13.8
576	13.09.10 08:49:11	82.2	76.0	2.9	0.015	0.012	3.6	822.0	13.9
577	13.09.10 08:49:26	79.6	75.9	2.9	0.015	0.012	3.6	743.0	13.8
578	13.09.10 08:49:41	47.7	44.3	6.5	0.055	0.027	13.2	1339.0	13.9
579	13.09.10 08:49:56	45.2	40.8	7.4	0.075	0.031	18.0	1512.0	14.0
580	13.09.10 08:50:11	45.0	41.1	7.8	0.075	0.031	18.0	1601.0	13.9
581	13.09.10 08:50:26	45.2	41.2	7.9	0.075	0.031	18.0	1607.0	14.0

Figure 1. Information from data recorder SKRT–21 Lite transferred into Microsoft Excel program (sample)

For the test, a level, nearly smooth-texture, loamy wheat-stubble field (after wheat harvesting) strip of 380 ± 10 m length was selected. Soil moisture at a depth of 10 cm was 17.6 ± 1.1%, soil hardness 1.14 ± 0.09 MPa, ambient temperature 19 ± 2 °C. The plough was set for 1.75 m working width and 0.20 m working depth. For processing of the results, a value of plough resistance to traction was accepted that prevailed during the tests. Only tests in which the traction force deviations from the prevailing (average) values did not exceed 5% were used in the results. Slip of tractor driving wheels during all tests was <15%. Tests were carried out with front driving axle of the tractor activated, driving differentials locked and the gear 3H engaged, while engine speed was 1620 ± 10 rpm. Tests were carried out by driving in one direction and then turning and driving in the opposite direction and the mean value was calculated from the obtained results. Tests were repeated 3 times. Tests were performed with 520, 360, 200, and 0 kg ballast mass in front of the tractor. The distribution of overall mass of tractor on the front and rear wheels, are presented in Table 4. Tests were carried out by making all combinations of selected ballast mass of 520, 360, 200 and 0 kg and inflation pressures in tractor tires of 240, 190, 150 and 100 kPa (16 combinations). Tractor tires: rear – ALLIANCE 18.4 R 38, 152 A8 ***, front – ALLIANCE 440/65 R 24, 149 A8. Wear of rear tires – 4.1%, front tires – 4.5%.

Table 4. Distributions of the overall tractor mass on front and rear axles

Ballast mass in front of the tractor, kg	On the front axles, kg	On the rear axles, kg
0	1920	3016
200	2180	2956
360	2389	2907
520	2597	2859

2.2 Calculations

The wheel slip (s) relates to the actual forward velocity (v_a) of the center of wheel and the angular velocity (ω) of the wheel as follows (Battiato & Diserens, 2013; Maclaurin, 2014; Moitzi et al., 2013; Senatore & Sandu, 2011):

$$s = \frac{\omega R_r - v_a}{\omega R_r} = 1 - \frac{v_a}{\omega R_r} \tag{1}$$

Where R_r is the rolling radius of the wheel.

Percent of slip of tractor was calculated according to the following equation:

$$s = \left(1 - \frac{v_a\, i_{tr}}{\omega_e\, R_{rd}}\right) 100 , \% \tag{2}$$

Where v_a is the actual speed of the tractor, ω_e is the angular velocity of the engine shaft, i_{tr} is the ratio of tractor transmission, and R_{rd} is the rolling radius of driving wheels of the tractor.

The tire rolling radius R_{rd} was determined according to the standard S296.2 of American Society of Agricultural Engineers (ASAE) as the distance travelled per revolution of the wheel divided by 2π, when operating at the specified zero conditions. Such conditions are assumed to be when the tractor is driving on a smooth road, and drawbar load is equal to zero (American Society of Agricultural Engineers 1983). Zoz and Grisso (2003) showed that the difference in measured rolling radiuses, when tractor is driving on a hard road compared to a test surface, is small under normal agricultural soil conditions (untilled soil), and thus has little impact on the final results. In our test the values of tractor driving wheel rolling radiuses (for each case of applied ballast mass and inflation pressure in the tires) were determined experimentally, i.e. by measuring the distance which driving wheels traveled per 10 revolutions. Tractor was not loaded by pulling force during the tests. The plough was set so that it would not till the soil ($h_p = 0$ m).

Rolling radius of tractor driving wheels was calculated according to the following equation:

$$R_{rd} = \frac{p_z}{2\pi z} , \text{m} \tag{3}$$

Where z is the count of driving wheels' revolutions, p_z is the distance traveled per count z of driving wheels' revolutions, π is the mathematical constant ($\pi = 3.14$).

Fuel consumption per hectare B_{ha} was calculated according to the following equation:

$$B_{ha} = \frac{B_h}{0.36\, v_a\, H} , \text{l ha}^{-1} \tag{4}$$

Where B_h is hourly fuel consumption, H is working width of the unit.

3. Results and Discussion

The present study was focused on the tire pressure and extra weight variation effect on fuel consumption at normal tractor wheels slippage (7-15%) during tillage operation. Figures 2, 3, 4 and 5 show tractor performance indicators' (such as ground speed, slip of driving wheels and fuel consumption) dependences on ballast mass at different inflation pressures in the tires. Figure 2 illustrates that when ballast mass was increased and inflation pressure in the tires was reduced, ground speed of tractor increased.

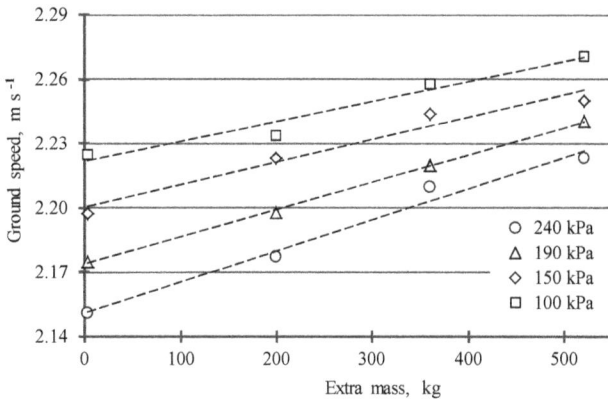

Figure 2. Ground speed dependences on the extra mass at different tire inflation pressures

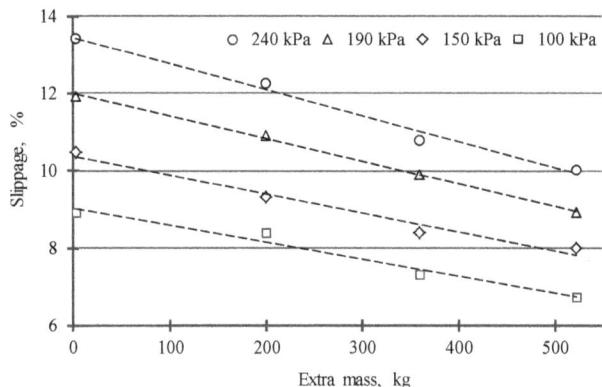

Figure 3. Driving wheels slippage dependences on the extra mass at different tire inflation pressures

Figure 3 illustrates that when ballast mass was increased and inflation pressure in the tires was reduced, slip of driving wheels decreased. When tilling work was performed with the front driving axle activated, increasing ballast mass in front of the tractor from 0 to 520 kg reduced slip on average by nearly 2.8%. When inflation pressure in the tires was 240 kPa, increasing ballast mass from 0 to 520 kg reduced slip from 13.5% to 10.2%. When inflation pressure in the tires was 190 kPa, slip decreased from 11.7% to 9.1%. When inflation pressure in the tires was 150 kPa, slip decreased from 10.3% to 7.8%. When inflation pressure in the tires was 100 kPa, slip decreased from 9.0% to 6.7%. After reducing inflation pressure in the tires from 240 kPa to 100 kPa, slip decreased on average by 3.8%. From the results presented in Figure 3 we can see that tractor slip varied in the range from 6.5 to 13.5% during all the tests. This corresponds to the tractor slip that is recommended for tillage works in a number of sources (Battiato & Diserens, 2013; Battiato, Diserens, Laloui, & Sartori, 2013; Moitzi et al., 2013; Srinivasa Rao et al., 2012; Šmerda & Čupera, 2010). In loam stubble of average moisture and hardness, such slip corresponds to adequate grip of driving wheels with the soil (Battiato & Diserens, 2013; Janulevičius & Giedra, 2008; Maclaurin, 2014; Zoz & Grisso, 2003).

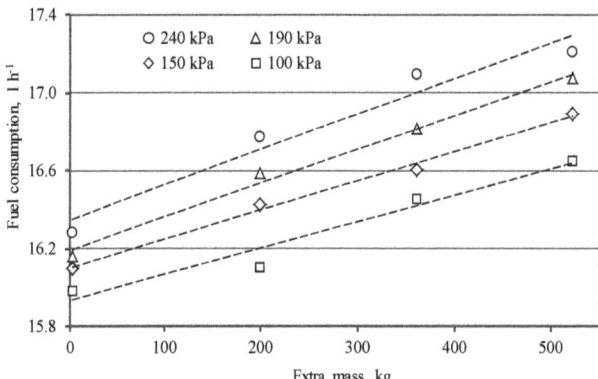

Figure 4. Tractor hourly fuel consumption dependences on the extra mass at different tire inflation pressures

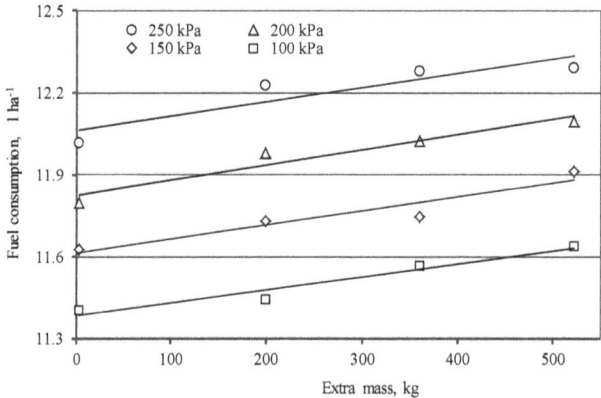

Figure 5. Tractor fuel consumption per hectare dependences on the extra mass at different tire inflation pressures

Figures 4 and 5 shows that when the ballast mass was increased, tractor fuel consumption also increased. When inflation pressure in the tires was 240 kPa, after increasing the ballast mass from 0 to 520 kg, hourly fuel consumption (Figures 4) increased from 16.3 l h^{-1} to 17.25 l h^{-1}, and fuel consumption per hectare (Figures 5) increased from 12.0 l ha^{-1} to 12.3 l ha^{-1}. When inflation pressure in the tires was 190 kPa, hourly fuel consumption increased from 16.2 l h^{-1} to 17.1 l h^{-1}, and fuel consumption per hectare increased from 11.8 l ha^{-1} to 12.1 l ha^{-1}. When inflation pressure in the tires was 150 kPa, hourly fuel consumption increased from 16.1 l h^{-1} to 16.9 l h^{-1}, and fuel consumption per hectare increased from 11.6 l ha^{-1} to 11.9 l ha^{-1}. When inflation pressure in the tires was 100 kPa, hourly fuel consumption increased from 16.0 kg h^{-1} to 16.7 kg h^{-1}, and fuel consumption per hectare increased from 11.4 l ha^{-1} to 11.6 l ha^{-1}. When loam stubble was tilled with tractor front driving axle activated, after increasing the front ballast mass from 0 to 520 kg, hourly fuel consumption increased on average by 0.8 l h^{-1}, and fuel consumption per hectare increased by 0.3 l ha^{-1}. These results do not include fuel consumption per hectare at the headlands. When tire inflation pressure was lowered from 240 kPa to 100 kPa, hourly fuel consumption fell on average by 0.5 l h^{-1}, and fuel consumption per hectare – by 0.7 l ha^{-1}. For the sufficient grip conditions, tractor ballast reduces driving wheels slip, but increases fuel consumption. When tire inflation pressure is reduced, the driving wheels slip, and tractor fuel consumption is reduced. Tractor work productivity, hourly fuel consumption and fuel consumption per hectare, dependence on the driving wheels slip, tires inflation pressure and the extra mass (ballast weights) is presented in Figures 6, 7 and 8, respectively.

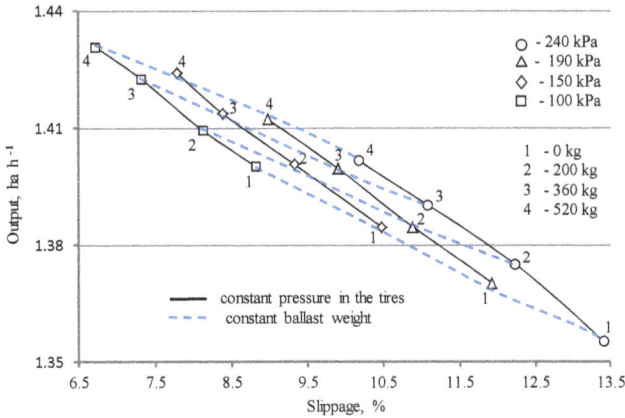

Figure 6. Work productivity dependences on driving wheel slippage, tire inflation pressures and extra mass (ballast weights)

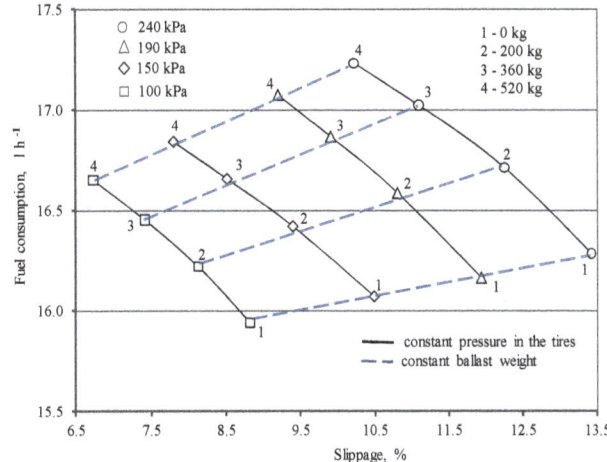

Figure 7. Hourly fuel consumption dependences on driving wheel slippage, tire inflation pressures and extra mass (ballast weights)

Figures 6, 7 and 8 illustrates that when extra mass (ballast weights) was added and inflation pressure in the tires was reduced, slip of driving wheels decreased. When extra mass (ballast weights) is added, slip of driving wheels decreases, and work productivity increases together with increased fuel consumption. Keeping tire inflation pressure constant and increasing the ballast mass from 0 to 520 kg, slip of driving wheels decreased on average 2.3%, work productivity increased by approximately 0.04 ha h^{-1}, hourly fuel consumption increased by nearly 0.9 l h^{-1}, and fuel consumption per hectare increased by nearly 0.3 l ha^{-1}. When tire inflation pressure is reduced, driving wheels slip as well as fuel consumption decreases, while productivity increases. Keeping tractor ballast mass constant and reducing the tire inflation pressure from 240 kPa to 100 kPa, driving wheel slip decreased on average 4.2%, hourly fuel consumption also decreased by approximately 0.5 l h^{-1}, fuel consumption per hectare decreased by approximately 0.3 l ha^{-1}, and work productivity increased by about 0.035 ha h^{-1}.

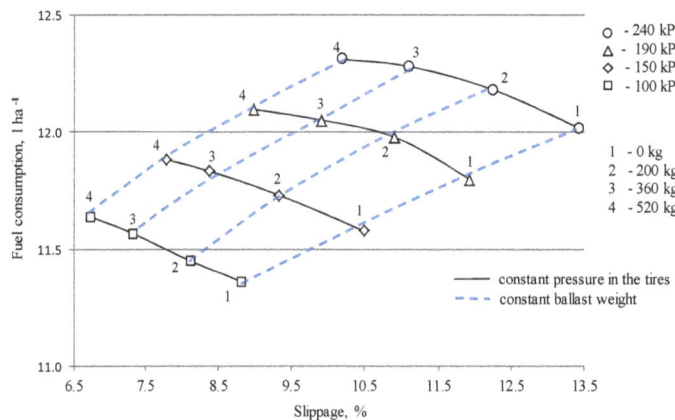

Figure 8. Fuel consumption per hectare dependences on driving wheel slippage, tire inflation pressures and extra mass (ballast weights)

Figure 6 shows that the largest slip of driving wheels (13.5%) and the lowest work productivity (1.36 ha h^{-1}) occurred when working without ballast weights and when tire inflation pressure was 240 kPa (maximum). Minimum slip of driving wheels (6.6%) and the highest productivity (1.43 ha h^{-1}) occurred when tire inflation pressure was 100 kPa (minimum) and ballast weight mounted in front of the tractor was 520 kg (maximum). Figures 7 and 8 show that the lowest hourly fuel consumption (16.0 l h^{-1}) and the lowest fuel consumption per hectare (11.4 l ha^{-1}) occurred when the tractor was working without ballast weights and tire inflation pressure was 100 kPa (minimum). When the work was being done in the mode of the lowest fuel consumption, driving wheels slip 8.8%, and the work productivity was 1.39 ha h^{-1}.

The highest hourly fuel consumption (17.2 l h^{-1}) and the highest fuel consumption per hectare (12.3 l ha^{-1}) occurred when the tractor was working with the biggest ballast weight (520 kg) and tire inflation pressure was 240 kPa (maximum). When working in a mode of maximum fuel consumption, the slip of driving wheels was not the greatest and work productivity was not minimal.

The dependencies of fuel consumption and productivity on the driving wheel slip, tire inflation pressure and extra mass (ballast weight) reveal preparation of the tractor for implementation of the desired operating parameters found in this study. The results of this study may provide helpful insights into a reasonable choice of tractor configuration as well as effective control of driving wheel slip, with a view to optimizing tractor performance parameters, thereby saving time and reducing the costs of tillage.

4. Conclusions

The aim of present study was to analyze the effect of variations in tire inflation pressures and extra mass on the fuel consumption and productivity of a 4 WD tractor when tire slip varied in the range from 7 to 15 percent (normal slip in the soil).

When normal slip occurs, reduced the inflation pressure decreases the driving wheel slip and fuel consumption, and increases work productivity. Added weight of the tractor decreases the driving wheel slip and increases work productivity, but also increases fuel consumption.

When tilling wheat stubble where soil moisture content in 10 cm depth was 17.6 ± 1.1% and hardness was 1.14 ± 0.09 MPa, reducing inflation pressure from 240 kPa to 100 kPa resulted in driving wheel slippage to decrease on average 4.2%, hourly fuel consumption also dropped by approximately 0.5 l h^{-1} work productivity increased by about 0.035 ha h^{-1}, and fuel consumption per hectare decreased by approximately 0.5 l ha^{-1}. Adding weight from 0 to 520 kg caused slip of driving wheels to decrease by average 2.3%, while hourly fuel consumption increased by nearly 0.9 l h^{-1}, work productivity increased by approximately 0.04 ha h^{-1}, and fuel consumption per hectare increased by nearly 0.3 l ha^{-1}.

In conclusion, the ballast weight and tire pressure reductions during normal slippage of the tractor wheels bring a positive effect on fuel consumption, maintaining the same slippage.

The results of this study may provide helpful insights into a reasonable choice of tractor configuration as well as effective control of driving wheel slip, with a view to optimizing tractor performance parameters, thereby saving time and reducing the costs of tillage.

Acknowledgements

The authors highly appreciate the comments and suggestions of the anonymous reviewers, which significantly contributed to improve the manuscript.

References

American Society of Agricultural Engineers. (1983). ASAE Standard S296.2 - Uniform terminology for traction of Agricultural Tractors, Self-Propelled Implements, and other Traction and Transport Devices. *Agricultural engineers yearbook of standards*. St. Joseph, Mich.: ASAE.

Backman, J., Oksanen, T., & Visala, A. (2013). Applicability of the ISO 11783 network in a distributed combined guidance system for agricultural machines. *Biosystems Engineering, 114*, 306-317. http://dx.doi.org/10.1016/j.biosystemseng.2012.12.017

Barzdžiukas, J., Augutis, S. V., & Žilinskas, R. P. (2012). The precise measurement of car velocity. *Transport, 27*(2), 138-142. http://dx.doi.org/10.3846/16484142.2012.690346

Battiato, A., & Diserens, E. (2013). Influence of Tyre Inflation Pressure and Wheel Load on the Traction Performance of a 65 kW MFWD Tractor on a Cohesive Soil. *Journal of Agricultural Science, 5*(8), 197-215. http://dx.doi.org/10.5539/jas.v5n8p197

Battiato, A., Diserens, E., Laloui, L., & Sartori, L. (2013). A Mechanistic Approach to Topsoil Damage due to Slip of Tractor Tyres. *Journal of Agricultural Science and Applications, 2*(3), 160-168. http://dx.doi.org/10.14511/jasa.2013.020305

Bručas, D., Šiaudinytė, L., Rybokas, M., & Grattan, K. (2014). Theoretical aspects of the calibration of geodetic angle measurement instrumentation. *Mechanika, 20*(1), 113-117. http://dx.doi.org/10.5755/j01.mech.20.1.6590

Dagiliūtė, R., & Juknys, R. (2012). Eco-efficiency: trends, goals and their implementation in Lithuania. *Journal of Environmental Engineering and Landscape Management, 20*(4), 265-272. http://dx.doi.org/10.3846/16486897. 2012. 661072

Grisso, R., Pitman, R., Perumpral, J. V., Vaughan, D., Roberson, G. T,. & Hoy, R. M. (2011). "Gear Up and Throttle Down" to Save Fuel. *VCE Publication*, 442-450. Retrieved from http://pubs.ext.vt.edu/442/442-450/442-450_pdf.pdf

Hashemi, A., Ahmad, D., Othman, J., & Sulaiman, S. (2012). Development and Testing of a New Tillage Apparatus. *Journal of Agricultural Science, 4*(7), 103-110. http://dx.doi.org/10.5539/jas.v4n7p103

Janulevičius, A., & Giedra, K. (2008). Tractor ballasting in field work. *Mechanika, 5*(73), 27-34. Retrieved from http://zurnalas.mechanika.ktu.lt/files/Janulevicius573.pdf

Janulevičius, A., Juostas, A., & Pupinis, G. (2013). Tractor's engine performance and emission characteristics in the process of ploughing. *Energy Conversion and Management, 75*, 498-508. http://dx.doi.org/10.1016/j.enconman.2013.06.052

Juostas, A., & Janulevičius, A. (2014). Tractor's engine efficiency and exhaust emissions' research in drilling work. *Journal of Environmental Engineering and Landscape Management, 22*(02), 141-150. http://dx.doi.org/10.3846/16486897.2013.852556

Keller, T. (2005). A model for the prediction of the contact area and the distribution of vertical stress below agricultural tyres from readily available tyre parameters. *Biosystems Engineering, 92*(1), 85-96. http://dx.doi.org/10.1016/j.biosystemseng.2005.05.012

Khambalkar, V., Pohare, J., Katkhede, S., Bunde, D., & Dahatonde, S. (2010). Energy and Economic Evaluation of Farm Operations in Crop Production. *Journal of Agricultural Science, 2*(4), 191-200. http://dx.doi.org/10.5539/jas.v2n4p191

Kheiralla, A. F., Azmi, Y., Zohadie, M., & Ishak, W. (2004) Modelling of power and energy requirements for tillage implements operating in Serdang sandy clay loam, Malaysia. *Soil & Tillage Research, 78*, 21-34. http://dx. doi:10.1016/j.still.2003.12.011

Lacour, S., Burgun, C., Perilhon, C., Descombes, G., & Doyen, V. (2014). A model to assess tractor operational efficiency from bench test data. *Journal of Terramechanics, 54*, 1-18. http://dx.doi.org/10.1016/j.jterra.2014.04.001

Lyasko, M. I. (2010). How to calculate the effect of soil conditions on tractive performance. *Journal of*

Terramechanics, 47, 423-445. http://dx.doi.org/10.1016/j.jterra.2010.04.003

Maclaurin, B. (2014). Using a modified version of the Magic Formula to describe the traction/slip relationships of tyres in soft cohesive soils. *Journal of Terramechanics, 52*, 1-7. http://dx.doi.org/10.1016/j.jterra.2013.11.005

Macor, A., & Rossetti, A. (2013). Fuel consumption reduction in urban buses by using power split transmissions. *Energy Conversion and Management, 71*, 159-171. http://dx.doi.org/10.1016/j.enconman.2013.03.019

Magalhães, A. C., Souza, J. M, Santana, M. A., & Sabbag, O. J. (2013). Analysis of the Mechanization Index of Wheel Tractors in Rural Farm Holdings. *Journal of Agricultural Science, 5*(44), 127-138. http://dx.doi.org/10.5539/jas.v5n11p127

Moitzi, G., Haas, M., Wagentristl, H., Boxberger, J., & Gronauer, A. (2013). Energy consumption in cultivating and ploughing with traction improvement system and consideration of the rear furrow wheel-load in ploughing. *Soil & Tillage Research, 134*, 56-60. http://dx.doi.org/10.1016/j.still.2013.07.006

Molari, G., Bellentani, L., Guarnieri, A, Walker, M., & Sedoni, E. (2012). Performance of an agricultural tractor fitted with rubber tracks. *Biosystems engineering, 111*, 57-63. http://dx.doi.org/10.1016/j.biosystemseng.2011.10.008

Naranjo, S. D., Sandu, C., Taheri, S., & Taheri, S. (2014). Experimental testing of an off-road instrumented tire on soft soil. *Journal of Terramechanics, 56*, 119-137. http://dx.doi.org/10.1016/j.jterra.2014.09.003

Patterson, M. S., Gray, J. P., Bortolin, G., & Vantsevich, V. V. (2013). Fusion of driving and braking tire operational modes and analysis of traction dynamics and energy efficiency of a 4 x 4 loader. *Journal of Terramechanics, 50*, 133-152. http://dx.doi.org/10.1016/j.jterra.2013.01.003

Peca, J. O., Serrano, J. M., Pinheiro, A., Carvalho, M., Nunes, M., Ribeiro, L., & Santos, F. (2010). Speed advice for power efficient drawbar work. *Journal of Terramechanics, 47*, 55-61. http://dx.doi.org/10.1016/j.jterra.2009.07.003

Pranav, P. K., & Pandey, K. P. (2008). Computer simulation of ballast management for agricultural tractors. *Journal of Terramechanics, 45*, 193-200. http://dx.doi.org/10.1016/j.jterra.2008.12.002

Saengprachatanarug, K., Ueno, M., Taira, E., & Okayasu, T. (2013). Modeling of soil displacement and soil strain distribution under a traveling wheel. *Journal of Terramechanics, 50*, 5-16. http://dx.doi.org/10.1016/j.jterra.2012.06.001

Senatore, C., & Sandu, C. (2011). Torque distribution influence on tractive efficiency and mobility of off-road wheeled vehicles. *Journal of Terramechanics, 48*, 372-383. http://dx.doi.org/10.1016/j.jterra.2011.06.008

Srinivasa Rao, S., Ramji, K., & Naidu, M. K. (2012). Analytical approach for the prediction of steady state tyre forces and moments under different normal pressure distributions. *Journal of Terramechanics, 49*, 281-289. http://dx.doi.org/10.1016/j.jterra.2012.10.002

Stoilov, S., & Kostadinov, G. D. (2009). Effect of weight distribution on the slip efficiency of a four-wheel-drive skidder. *Biosystems engineering, 104*, 486-492. http://dx.doi.org/10.1016/j.biosystemseng.2009.08.011

Szendrő, G., & Török, Á. (2014). Theoretical investigation of environmental development pathways in the road transport sector in the European Region. *Transport, 29*(1), 12-17. http://dx.doi.org/10.3846/16484142.2014.893538

Šmerda, T., & Čupera, J. (2010). Tire inflation and its influence on drawbar characteristics and performance – Energetic indicators of a tractor set. *Journal of Terramechanics, 47*, 395-400. http://dx.doi.org/10.1016/j.jterra.2010.02.005

Taghavifar, H., & Mardani, A. (2014). On the modeling of energy indices of agricultural tractor driving wheels applying adaptive neuro-fuzzy inference system. *Journal of Terramechanics, 56*, 37-47. http://dx.doi.org/10.1016/j.jterra.2014.08.002

Taghavifar, H., & Mardani, A. (2013). Investigating the effect of velocity, inflation pressure, and vertical load on rolling resistance of a radial ply tire. *Journal of Terramechanics, 50*, 99-106. http://dx.doi.org/10.1016/j.jterra.2013.01.005

Vantsevich, V. V. (2008). Power losses and energy efficiency of multi-wheel driver vehicles: A method for evaluation. *Journal of Terramechanics, 45*, 89-101. http://dx.doi.org/10.1016/j.jterra.2008.08.001

Way, T. R, & Kishimoto, T. (2004). Interface pressures of a tractor drive tyre on structured and loose soils.

Biosystems Engineering, 87(3), 375-386. http://dx.doi.org/10.1016/j.biosystemseng.2003.12.001

Wong, J. Y. (2010). *Terramechanics and off-road vehicle engineering* (2nd ed.). Elsevier.

Xia, K. (2011). Finite element modeling of tire/terrain interaction: Application to predicting soil compaction and tire mobility. *Journal of Terramechanics, 48*, 113-123. http://dx.doi.org/10.1016/j.jterra.2010.05.001

Zoz, F. M., & Grisso, R. D. (2003). Traction and tractor performance. *ASAE distinguished lecture series* (Tractor design No. 27). ASAE publication No. 913C0403. St. Joseph, Mich.: ASAE. Retrieved from http://bsesrv214.bse.vt.edu/Dist_Lecture_27/Resources/Traction_Tractor_Performance.PDF

Appendix

Appendix 1

4WD: four-wheel drive;

B_{ha}: fuel consumption per hectare (l ha^{-1});

B_h: hourly fuel consumption (l h^{-1});

H: working width of the unit (m);

h_p: ploughing depth (m);

s: wheel slip coefficients;

π: mathematical constant ($\pi \approx 3.14$);

z: revolutions of the wheel;

i_{tr}: gear-ratio of the tractor transmission;

R_r: rolling radius of the wheel (m);

R_{rd}: rolling radius of the drive wheels of tractor (m);

p_z: distance of wheels travel during z rotations (m);

v_a: actual speed (m s^{-1});

ω: angular velocity of the wheel (s^{-1});

ω_e: angular velocity of the engine (s^{-1}).

12

Effect of Trap Orientation and Interval Distance on Captures of *Isoceras sibirica* Alpheraky (Lepidoptera: Cossidae)

Hongxia Liu[1], Zhixiong Liu[1], Haixia Zheng[1], Meihong Yang[1], Jinlong Liu[1] & Jintong Zhang[1]

[1] Institute of Chemical Ecology, Shanxi Agricultural University, Shanxi, China

Correspondence: Jintong Zhang, Institute of Chemical Ecology, Shanxi Agricultural University, Shanxi, 030801, China. E-mail: sxaulhx@163.com

Abstract

Studies were conducted in an asparagus field in taigu (37°18′N, 112°29′E, 824 m above sea level), Shanxi province, China, May to June, 2009 to 2011, to evaluate the influence of interval distance and orientation on catches of the carpenterworm, *Isoceras sibirica* Alpheraky (Lepidoptera: Cossidae) in pheromone-baited traps. The results showed that catches of male *I. sibirica* moths in upwind were higher than in other traps for any intertrap distances. When intertrap distances were shorter than 30 m, interference between traps occurred. These results reveal the effective trap orientation and interval distance for *I. sibirica* and thus provide guidelines for improving the effectiveness of traps in monitoring and controlling *I. sibirica* in fields.

Keywords: carpenterworm, monitoring and controlling, sex pheromone, trap orientations, trap interval distances

1. Introduction

The carpenterworm, *Isoceras sibirica* Alpheraky (Lepidoptera: Cossidae), is one of the major pest of *Asparagus officinalis* Linn. in China, causing significant yield losses (Duan et al., 2008). To date, no effective measures for controlling *I. sibirica* are available; this is due to the root-boring habit of the larvae. The use of synthetic sex pheromones to interfere with reproduction offers an environmental friendly measure to control the pest. Sex pheromones are species-specific and highly selective. They are valuable tools for use in integrated pest management as they are non-toxic and do not represent a health risk to humans and animals. Indeed, the use of pheromones has been reported for a number of insect species, for such purposes as monitoring emergence patterns (Patricia et al., 2008), monitoring pest populations for management decisions (Kehat et al., 1992), assessing the levels of insecticide resistance in pest populations (Haynes et al., 1986, 1987), luring and trapping adult males to suppress pest populations (Zhang et al., 2002), and for mating disruption (Higbee et al., 2008; Il'Ichev et al., 2006; Stelinski et al., 2007).

The effectiveness of pheromone-baited traps in capturing insects may be influenced by many factors, including distance of traps (Schlyter, 1992; Byers, 1999; Laboke et al., 2000), environment (Jansson et al., 1989; Laboke et al., 2000; Sappington & Spurgeon, 2000), trap type and trap height (Murad, 2001; Sarzynski, 2004). One of these factors is the density of traps, intertrap distances may affect the number of males captured because of interference between traps. When traps are placed close to each other, their radius of attraction may overlap. Furthermore, the interaction of traps is not constant with distance, but varies with lure concentration and wind conditions. By reducing the pheromone release, the active space of a trap decreases because the average concentration of pheromone downwind from the traps decreases (Bradshaw et al., 1989; Judd & Borden, 1989). In *I. sibirica*, active components present within the extract from the female sex pheromone gland are (Z)-9-tetradecenyl acetate (Z9-14:Ac), (Z)-7-tetradecenyl acetate (Z7-14:Ac), and (Z)-9-hexadecadecenyl acetate (Z9-16:Ac) (Zhang et al., 2011). However, factors relative with traps of *I. sibirica* in field remains unknown, so examination of trap orientation and interval distance will be required for successful implementation of pheromone-based trapping for this insect. Our goals were to determine the effects of trap location and interval distance on efficiency of capture of *I. sibirica* in pheromone-baited traps in Asparagus growing areas. Based on our results, we recommend steps for implementing an effective pheromone-baited trapping program against *I. sibirica* in field.

2. Materials and Methods

2.1 Experimental Setup

The study was conducted in an asparagus field in taigu (37°18′N, 112°29′E, 824 m above sea level), Shanxi province, China, May to June, 2009 to 2011, during the main peak periods of *I. sibirica* adult moth emergence and flight in this region. Traps were hung on wooden supports, 0.5 to 1 m in height, and set at 30 m intervals for the trap orientation and intertrap distance tests. Traps were deployed in arrays of nine traps in a 3 × 3 grid pattern with 10, 20, 30, 40, or 50 m between traps (Figure 1). Each treatment was replicated four times. The minimum distance between neighboring plots was 50 m. This experiment used 180 traps. The prevailing winds were from the south-east (local meteorological data for the past ten years). Trap catches were checked every morning and captured moths were recorded and removed daily.

2.2 Statistical Analysis

An assumption of normality was tested for all data sets with the Shapiro-Wilk test. If the null hypothesis that data were normally distributed was rejected, count data were transformed with a square-root transformation to stabilize variances (Snedecor & Cochran, 1967). If the transformed data were normally distributed (Shapiro-Wilk test), analysis of variance (ANOVA) was used and means with significant differences were separated using Tukey's studentized range test. Otherwise, the data were analyzed with a Kruskal-Wallis parametric ANOVA of mean ranks.

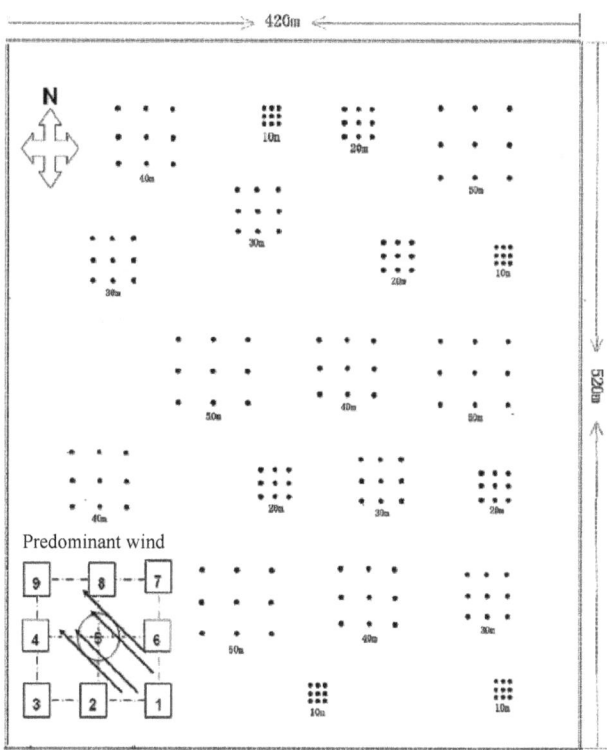

Figure 1. Schematic representation of locations of the *I. sibirica* pheromone-baited traps, showing arrays of 10, 20, 30, 40 and 50 m distance between traps, and the position relative to the cardinal directions and prevailing wind. Arabic numerals in panels expressed the serial numbers of trap position. •: *I. sibirica* Sex Pheromone-Baited Trap

The male capture per square metre as a function of grid distances was estimated using the non-linear regression function y = axb, where y is the density of male *I. sibirica* captured per square metre, x is the grid distance, and a and b are constants obtained as fitted parameters. The number of males captured per square meter for each inter-trap distance was estimated by the ratio of the total number of males captured in nine traps to the total area within the trap array for each inter-trap distance (Bacca et al., 2006). All analyses were done with SPSS for Windows Version 16.0 software.

3. Results

3.1 Effect of Trap Orientation

We recorded significantly higher male moth captures on pheromone-baited traps located the upwind edge of grids than the downwind edge, but differences were not significant for any intertrap distance (10 m: H = 1.5, df = 8, P = 0.9 > 0.05; 20 m: H = 5.4, df = 8, P = 0.7 > 0.05; 30 m: H = 5.8, df = 8, P = 0.7 > 0.05; 40 m: H = 4.3, df = 8, P = 0.8 > 0.05; 50 m: H = 2.4, df = 8, P = 0.9 > 0.05) (Figure 2).

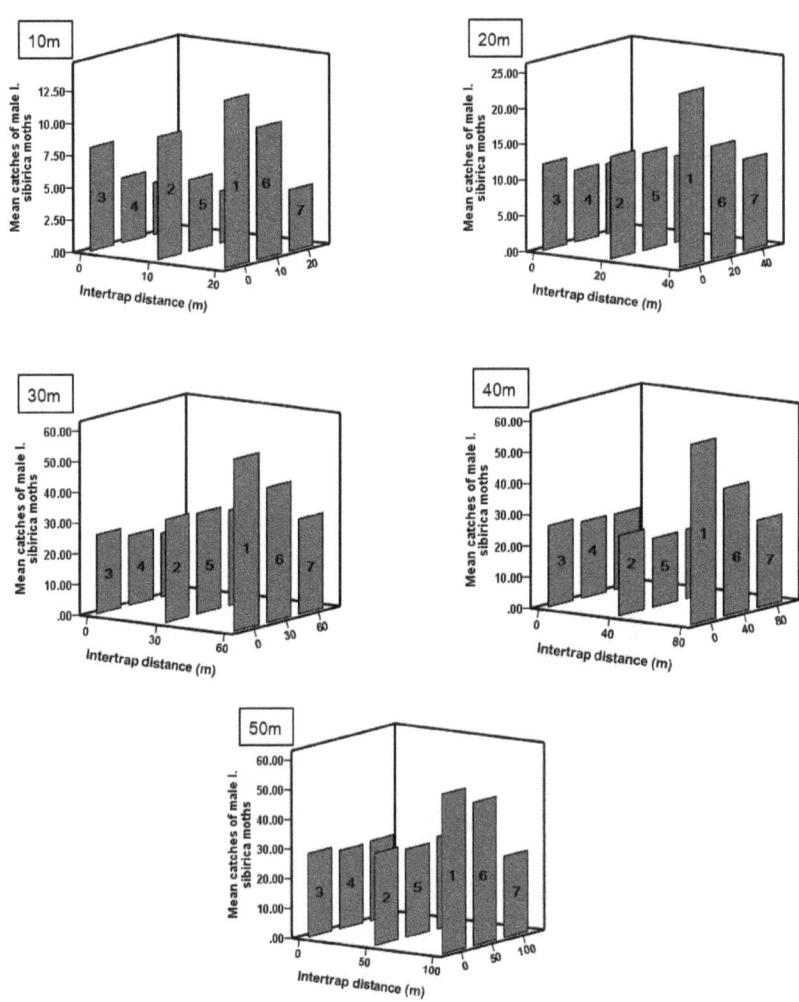

Figure 2. Effect of trap position within grids with different intertrap distances (shown in the upper left of each graph) on mean number of male *I. sibirica* moths captured per trap during the entire flight season. Trap position had no significant effect on mean catch (10 m: H = 1.5, df = 8, P = 0.9 > 0.05; 20 m: H = 5.4, df = 8, P = 0.7 > 0.05 ; 30 m: H = 5.8, df = 8, P = 0.7 > 0.05; 40 m: H = 4.3, df = 8, P = 0.8 > 0.05; 50 m: H = 2.4, df = 8, P = 0.9 > 0.05;). The prevailing winds were from the south-east

3.2 Effect of Distances between Traps

The largest numbers of *I. sibirica* were captured in the traps placed 30m apart (H = 116.6, df = 4, P = 0 < 0.05) (Figure 3), but no significant difference in captures among traps placed more than 30 m apart (H=1.6, df=2, P = 0.44 > 0.05) (Figure 3). The capture density (males captured per square meter) depended on the distance among traps: the greater the distance, the smaller the capture density (Figure 4). At intertrap distance of 30 m or greater, capture density stabilized and traps did not compete with each other (Figure 4).

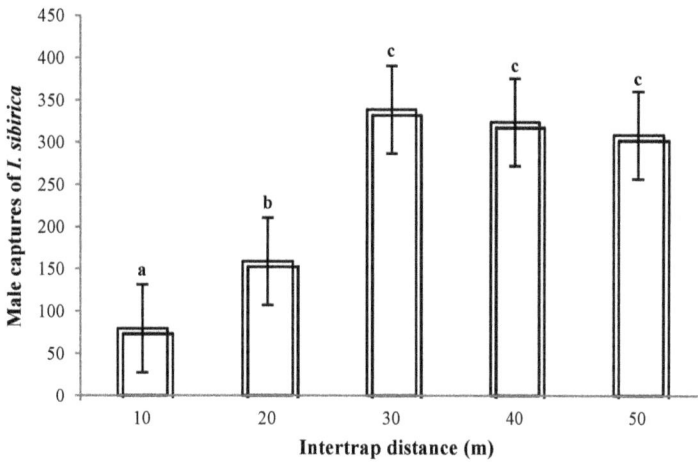

Figure 3. Effect of intertrap distance among traps and mean number (±SE) of male *I. sibirica* moths captured in the whole period corresponding to peak moth flight (H = 116.6, df = 4, P = 0 < 0.05; n = 1800 traps)

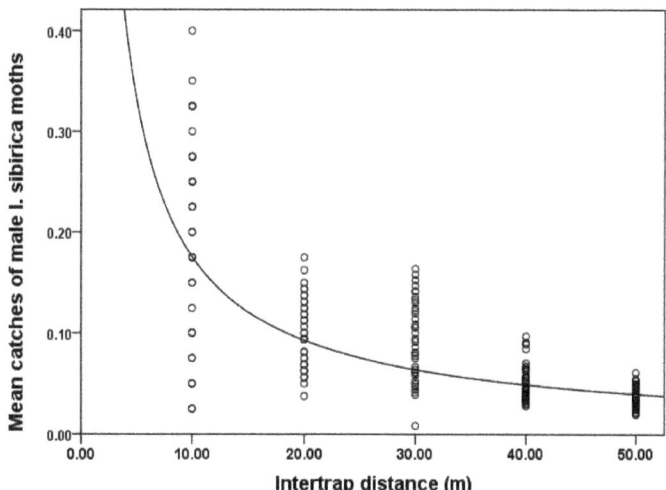

Figure 4. Relationship between capture of male *I. sibirica* moths per square meter and distance among traps
($y = 3.931x^{-0.50}$, $R^2 = 0.930$, $F = 54.2$, d.f. = 4, 179, $P < 0.0001$; n = 1800)

4. Discussion

For trap spacing, the present results show that capture density stabilized and traps did not compete with each other at intertrap distance of 30 m or greater, suggesting 30 m should be the minimum intertrap distance for monitoring *I. sibirica*. The interference between traps occurred when intertrap distances were shorter than 30 m, the reason is that the short distances among traps can lead to competition among the pheromone plumes of neighboring traps (Wedding et al., 1995; Wall & Perry, 1982; Bacca et al., 2006; Elkinton & Cardé, 1988).

For trap orientation, we found higher catches of male *I. sibirica* moths in upwind than in other traps regardless of intertrap distances but differences were not significant. Pheromone plumes from traps deployed upwind apparently prevented captures in traps deployed downwind (Wall & Perry, 1982; Knight et al., 2007).

Our results indicate that trap orientation and intertrap distance significantly affect mean catch of *I. sibirica* in pheromone-baited traps and should thus be considered when using traps to monitor or control *I. sibirica* in Asparagus growing areas. Trap interval distances should be adjusted depending on the circumstances. An intertrap distance of over 30m would be suitable for monitoring and mass trapping of *I. sibirica*, respectively.

Acknowledgements

This research was supported by the National 12th Five-year Science and Technology Support Plan of China (Grant no.2012BAD19B07) and the Research Foundation of the Introduction of Talents of Shanxi Agricultural University (Grant no.2014ZZ09).

References

Bacca, T., Lima, E. R., Picanço, M. C., Guedes, R. N. C., & Viana, J. H. M. (2006). Optimum spacing of pheromone traps for monitoring the coffee leaf miner *Leucoptera coffeella*. *Entomologia Experimentalis et Applicata, 119*, 39-45. http://dx.doi.org/10.1111/j.1570-7458.2006.00389.x

Byers, J. A. (1999). Effects of attraction radius and flight paths on catch of scolytid beetles dispersing outward through rings of pheromone traps. *Journal of Chemical Ecology, 25*, 985-1005. http://dx.doi.org/10.1023/A:1020869422943

Bradshaw, J. W. S., Ellis, N. W., Hand, S. C., & Stoakley, J. T. (1989). Interactions between pheromone traps with different strength lures for the pine beauty moth, *Panolis flammea* (Lepidoptera: Noctuidae). *Journal of Chemical Ecology, 15*, 2485-2494. http://dx.doi.org/10.1007/BF01020378.

Duan, G. Q., Zhang, Z. B., Zhang, H. J., Du, Y. Q., Hua, B. Z., & Ma, L. P. (2008). The bionomics of *Isoceras sibirica*. *Chinese Bull Entomol, 45*(3), 397-400. http://dx.doi.org/10.3969/j.issn.0452-8255.2008.03.012

Elkinton, J. S., & Cardé, R. T. (1988). Effects of intertrap distance and wind direction on the interaction of gypsy moth (Lepidoptera: Lymantriidae) pheromone-baited traps. *Environmental Entomology, 17*, 764-769. http://dx.doi.org/10.1093/ee/17.5.764

Haynes, K. F., Miller, T. A., Staten, R. T., Li, W. G., & Baker, T. C. (1986). Monitoring insecticide resistance with insect pheromones. *Experienlia, 42*, 1293-95. http://dx.doi.org/10.1007/BF01946429

Haynes, K. F., Miller, T. A., Staten, R. T., Li, W. G., & Baker, T. C. (1987). Pheromone trap for monitoring insecticide resistance in the pink bollworm moth (Lepidoptera: Gelechiidae): New tool for resistance management. *Environmental Entomology, 16*, 84-89. http://dx.doi.org/10.1093/ee/16.1.84

Higbee, B. S., & Burks, C. S. (2008). Effects of mating disruption treatments on Navel orangeworm (Lepidoptera: Pyralidae) sexual communication and damage in almonds and pistachios. *Journal of Chemical Ecology, 101*, 1633-1642. http://dx.doi.org/10.1603/0022-0493(2008)101[1633:EOMDTO] 2.0.CO;2

Il'Ichev, A. L., Stelinski, L. L., Williams, D. G., & Gut, L. J. (2006). Sprayable microencapsulated sex pheromone formulation for mating disruption of Oriental fruit moth (Lepidoptera: Tortricidae) in Australian peach and pear orchards. *Journal of Chemical Ecology, 99*, 2048-2054.

Judd, G. J. R., & Borden, J. H. (1989). Distant olfactory response of the onion fly, *Delia antiqua* to host-plant odour in the field. *Physiological Entomology, 14*, 429-441. http://dx.doi.org/10.1111/j.1365-3032.1989.tb01112.x

Jansson, R. K., Heath, R. R., & Coffelt, J. A. (1989). Temporal and spatial patterns of sweet potato weevil (Coleoptera: Curculionidae) counts in pheromone-baited traps in sweet potato fields in southern Florida. *Environmental Entomology, 18*, 691-697. http://dx.doi.org/10.1093/ee/18.4.691

Kehat, M., Eitam, A., Blumberg, D., Dunkelblum, E., & Anshelevish, L. (1992). Sex pheromone traps for detecting and monitoring the raisin moth, *Cadra figulilella*, in date palm plantations. *Phytoparasitica, 20*(2), 99-106. http://dx.doi.org/10.1007/BF02981275

Knight, A. L. (2007). Influence of within-orchard trap placement on catch of codling moth (Lepidoptera: Tortricidae) in sex pheromone-treated orchards. *Environmental Entomology, 36*, 425-432. http://dx.doi.org/10.1603/0046-225X(2007)36[425:IOWTPO]2.0.CO;2

Laboke P. O., Ogenga-Latigo, M. W., Smit, N. E. J., Downham, M. C. A., Odongo, B., Hall, D. R., & Farman, D. (2000). Environmental factors affecting catches of sweat potato weevils, *Cylas brunneus* (Fabricius) and *C. puncticollis* (Boheman) in pheromone traps. *African Potato Association Conference Proceedings, 5*, 217-227.

Murad, Z. (2001). Using pheromones to trap banana weevil borers. *Banana Topics*, DPI South Johnstone, Australia.

Patricia, L. S., Mochel, G., & Julio, K. (2008). Effect of pheromone trap density on mass trapping of male potato tuber moth *Phthorimaea operculella* (Zeller) (Lepidoptera: Gelechiidae), and level of damage on potato tubers. *Chilean Journal of Agricultural Research, 69*(2), 281-285.

Stelinski, L. L., McGhee, P., Haas, M., & Il'ichev, A. L. (2007). Sprayable microencapsulated sex pheromone formulations for mating disruption of four tortricid species: Effects of application height, rate, frequency, and sticker adjuvant. *Journal of Economic Entomology, 100,* 1360-1369. http://dx.doi.org/10.1603/0022-0493(2007)100[1360:SMSPFF]2.0.CO;2

Schlyter, F. (1992). Sampling range, attraction range and effective attraction radius: estimates of trap efficiency and communication distance in coleopteran pheromone and host attractant systems. *Journal of Applied Entomology, 114,* 439-454. http://dx.doi.org/10.1111/j.1439-0418.1992.tb01150.x

Sappington, T. W., & Spurgeon, D. W. (2000). Variation in boll weevil (Coleoptera: Curculionidae) captures in pheromone traps arising from wind speed moderation by brush lines. *Environmental Entomology, 29,* 807-814. http://dx.doi.org/10.1603/0046-225X-29.4.807

Sarzynski, E. M., & Liburd, O. E. (2004). Effect of trap height and within‑planting location on captures of cranberry fruitworm (Lepidoptera: Pyralidae) in highbush blueberries. *Agricultural and Forest Entomology, 6*(3), 199-204. http://dx.doi.org/10.1111/j.1461-9555.2004.00222.x

Schlyte, R. F. (1992). Sampling range, attraction range and effective attraction radius: estimates of trap efficiency and communication distance in coleopteran pheromone and host attractant systems. *Journal of Applied Entomology, 114,* 439-454. http://dx.doi.org/10.1111/j.1439-0418.1992.tb01150.x

Sappington, T. W., & Spurgeon, D. W. (2000). Variation in boll weevil (Coleoptera: Curculionidae) captures in pheromone traps arising from wind speed moderation by brush lines. *Environmental Entomology, 29,* 807-814. http://dx.doi.org/10.1603/0046-225X-29.4.807

Wall, C., & Perry, J. N. (1982). The behaviour of moths responding to pheromone sources in the field: A basis for discussion. *Les Colloques de I'INRA, 7,* 169-186.

Snedecor, G. W., & Cochran, W. G. (1967). *Statistical methods.* Iowa State University Press, Ames, IA.

Wedding, R., Anderbrant, O., & Jönsson, P. (1995). Influence of wind conditions and intertrap spacing on pheromone trap catches of European pine sawfly, *Neodiprion sertifer. Entomologia Experimentalis et Applicata, 77,* 223-232. http://dx.doi.org/10.1111/j.1570-7458.1995.tb02005.x

Zhang, G. F., Meng, X. Z., Han, Y., & Sheng, C. F. (2002). Chinese tortrix *Cydia trasias* (Lepidoptera: Olethreutidae): Suppression on street-planting trees by mass trapping with sex pheromone traps. *Environmental Entomology.* http://dx.doi.org/10.1603/0046-225X-31.4.602

Zhang, J. T., & Liu, H. X. (2011). Identification of the Sex Pheromone of the Asparagus Carpenterworm, *Isoceras sibirica* Alpheraky (Lepidoptera: Cossidae). *Zeischrift Fur Naturforschung Section, 66c,* 527-533.

13

Physicochemical Characteristics of the Maltese Grapevine Varieties – Ġellewża and Girgentina

Marilyn Theuma[1], Claudette Gambin[2] & Everaldo Attard[1]

[1] Institute of Earth Systems, Division of Rural Sciences and Food Systems, University of Malta, Msida, MSD, Malta

[2] Permanent Crops - Agriculture Directorate, Department for Rural Affairs and Aquaculture, Agricultural Research and Development Centre, Għammieri, Malta

Correspondence: Everaldo Attard, University of Malta, Institute of Earth Systems, Division of Rural Sciences and Food Systems, Msida, MSD 2080, Malta. E-mail: everaldo.attard@um.edu.mt

Abstract

Two indigenous Maltese grape vine varieties (cv. Ġellewża and Girgentina) juice extracts were studied for their physicochemical properties at three different locations on the Island of Malta. The mean acidity for Ġellewża and Girgentina was 2.729±0.088 and 3.971±0.179 g/L, pH was 4.026±0.039 and 3.704±0.042 and a %Brix was 17.913±0.364 and 17.531±0.189, indicating similarities between the variety-location combinations. Spectroscopic analysis revealed significant difference between the two varieties. The Ġellewża variety exhibited a high colour index (3.055-10.774) while the Girgentina variety showed a high tonality ratio (2.656-3.111). Although, the total polyphenolic content of the two varieties was not significant differently in most cases (754.771-2643.552 mg/kg), the red grape Ġellewża had significantly higher anthocyanin content (708.236±68.451 mg/kg) compared to the white grape Girgentina (14.412±1.119 mg/kg). Principal component analysis confirmed the differences between the varieties and also exhibited distinctive location differences, based on their physicochemical characteristics.

Keywords: *Vitis vinifera* varieties, Ġellewża, Girgentina, physicochemical characteristics

1. Introduction

The dark-skinned indigenous Ġellewża grape variety has been utilised by traditional wine makers to produce a red wine. It has been also transformed into still and semi-sparkling rosé wines. Wine-makers blend this variety with Syrah, softening the spiciness of the latter wine and at the same time adding a cherry flavour. A typical *passito* is produced by drying the berries in the sun, to intensify the sugars, prior to vinification. The resultant wine is spicier with a more pungent cherry flavour and earthy undertones. On the other hand, the white indigenous Girgentina grape variety produces a crispy and fruity wine. It is usually blended with Chardonnay, to produce a smooth wine with a fruity aroma and buttery undertones (Marsovin, 2008; Delicata, 2013).

The berries of international varieties have been distinctively studied for their physicochemical characteristics. Such characteristics include proteins (Sarry et al., 2004), proanthocyanidins (Czochanska et al., 1979; Souquet et al., 1996; Sun et al., 1998), anthocyanins (Revilla et al., 2001), organic acids (Conde et al., 2007) and sugars (Liu et al., 2006). Attention has also been drawn towards the health benefits derived from the consumption of grapes and their products. These phytochemicals contribute to the antioxidant (Tamura & Yamagami, 1994; Wang et al., 1997; Orak, 2007), cardioprotective (Sato et al., 1999; Xia et al., 2010), and anticancer (Ye et al., 1999; Singletary & Meline, 2001) activities amongst others.

In this study, we attempted to determine the quality of the Maltese grapevine varieties Ġellewża and Girgentina, grown in different regions of Malta, through physicochemical characteristics, including the total acids, sugar content, pH and polyphenolic parameters.

2. Materials and Methods

2.1 Materials

The berries of two Maltese indigenous varieties, Ġellewża and Girgentina, were collected between 30th August and 21st September 2011 every two days, from different vineyards. Ġellewża grape samples were collected from Mġarr (two sites) and Burmarrad while Girgentina grape samples were collected from Siġġiewi and Mġarr (two sites).

2.2 Extraction of Grape Samples

For the total acidity, Brix and pH analyses, the fresh grape samples were crushed in a large beaker, and the filtered juice was then analysed. For the polyphenolic analyses, approximately 5 g fresh grape aliquots were extracted with 35 ml methanol:HCl (1M) (95:5). The samples were placed in an ultrasonicator for 15 minutes followed by 15 minutes centrifugation at 3000rpm. The samples (supernatants) were ready for analysis. All samples were collected in triplicates and all tests were conducted in triplicates, for a total replicate number of nine.

2.3 Titratable Acidity, Brix and pH Determination

The titratable acidity was measured by using NaOH 0.1 N and bromothymol blue as indicator. The volume of titrant was then multiplied by 0.75 and then divided by two to determine the amount of tartaric acid in g/L as titratable acidity. The pH was measured using a Thermo Scientific Orion 4-Star pH meter (USA), while the %Brix content was determined using a Hanna Instruments HI 96814 wine refractometer (USA).

2.4 UV-Vis Analysis and Folin-Ciocalteu Colorimetric Method

Each grape extract was analysed with a Lightwave II – WPA UV-Vis spectrophotometer (UK), between 200 and 800nm (wave scans) and specifically at 280 nm, 420 nm, 520 nm and 620 nm. The Ġellewża and Girgentina extracts were generally in a 1:9 and 1:4 ratio with solvent, respectively. The colour intensity, % tint, tonality ratio (Glories, 1984) and anthocyanin content were determined according to the following equations

$$\text{Colour intensity} = (A_{420}.DF) + (A_{520}.DF) + (A_{620}.DF) \tag{1}$$

$$\text{Percentage Tint} = \frac{A_{420}}{A_{520}} \times 100 \tag{2}$$

$$\text{Tonality Ratio} = \frac{A_{420}}{A_{520}} \tag{3}$$

$$\text{Anthocyanin content(mg/kg)} = \frac{1000.V_S.DF.A_{520}}{\epsilon} \tag{4}$$

Where, V_S = volume of extracted sample per gramme of grape; DF = dilution factor; ϵ = extinction coefficient [58.3 ml (mg.cm)]; A420, A520, A620 = absorbance values at 420, 520 and 620 nm.

The test for polyphenols, using the Folin-Ciocalteu reagent, was conducted according to Attard (2013). Briefly, to 5 μl of each grape extract, 5 μl of distilled water, 100 μl of Folin-Ciocalteu reagent (Sigma-Aldrich, pre-diluted with distilled water (1:9)) and 80 μl of sodium carbonate Na_2CO_3 (Sigma-Aldrich, 1M) were added. The absorbance was read after 20 min at room temperature at 630 nm using a micro-plate reader (BioTek ELx800, Winooski, VT, USA). Tannic acid (Sigma-Aldrich) was used as standard (60-960 mg/ml; $r^2 = 0.9940$) to prepare the calibration curve in order to determine the total amount of phenols in each grape extract (mgTAE/kg of grape extract).

2.5 Data Analysis

The results for the samples, collected over the three-week period, were averaged. The titratable acidity, pH, %Brix, colour intensity (CI), tonality ratio, anthocyanin content, polyphenolic content, anthocyanin to polyphenolic ratio were investigated with multivariate analysis for all berry samples. The correlation matrix was calculated, giving the correlation coefficients between each pair of variables tested. To identify variability and to reduce the dimensions of the data set, principal component analysis (PCA) was performed, using the XLSTAT Version 2011.5.01 software (Addinsoft, USA).

3. Results and Discussion

3.1 Titratable Acidity, pH and %Brix

In previous studies, it has been observed that different physicochemical parameters are affected by genetic variability and environmental factors (Serrano-Megias et al., 2006; Makris et al., 2006). In this study we report the differences observed for the physicochemical characteritics for the two grapevine varieties wth different environmental conditions. The results for the titratable acidity, pH and %Brix are exhibited in Table 1. In particular, acidity values for Girgentina were superior to those of Ġellewża and while the opposite was observed for the pH values. Furthermore, samples collected from Mġarr have shown that grapes from the old vines have a higher acidity compared to the grapes of the new vines (p < 0.05). The total acidity results of the Ġellewża and Girgentina also showed to be much lower compared to Cabernet Sauvignon, Malbec, Bonarda, Merlot and Tempranillo grape varieties (Fanzone et al., 2012), i.e. less than 4.63 g/L as opposed to values greater than 5.3 g/L, respectively. Results also showed that the new Ġellewża and Girgentina grapevines in Mġarr have a higher pH than the old grapevines. Comparing the pH of Cabernet Franc, Merlot, Sangiovese and Syrah grape varieties (Gris et al., 2010) to the Ġellewża and Girgentina, it was observed that the pH of the local varieties is quite higher. In fact the pH of the international varieties were lower than 3.58 while those for the Maltese varieties the values ranged between 3.507 and 4.205. The Ġellewża grapevines located at Burmarrad exhibited the lowest sugar content (15.181±0.274%), while the new Ġellewża and Girgentina grapevines in Mġarr have shown the highest sugar content (19.458±0.347 and 18.801±0.607%, respectively). The old Ġellewża grapevines in Mġarr showed a higher sugar content than the old Girgentina grapevine (18.526±0.468 and 17.151±0.372%, respectively), while the Girgentina grapevines at Siġġiewi (17.033±0.345%) showed a similar sugar content to the old Girgentina grapevine at Mġarr. Compared to other grape varieties, including Alicante, Cabernet Franc, Cabernet Sauvignon, Carignan, Cinsault, Grenache, Merlot, Mourvedre, Sangiovese and Syrah (Gris et al., 2010; Jensen et al., 2008) both the Ġellewża and Girgentina varieties have a much lower sugar content. This was always considered as an important issue with local grapevine varieties.

Table 1. Total acidity, pH and %Brix for the six variety-location combinations

		Titratable Acidity (g/L)	pH	%Brix
Ġellewża	Burmarrad	2.721±0.166	3.941±0.073	15.181±0.274
	Mġarr-New	2.536±0.187	4.205±0.108	19.458±0.347*
	Mġarr-Old	2.874±0.329	3.949±0.100	18.526±0.468*
Girgentina	Siġġiewi	3.700±0.351	3.716±0.104*	17.033±0.345
	Mġarr-New	3.445±0.280	3.956±0.091	18.801±0.607*
	Mġarr-Old	4.650±0.732*	3.507±0.084*	17.151±0.372

Note. Data are means±S.E.M. of three independent determinations. Means within an asterisk show a significant difference between variety-location for the individual parameters, at $P < 0.05$ by Student's *t*-test.

3.2 UV-Vis Analysis and Absorbance Values

Colour analysis was carried out by spectrophotometric methods (Gris et al., 2010). Although these parameters are usually studied in red grape varieties, similar to Ġellewża, these were also considered for Girgentina as a means to show distinctive differences between the two varieties. Percentage tint is the percentage ratio of the yellow components (proanthocyanidins) at 420 nm and the red components (anthocyanins) at 520 nm. The results for CI and Tonality are illustrated in Table 2. Results showed that the percentage tint is lower in Ġellewża grapes than in Girgentina. It is clearly indicated that although the anthocyanin content of Girgentina white grapes is low, the proanthocyanidin content is high. Consequently, the proanthocyanidin to anthocyanin percentage ratio of Ġellewża red grapes is lower than the Girgentina type. Another parameter related to percentage tint is the tonality. Tonality ratio is the classical ratio between the two wavelengths. As a result, Girgentina samples resulted in a higher tonality ratio compared to the Ġellewża samples. Colour intensity depends on pH and the content of anthocyanins. This reflects the content and structure of the anthocyanins present in both grapes and wine. In fact, Colour intensity featured distinctively with Ġellewża samples. Colour intensity is obtained by the addition of the absorbances at three wavelengths, i.e. 420, 520 and 620 nm (Glories, 1984). When comparing red grape varieties or wines, the more acidic the grape juice or wine, the brighter red the colour is (Mollah, 2011). Although Ġellewża samples showed higher colour intensities than Girgentina samples, the important parameter, in this study, was the content of anthocyanins. This is because the Ġellewża samples exihibited higher pH values than Girgentina samples, even though their colour intensities were higher. This may be due to the adaptability of the local varieties to the highly alkaline and calcareous Maltese soil.

In the measurement of the anthocyanin content, the absorbance value at 520nm is taken into consideration. Since anthocyanins have a direct influence on the red colour of the grapes, total anthocyanin content in Ġellewża samples were higher than in Girgentina (Figure 1). The Mġarr Ġellewża anthocyanin content compares well with the Syrah anthocyanin content (808-1112 mg/kg) obtained by Barbagallo et al. (2011). Ġellewża grapes compared to other varieties such as Merlot, Cabernet Sauvignon and Vranec (Ivanova et al., 2010) possess a lower tint but a higher colour intensity, while on the other hand, Girgentina grapes showed a lower colour intensity with a higher tint.

Table 2. Colour Intensity and Tonality for the six variety-location combinations

		Colour Intensity(Abs)	Percentage Tint (%)	Tonality (ratio)
Ġellewża	Burmarrad	3.055±0.689*	38.7±4.5	0.387±0.045
	Mġarr-New	9.488±0.670*	26.2±0.8	0.262±0.008
	Mġarr-Old	10.774±1.298*	26.2±0.8	0.262±0.008
Girgentina	Siġġiewi	0.447±0.088	311.1±35.5*	3.111±0.355*
	Mġarr-New	0.467±0.056	304.5±24.3*	3.045±0.243*
	Mġarr-Old	0.663±0.141	265.6±25.1*	2.656±0.251*

Note. Data are means±S.E.M. of three independent determinations. Means within an asterisk show a significant difference between variety-location for the individual parameters, at $P < 0.05$ by Student's *t*-test.

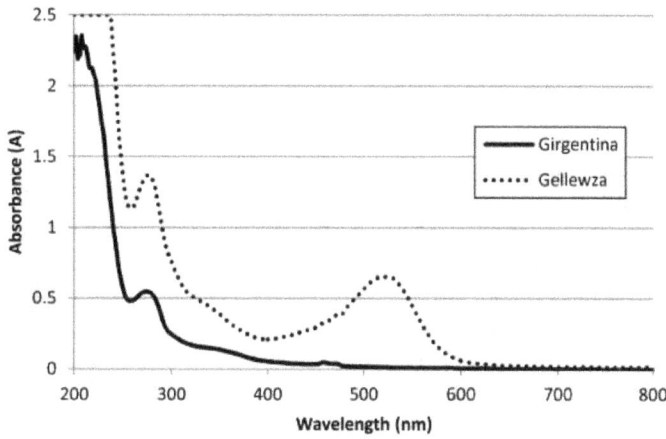

Figure 1. The UV-Vis profile for the Ġellewża and Girgentina grape varieties between 200 and 800 nm

3.3 Folin-Ciocalteu Test for Polyphenols

The determination of the polyphenolic content of Ġellewża and Girgentina grapes was carried out by using the Folin-Ciocalteu test, a commonly used spectrophotometric method in viticulture and oenology. The results for anthocyanins, polyphenols and their ratio are exhibited in Table 3. These have showed that Ġellewża and Girgentina samples collected from the old grapevines in Mġarr had the highest content of polyphenols. Compared to the other samples, grapes collected from Burmarrad possessed the lowest amount of polyphenols. Samples collected from the new grapevines in Mġarr showed that Ġellewża grapes contained a higher amount of polyphenolic compounds compared to Girgentina samples. The results also showed Girgentina from Siġġiewi to have less polyphenols than the other Girgentina grapes sampled from Mġarr, though the difference was not statistically significant.

Table 3. Anthocyanins, Polyphenols and their Ratio for the six variety-location combinations

		Anthocyanin content (mg/kg)	Polyphenolic content (mg/kg)	Anthocyanin:Polyphenols ratio
Ġellewża	Burmarrad	216.612±46.882*	754.771±207.066	0.287*
	Mġarr-New	803.092±78.727*	1853.426±178.299	0.433*
	Mġarr-Old	976.487±100.111*	2643.552±193.592*	0.369*
Girgentina	Siġġiewi	11.662±3.335	1021.562±276.026	0.011
	Mġarr-New	12.185±1.841	1447.374±286.619	0.008
	Mġarr-Old	19.032±3.468	1818.667±304.349	0.010

Note. Data are means±S.E.M. of three independent determinations. Means within an asterisk show a significant difference between variety-location for the individual parameters, at $P < 0.05$ by Student's *t*-test.

The amount of polyphenols found in Ġellewża samples collected from Burmarrad was compared to the Cinsault variety with a value of 876 mg/kg. The polyphenolic content of both the Ġellewża and Girgentina collected from the new grapevines in Mġarr, and the Ġellewża collected from the old grapevines resulted in a much higher concentration compared to the popular varieties Cabernet Sauvignon, Grenache, Merlot and Syrah. Girgetina sampled from Siġġiewi had similar polyphenolic content to the Carignan variety with a value of 1210 mg/kg (Jensen et al., 2008).

It was observed from the scree plot (data not shown) that the first three principal components accounted for 98.68 % of the total variance. This indicates that little numerical noise and/or experimental error is observed. However, the parameters studied fall within the first two principal components, which in fact contributed significantly to the total variance. The loadings plot provides the direction of each original variable, and the scores plot, indicates the position of each variety/location combination. The first factor is loaded heavily on titratable acidity, pH, colour intensity, tonality, anthocyanin content and anthocyanin/polyphenolic ratio, representing a mix of physical and chemical parameters. The second factor is heavily loaded on %brix and polyphenlic content, representing the main categories of phytochemicals in grapes. The titratable acidity and tonality showed an inverse correlation with anthocyanin/polyphenol ratio and pH. There is no direct relationship between titratable acidity and pH for grape extracts, although generally the pH decreases as the acid increases and vice-versa. The latter two parameters did not show any strong correlation with the other parameters (Figure 2A).

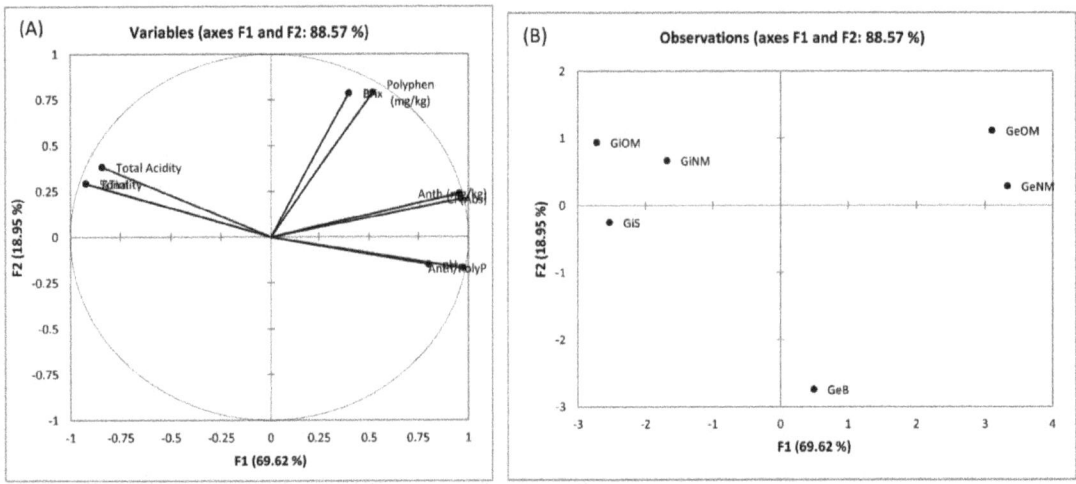

Figure 2. Graph of loading plot (A) of physicochemical parameters and scores plot (B) of six variety-location combinations: *Ġellewża* Burmarrad, Mġarr-New, Mġarr-Old (GeB, GeNM and GeOM, respectively) and *Girgentina* Siġġiewi, Mġarr-New, Mġarr-Old (GiS, GiNM and GiOM, respectively

The score plot reported in Figure 2B shows the physicochemical parameters of grape extracts in the space of the two new variables F1 and F2. Moving along F1 from left to right in the graph, we observed that the Girgentina grape extracts exhibited siginificantly different physicochemical characteristics from the Ġellewża grape extracts. Moving along F2 from top to bottom, the Mġarr-grown Ġellewża and Girgentina vines exhibited superior sugar and polyphenolic contents, as compared to those grown at Burmarrad and Siġġiewi.

4. Conclusion

These results show that the two Maltese grapevine varieties, Ġellewża and Girgentina, are distinct from other international grapevine varieties, based on their physicochemical characteristics. There were also remarkable differences between the Ġellewża and Girgentina, which also are reflected in the locality, hence environmental differences. The high polyphenolic content in both grape varieties, indicates that the grape products may be valued for their nutritional properties, apart from the production of wine.

References

Attard, E. (2013). A rapid microtitre plate Folin-Ciocalteu method for the assessment of polyphenols. *Central European Journal of Biology, 8*, 48-53. http://dx.doi.org/10.2478/s11535-012-0107-3

Barbagallo, M. G., Guidoni, S., & Hunter, J. J. (2011). Berry Size and Qualitative Characteristics of *Vitis vinifera* L. cv. Syrah. *South African Journal of Enology and Viticulture, 32*(1), 129-136.

Conde, C., Silva, P., Fontes, N., Dias, A. C. P., Tavares, R. M., Sousa, M. J., ... Hernani, G. (2007). Biochemical Changes throughout Grape Berry Development and Fruit and Wine Quality. *Food, 1*(1), 1-22.

Czochanska, Z., Yeap Foo, L., & Porter, L. J. (1979). Compositional changes in lower molecular weight flavans during grape maturation. *Phytochemistry, 18*, 1819-1822, http://dx.doi.org/10.1016/0031-9422(79)83060-5

Delicata. (2013). *Our Wines*. Retrieved from http://www.delicata.com

Fanzone, M., Zamora, F., Jofré, V., Assof, M., Gómez-Cordovés, C., & Peña-Neira, A. (2012). Phenolic characterisation of red wines from different grape varieties cultivated in Mendoza province, Argentina. *Journal of the Science of Food and Agriculture, 92*, 704-718. http://dx.doi.org/10.1002/jsfa.4638

Glories, Y. (1984). La couleur des vins rouges. 2éme partie. Mesure, origineet interpretation. *Connaissance de la Vigne et du Vin, 18*, 253-271.

Gris, E. F., Burin, V. M., Brighenti, E., Vieira, H., & Bordignon-Luiz, M. T. (2010). Phenology and ripening of Vitis vinifera L. grape varieties in São Joaquim, southern Brazil: a new South American wine growing region. *Ciencia e investigación agraria, 37*, 61-75. http://dx.doi.org/10.4067/s0718-16202010000200007

Ivanova, V., Stefova, M., & Chinnici, F. (2010). Determination of the polyphenol contents in Macedonian grapes and wines by standardized spectrophotometric methods. *Journal of the Serbian Chemical Society, 7*, 45-59. http://dx.doi.org/10.2298/JSC1001045I

Jensen, J. S., Demiray, S., Egebo, M., & Meyer, A. S. (2008). Prediction of Wine Color Attributes from the Phenolic Profiles of Red Grapes. *Vitis vinifera. Journal of the Science of Food and Agriculture, 56*, 1105-1115. http://dx.doi.org/10.1021/jf072541

Liu, H. F., Wu, B. H., Fan, P. G., Li, S. H., & Li, L. S. (2006). Sugar and acid concentrations in 98 grape cultivars analyzed by principal component analysis. *Journal of the Science of Food and Agriculture, 86*, 1526-1536. http://dx.doi.org/10.1002/jsfa.2541

Makris, D. P., Kallithraka, S., & Kefalas, P. (2006). Flavonols in grapes, grape products and wines: Burden, profile and influential parameters. *Journal of Food Composition and Analysis, 19*, 396-404. http://dx.doi.org/10.1016/j.jfca.2005.10.003

Marsovin. (2008). *Wines From Malta - I.G.T Maltese Islands Wines*. Retrieved from http://www.marsovin.com/wineproducer/WinesFromMalta/

Mollah, M. (2011). *Colour Production and Measurement in the Berry, Cooperative Research Centre for Viticulture 2004*. Retrieved from http://www.crcv.com.au/resources/Grape%20and%20Wine%20Quality/Workshop%20Notes/Colour%20Measurement.pdf

Orak, H. H. (2007). Total antioxidant activities, phenolics, anthocyanins, polyphenoloxidase activities of selected red grape cultivars and their correlations. *Scientia Horticulturae, 111*, 235-241. http://dx.doi.org/10.1016/j.scienta.2006.10.019

Revilla, E., García-Beneytez, E., Cabello, F., Martín-Ortega, G., & Ryan, J. M. (2001). Value of high-performance liquid chromatographic analysis of anthocyanins in the differentiation of red grape cultivars and red wines made from them. *Journal of Chromatography A, 915*, 53-60.

Sarry, J. E., Sommerer, N., Sauvage, F. X., Bergoin, A., Rossignol, M., Albagnac, G., & Romieu, C. (2004). Grape berry biochemistry revisited upon proteomic analysis of the mesocarp. *Proteomics, 4*, 201-215. http://dx.doi.org/10.1002/pmic.200300499

Sato, M., Maulik, G., Ray, P. S., Bagchi, D., & Das, D. K. (1999). Cardioprotective effects of grape seed proanthocyanidin against ischemic reperfusion injury. *The Journal of Molecular and Cellular Cardiology, 31*, 1289-1297. http://dx.doi.org/10.1006/jmcc.1999.0961

Serrano-Megías, M., Núñez-Delicado, E., Pérez-López, A. J., & López-Nicolás, J. M. (2006). Study of the effect of ripening stages and climatic conditions on the physicochemical and sensorial parameters of two varieties of *Vitis vinifera* L. by principal component analysis: Influence on enzymatic browning. *Journal of the Science of Food and Agriculture, 86*, 592-599. http://dx.doi.org/10.1002/jsfa.2375

Singletary, K. W., & Meline, B. (2001). Effect of Grape Seed Proanthocyanidins on Colon Aberrant Crypts and Breast Tumors in a Rat Dual-Organ Tumor Model. *Nutrition and Cancer, 39*, 252-258. http://dx.doi.org/10.1207/S15327914nc392_15

Souquet, J. M., Cheynier, V., Brossaud, F., & Moutounet, M. (1996). Polymeric proanthocyanidins from grape skins. *Phytochemistry, 43*, 509-512. http://dx.doi.org/10.1016/0031-9422(96)00301-9

Sun, B., Leandro, C., da Silva, J. M. R., & Spranger, I. (1998). Separation of Grape and Wine Proanthocyanidins According to Their Degree of Polymerization. *Journal of Agricultural and Food Chemistry, 46*, 1390-1396. http://dx.doi.org/10.1021/jf970753d

Tamura, H., & Yamagami, A. (1994). Antioxidative activity of monoacylatedanthocyanins isolated from Muscat Bailey A grape. *Journal of Agricultural and Food Chemistry, 42*, 1612-1615. http://dx.doi.org/10.1021/jf00044a005

Wang, H., Cao, G., & Ronald, L. (1997).Oxygen radical absorbing capacity of anthocyanins. *Journal of Agricultural and Food Chemistry, 45*, 304-309. http://dx.doi.org/10.1021/jf960421t

Xia, E. Q., Deng, G. F., Guo, Y. J., & Li, H. B. (2010). Biological Activities of Polyphenols from Grapes. *International Journal of Molecular Sciences, 11*, 622-646. http://dx.doi.org/10.3390/ijms11020622

Ye, X., Krohn, R. L., Liu, W., Joshi, S. S., Kuszynski, C. A., McGinn, T. R., … Bagchi, D. (1999). The cytotoxic effects of a novel IH636 grape seed proanthocyanidin extract on cultured human cancer cells. *Molecular and Cellular Biochemistry, 196*, 99-108. http://dx.doi.org/10.1023/A:1006926414683

14

Jasmine 85 from Seven Rice Seed Production Sources in Ghana Are Genetically Different

O. T. Akintayo[1], B. K. Maalekuu[1] & J. K. Saajah[2]

[1] Kwame Nkrumah University of Science and Technology, Kumasi, Ghana

[2] Ghana Cocoa Board/Newmont Mining Cooperation, Ghana

Correspondence: B. K. Maalekuu, Kwame Nkrumah University of Science and Technology, Kumasi, Ghana. E-mail: kbmaalekuu.agric@knust.edu.gh

Abstract

Jasmine 85 is the most popular and widely cultivated rice variety in Ghana. Samples of Jasmine 85 were collected from seven seed sources with a reference sample obtained from Africa Rice Center, Senegal. Morphological evaluation involved qualitative and quantitative data. Jasmine 85 from all sources were similar in terms of aroma, anthocyanin coloration, leaf pubescence, and ligule shape. The sources showed significant differences to pericarp colour, days to 50% heading, plant height, seed length and seed width.

Physico-chemical analyses showed grain size and shape, grain chalkiness; cooking time, head rice yield, gelatinisation temperature, amylose content and viscosity properties differed significantly among the sources whilst grain hardness was not significantly different.

Molecular characterisation using 15 SSR markers showed that although closely related, the sources differed significantly. None of the sources in Ghana was genetically identical to the reference sample. Seeds from different sources should not be mixed for sale or production, and these varieties should be treated separately in future evaluations.

Keywords: chalkiness, physic-chemical, gelatinisation temperature, anthocyanin coloration, viscosity, amylose content, significant, clustering

1. Introduction

The rice import bill of Ghana is about US$300 million accounting for 70% of local demand (MoFA, 2009). As rice imports surge ahead of production in Ghana, increasing rice production and yields has become a priority. Annual per capita consumption of rice in Ghana grew from 17.5 kg during 1999–2001 to 24 kg during 2010–2011. As only 5 per cent of global production is traded, local production would also protect consumers from price shocks in the world rice market (World Bank, 2013). While substantial investments in national rice production have been made, local production is still not able to keep up with growing demand for rice in Ghana. Although local production of milled rice recently has grown by 10.5 per cent annually, from 242,000 metric tons (MT) in 2004 to 481,000 MT in 2012, most of this growth in production has come from area expansion (7.5 per cent), with the remaining 3.0 per cent coming from productivity improvements. Despite these efforts, Ghana imported 640,000 MT of rice in 2013 (IFPRI, 2014)

Ghana rice market places premium on long slender intermediate amylose aromatic grains. Thus Jasmine 85, market quality rice variety developed by the International Rice Research Institute (IRRI) from the cross IR262-43-8-11/KHAO DAWK MALI 4-2-105 (Marco et al., 1997) and released 1998 in Ghana as Gbewaa rice by CSIR-Savannah Agricultural Research Institute (CSIR-SARI) (Diako et al., 2011) is grown all over the country. Dartey (unpublished) found that Jasmine 85 grown by Prairie Volta Co. Ltd. was morphologically different from that grown by CSIR-Crops Research Institute (CSIR-CRI), and both were different from that obtained from CSIR-SARI. This study sought to use morphology, biochemical and molecular tools to determine differences among Jasmine 85 from different sources so as to improve seed quality.

2. Materials and Methods

Two samples Jasmine 85 were collected from SARI, Tamale (denoted GBEWAA and SARI) and from CSIR-CRI in Nobewam (denoted DARTEY) and Kumasi (denoted CRI). Samples were also collected from Kpong

Irrigation Project, Asutsuare (denoted KIP), Prairie Volta Co Ltd, Aveyime (denoted PV), Tono Irrigation Project, Tono (denoted TONO) with the reference sample from AfricaRice, Senegal (denoted ARI).

The varieties were evaluated under irrigation in a randomized complete block design with four replicates in Nobewam (North 6°37′6.2″, West 1°17′0.4″, 200 m above the sea level) on 4 m² plots.

After harvest, visibly off-types were disregarded and seed from true to type plants were used to carry out the laboratory analyses. Complete randomized design of ... replicates was used for analyses of the physico-chemical and molecular properties.

To detect aroma, young leaf samples were put in an eppendorf tube containing alcohol (90% ethanol), closed for 24 hours and smelt by a panel of 4 persons.

All other morphological data were taken following the Standard Evaluation System for rice.

2.1 Moisture Content

Moisture content of samples were taken on two replicate samples of 50grains using PQ-510 Single Kernel Moisture Meter (Kett, Japan).

2.2 Milling Recoveries

Rice samples were dehusked in a THU-34A Satake Testing Rice Husker (Satake, Japan). The brown rice obtained was polished in a Rice pal 32 (Yamamoto Co., Japan) rice polisher. Milled rice was separated into whole and broken grains using a Test rice grader. The milling recoveries were then estimated using the following equations:

$$\text{Brown rice yield}(\%) = 100 \times \frac{\text{weight of dehusked rice}}{\text{weight of paddy}} \tag{1}$$

$$\text{Total milling yield}(\%) = 100 \times \frac{\text{weight of polished rice}}{\text{weight of paddy}} \tag{2}$$

$$\text{Head rice yield}(\%) = 100 \times \frac{\text{weight of whole grains}}{\text{weight of paddy}} \tag{3}$$

2.3 Raw Grain Hardness

Grain hardness was measured using a grain hardness tester (Fujihara Seisakusho LDT, Japan). Ten grains were used for each sample. The handle of the equipment was initially turned anti-clockwise to make room to place a grain on the sample table. Consequently, the handle was turned clockwise until a cracking sound was heard. At this time, the black pointer returns to the zero point and the red pointer remained. The reading of the red pointer (kg) indicated the hardness of the grain.

2.4 Grain Dimension and Chalkiness

Grain dimensions and chalkiness were estimated using the S21 Rice Statistic Analyzer, (LKL Technologia, Brazil). The equipment is run with a Classficador S21 version 4.05 software. The S21 was calibrated using a reference sample supplied by the manufacturer. The fluorescent light on the S21 was turned on and approximately 50 g of whole grains weighed and emptied into its sample receiver. The "long white" classification set up was opened in the capture mode on the software. The equipment was then switched on to vibrate and cause the release of individual grains from the receiver to slide on a blue tile background and pass beneath the attached camera that captured images of the grains. When all the grains had exited the receiver the image capturing mode was stopped. The grain dimensions were determined by processing the captured images and applying the "advanced filter-length distribution" on the software. The grain length and width were then recorded and the length/width ratio calculated. To determine chalkiness, the "basic filter – chalky distribution" was applied. The % total chalky area for the samples were recorded and reported as the percentage chalkiness of the samples.

2.5 Alkali Spreading Value (ASV)

ASV was determined using the method developed by Little et al. (1958) which involves visual observation of the degree of dispersion of grains of the milled rice after their immersion in 1.7% potassium hydroxide solution (KOH). Approximately10 ml of 1.7% KOH solution was poured on 6 rice grains placed in a transparent petri dish and incubated at room temperature for 23 hours after which the samples were observed and visual scores (Jennings et al., 1979) assigned. Based on the alkali spreading value, the gelatinization temperature (GT) of samples was indirectly determined.

2.6 Pasting Properties

The pasting properties of rice flour samples were measured using a Rapid Visco Analyzer (RVA) model - Super 4 (Newport Scientific, Warriewood, Australia) and Thermocline for Windows (TCW3) software. The general pasting method 162 (ICC, 2004) for flour samples was used. Rice flour (3 g sample) was weighed directly into the RVA canister, 25 ml of distilled water was added and mixed with the rice flour. The canister and its content were then placed in the RVA and run using the following RVA test profile:

- Rotating paddle speed – 160 rpm,

- Heating the rice-water mixture to 95 °C at a rate of 12 °C/min (i.e. in 3.75 min),

- Holding at 95 °C for 2.5 min,

- Cooling to 50 °C at a rate of 12 °C/min (i.e. in 3.75 min), and

- Holding at 50 °C for 2.5 min.

2.7 Cooking Time

Cooking time was determined using the method described by Fofana et al. (2011). Five grams of milled rice of each sample were weighed in duplicate and poured into 135 ml of vigorously boiling distilled water in a 400 ml beaker and covered with a watch glass. After 10 minutes of further boiling, 10 grains were taken out every minute with a perforated ladle. The grains were pressed between two petri dishes and were considered cooked when at least 9 out of the 10 grains no longer had opaque centers. The time it took for this to happen was then recorded as the cooking time for the sample.

2.8 Amylose Content

Amylose content is measured using the standard iodine colorimetric method ISO 6647-2-2011. Ethanol (1 mL, 95%) and 1 M sodium hydroxide (9 mL) is added to rice flour (100 mg) and this is heated in a boiling water bath until gelatinization of the starch occurred. After cooling, 1 M acetic acid (1 mL) and iodine solution (2 mL) are added and the volume is made up to 100 mL with Millipore water. The iodine solution is prepared by dissolving 0.2 g iodine and 2.0 g potassium iodide in 100 mL Millipore water. Absorbance of the solution is measured using an Auto Analyzer 3 (Seal Analytical, Germany) at 600 nm. Amylose content is quantified from a standard curve generated from absorbance values of 4 well-known standard rice varieties (IR65, IR24, IR64 and IR8).

Molecular tests were done using fresh leaves from 3 weeks old rice seedlings. The CTAB method was used for DNA extraction. The DNA was suspended in 50 µl 1X TE buffer and centrifuged at high speed for 30 sec to remove all insolubles. During that time 1.0% of agarose gel was prepared with (3 µl) 0.003% Ethidium bromide. 5 µl of the sample was taken using pipette, and added to 1 µl loading buffer. The sample was loaded in the wells on gel submerged in 1X TAE buffer and the sample was run at (90 to 120) volts for 45 minutes. The computer connected to the machine recorded the data and pictured the gel.Fifteen SSR rice primers (RM252, RM511, RM316, RM8236, RM224, RM19, RM452, RM144, RM514, RM105, RM277, RM25, RM178, RM11& RM55) were used.

3. Results and Discussion

3.1 Qualitative Traits

Data for qualitative traits considered for the morphological evaluation is presented in Table 1.

Table 1. Aroma, anthocyanin coloration, leaf pubescence, pericarp colour and ligule shape of the various seed sources

Treatment	Aroma	Anthocyanin coloration	Leaf pubescence	Pericarp colour	Ligule Shape
ARI	Lightly scented	Absent	Pubescent	Light brown	2-cleft
CRI	Lightly scented	Absent	Pubescent	Light brown	2-cleft
DARTEY	Lightly scented	Absent	Pubescent	Light brown	2-cleft
GBEWAA	Lightly scented	Absent	Pubescent	Light brown	2-cleft
KIP	Lightly scented	Absent	Pubescent	Light brown	2-cleft
PV	Lightly scented	Absent	Pubescent	Light brown	2-cleft
SARI	Lightly scented	Absent	Pubescent	Red	2-cleft
TONO	Lightly scented	Absent	Pubescent	Light brown	2-cleft

With the exception of pericarp colour where SARI had red whilst the others had light brown, no difference was observed with respect to the qualitative data. According to IRRI (1998), Jasmine 85 has pubescent leaves, no anthocyanin colouration, light brown pericarp and is aromatic. With respect to the above description, seeds from SARI do not match a true Jasmine 85 variety.

3.2 Quantitative Data

Data for the quantitative traits are presented in Table 2.

Table 2. Means of 50% heading (days); plant height (m); number of tiller; seed length (mm) and seed width (mm) for the various seed sources

Treatment	Means of 50% heading (days)	Plant height (m)	Number of tiller	Seed length (mm)	Seed width (mm)
ARI	69.25^c	95.1^e	7.62^{cd}	10.57^a	2.25^a
CRI	84^a	111.4^{abc}	10.75^b	10.53^a	2.22^a
DARTEY	85^a	113.7^{ab}	6.84^{cd}	10.32^b	2.213^{ab}
GBEWAA	74.25^b	106.9^{cd}	8.40^{bcd}	9.56^f	2.213^{ab}
KIP	71.50^{bc}	97.1^e	7.23^{cd}	10.24^{bc}	2.28^a
PV	83.50^a	115.7^a	16.34^a	10.12^{cd}	2.05^c
SARI	84.25^a	109^{bcd}	9.18^{bc}	9.94^{de}	2.123^{bc}
TONO	71.50^{bc}	105.2^d	6.28^d	9.86^e	2.205^{ab}
LSD	3.02	5.65	2.68	0.18	0.094
CV (%)	2.6	3.6	20.1	1.3	2.9

Note. Means with same alphabet within column are not significantly different at 5%.

The average heading days ranged from 85 days for DARTEY to 69 days for ARI. Significant differences were observed in all the sources implying that the materials differ in terms of maturity days. Significant differences were also observed in number of tillers, seed length and seed weight. Although the study recorded a plant height range of 115.7 cm to 95 cm which is in agreement with IRRI's (1998) report for Jasmine 85, significant variations recorded for other traits undermine the genetic purity of this popular variety in Ghana. Ideally, a single variety from different sources should not vary significantly in these traits.

3.3 Physico-Chemical Analysis

The physico-chemical variables measured on the grain helped reveal differences in the Jasmine 85 from the various sources. The analysis of variance (ANOVA) showed significant difference for all parameters measured except for Grain hardness (Table 3).

Table 3. Means of Grain length (mm), grain width (mm), grain length to the width ratio, grain size, grain shape, total chalky area (%), and grain hardness (kg) of Jasmine 85 from various sources

Treatment	Length (mm)	Width (mm)	L/W	Size	Shape	Total chalky area (%)	Grain hardness (kg)
ARI	6.61^a	2.208^{ab}	3.0	Long	Medium	9.24^e	9.02
CRI	6.505^{ab}	2.198^b	3.0	Medium	Medium	15.04^b	8.13
DARTEY	6.38^{bc}	2.035^f	3.1	Medium	Slender	12.41^c	7.83
GBEWAA	6.308^c	2.127^d	3.0	Medium	Medium	20.80^a	8.89
KIP	6.367^{bc}	2.225^a	2.9	Medium	Medium	10.99^d	8.09
PV	6.365^{bc}	2.035^f	3.1	Medium	Slender	13.16^c	8.25
SARI	6.078^d	2.083^e	2.9	Medium	Medium	7.34^f	8.96
TONO	6.44^{bc}	2.155^c	3.0	Medium	Medium	14.30^b	8.43
LSD	0.156	0.021	-	-	-	1.019	1.22
CV (%)	1.7	0.7	-	-	-	5.4	10

Note. Means with same alphabet within columns are not significantly different at 5%.

The alkaline spreading value gave three classes of gelatinisation temperature: low (< 70 °C), intermediate (70-74 °C) and high (74.5-80 °C) (Table 4). The amylose content was also significantly different for all seed sources except between PV and DARTEY and ARI and DARTEY. GBEWAA recorded the highest amylose content of 20.64% while SARI recorded 10.63% as the lowest. IRRI (1998) report indicates that Jasmine 85 has low gelatinisation temperature (65 to 68%) and low amylose. Diako et al. (2011) also reported in their study that Jasmine 85 has amylose content of 20.2%.

Table 4. Alkaline spreading value, gelatinisation temperature (°C) and its classification (°C), and means of the amylose content (%) for the various seed sources

Treatment	Alkaline spreading value	Gelatinisation temperature (°C)	Temperature classification	Amylose content (%)
PV	7	< 70°C	Low	14.51^e
GBEWAA	5	70-74°C	Intermediate	20.64^a
ARI	7	< 70°C	Low	14.85^d
TONO	6	< 70°C	Low	18.07^b
DARTEY	7	< 70°C	Low	14.65^{de}
SARI	2	74.5-80°C	High	10.63^g
KIP	7	< 70°C	Low	14.13^f
CRI	7	< 70°C	Low	15.79^c
LSD	-	-	-	0.31
CV (%)	-	-	-	1.4

Note. The means with the same alphabet within columns are not significantly different at 5%.

The peak obtained from the viscosity test (Figure 1) gave information on treatment viscosity. The Treatment with the highest peak viscosity was SARI (3537cP). GBEWAA (2791cP) had the lowest peak viscosity.Pasting properties is influenced by amylose content (Tester & Morrison, 1990); the higher the amylose content the less expansion potential and the lower the gel strength for the same starch concentration (Li & Yeh, 2001; Singh et al., 2003). The lower the amylose content the starch granules swell much more, and hence a thicker paste is produced (Diako et al., 2011). The viscosity graph gave significant difference among the sources especially with the peak and the final viscosities. SARI had the highest peak viscosity which can be explained by its lowest amylose content; GBEWAA on the other hand expressed the lowest peak viscosity and high amylose content.

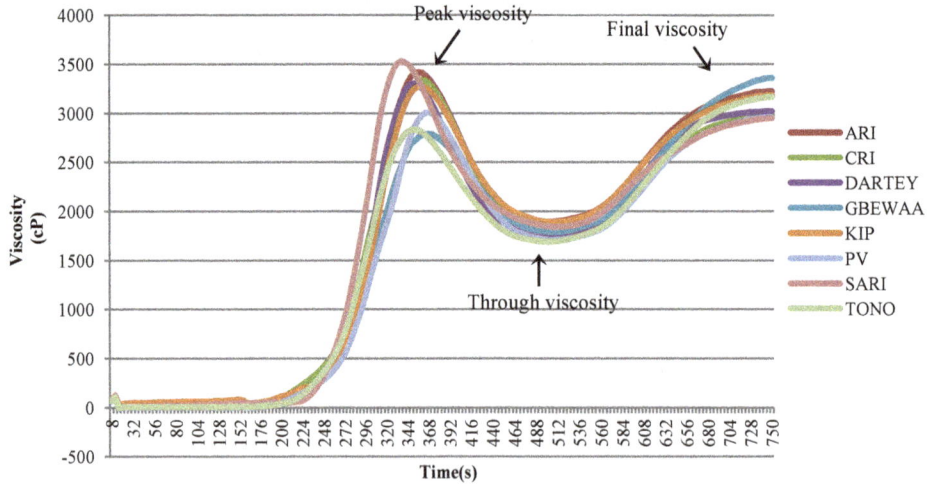

Figure 1. Means of the Viscosity properties of the various sources plotted against time (Centipoise)

Figure 2 represents the means of the cooking times; it varies from 23 minutes to 15.75 minutes respectively for GBEWAA and for PV. The source which has the shortest cooking time is PV and the source which has the longest cooking time is GBEWAA. The significant difference among treatments implies that mixing grains from two or more of the different sources for boiling would result in inconsistent texture of the meal. The physical appearance of the milled rice is also important in terms of consumer preference because most consumers are selective in what they eat so a mixture of different varieties would not be palatable to most consumers.

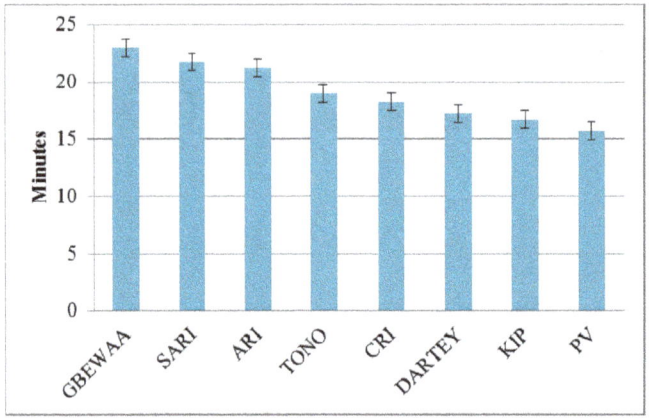

Figure 2. Means of the cooking time of the various sources

Note. LSD (1.53) is used as standard error bar to show the differences between treatments at 5%.

3.4 Source Clustering

To categorize the seeds from the eight sources, clustering analysis was performed with morphological and grain quality characteristics. The cluster is presented in Figure 3. The hierarchical tree can be cut into 4 groups. The first group comprised CRI, DARTEY, ARI and KIP; GBEWAA and TONO formed the second group; the remaining two classes were SARI and PV in that order. For the study, ARI obtained from Africa Rice Centre in Senegal with IRRI as its original source, was used as the check. Based on the clustering, CRI, KIP and DARTEY resembles ARI and might be the true Jasmine 85.

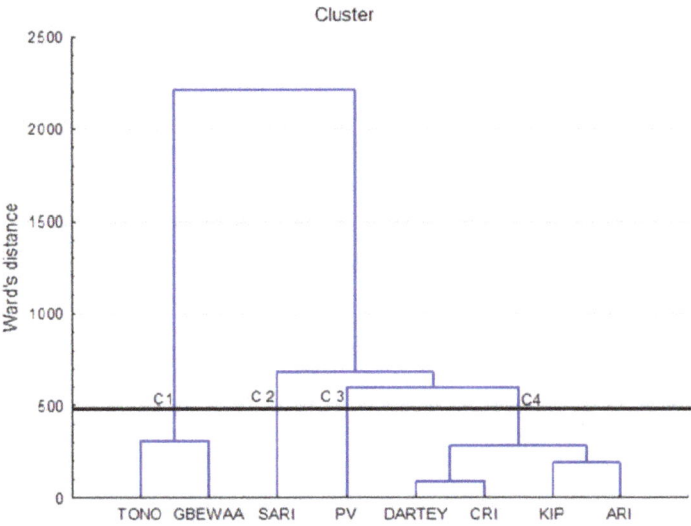

Figure 3. Cluster of the various sources using field and physico-chemical data

3.5 Molecular Analysis

The molecular analysis revealed GBEWAA as the most distant relative to the other seed sources with genetic distance of 0.3 cM between (GBEWAA and CRI). The least genetic distance (0.03 cM) was recorded between ARI and TONO; this means that ARI and TONO are closely related genetically than the other sources. The average distance between the various seed sources was 0.16 cM.The results of the similarity and dissimilarity between sources is presented in Table 5.

Table 5. Dissimilarity between and within sources (centimogan: cM)

	ARI	DARTEY	CRI	SARI	TONO	PV	GBEWAA
DARTEY	0.17						
CRI	0.13	0.17					
SARI	0.17	6.67E-02	0.17				
TONO	3.33E-02	0.13	0.1	0.13			
PV	0.1	6.67E-02	0.17	0.13	6.67E-02		
GBEWAA	0.23	0.267	0.3	0.2	0.2	0.2	
KIP	0.1	0.13	0.17	6.67E-02	6.67E-02	6.67E-02	0.13

Compared to the morpho-physico chemical cluster (Figure 3), the molecular clustering showed three classes of cluster (Figure 4). Cluster 1 composed of PV; DARTEY; and SARI, within this cluster PV is distinguished from the two other sources. Cluster 2 composed CRI, ARI, and TONO. Cluster 3 also composed of KIP and GBEWAA, GBEWAA also is distinguished from KIP in this cluster.

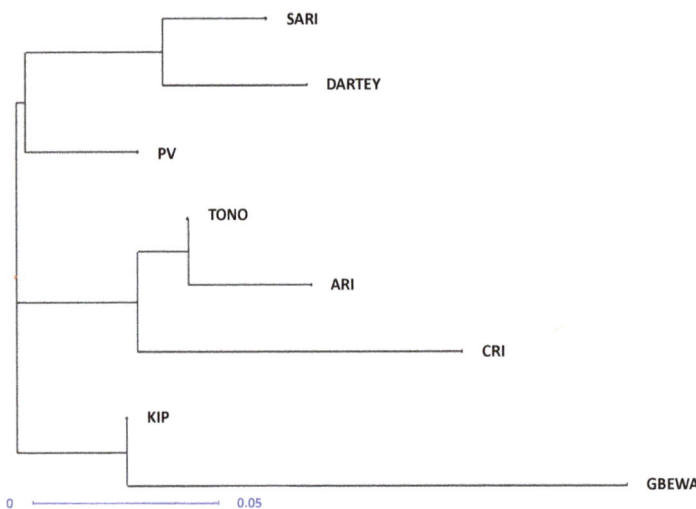

Figure 4. Molecular cluster of the sources

4. Conclusion

Significant genetic differences were observed at the morphological, physico-chemical and molecular levels. The seeds from the various sources are therefore genetically different and should not be mixed for production or sale.

Acknowledgements

The contributions of the various authors (O. T. Akintayo, P. K. A. Dartey, B. K. Maalekuu, J. Manful, M. Sow, M. Ndjiondjop, S. Graham-Acquaah, S. O. Abebrese, and J. K. Saajah) have been duly acknowledged for their respective roles in the completion of this work. Special thanks also go to God Almighty, for granting the needed health, understanding and guidance for towards the completion of the work. Other staff of Kwame Nkrumah University of Science and Technology, Kumasi, Ghana; Africa Rice Center, Suakoko, Liberia/CSIR-Crops Research Institute, Kumasi, Africa Rice Center, Cotonou, Benin, CSIR-Savannah Agricultural Research Institute, Tamale, Ghana and Ghana Cocoa Board/Newmont Mining Cooperation-Ghana who in diverse ways contributed to the completing of this work are all duly acknowledged.

The special roles played by the publishers of this work, who provided a platform for the work to get to the reading public are also duly acknowledged for their support.

References

Diako, C., Manful, J. T., Johnson, E. P. N. T., Dawson, S., Bediako-Amoa, B., & Saalia, F. K. (2011). Physicochemical characterisation of four commercial Rice varieties in Ghana. *Advance Journal of Food Science and Technology* (3rd ed., Vol. 3, pp. 196-202). Department of Nutrition and Food Science, University of Ghana.

Fofana, M., Futakuchi, K., Manful, J. T., Yaou, I. B., Dossou, J., & Bleoussi, R. T. M. (2011). Rice grain quality: A comparison of imported varieties, local varieties with new varieties adopted in Benin. *Food Control, 22*(12), 1821-1825. http://dx.doi.org/10.1016/j.foodcont.2011.04.016

International Rice Research Institute (IRRI). (1998). Registration of Jasmine 85. *Crop Science, 38*, 895-896.

ICC. (2004). Standard Methods of the International Association for Cereal Science and Technology. *6th Supplement. Methods No. 162 and No. 164, approved 1996.* The Association: Vienna.

International Organisation for Standardisation. (2011). ISO/DIS 6647-2-Rice-Determination of amylose content. *Part 2: Routine methods 10.*

International Food Policy Research Institute. (2014). Substituting Rice Import in Ghana. *Ghana Strategy Support Program* (Policy No. 6).

Jennings, P. R., Coffmanm, W. R., & Kauffmanm, H. E. (1979). *Rice improvement.* International Rice Research Institute Los Banos, Laguna, Philippines.

Li, J. Y., & Yeh, A. L. (2001). Relationships between thermal, Rheological characteristics and swelling power for various   starches.   *Journal   of   Food   Engineering,   50,   141-148.* http://dx.doi.org/10.1016/S0260-8774(00)00236-3

Little, R. R., Hilder, G. B., & Dawson, E. H. (1958). Differential effect of dilute alkali on 25 varieties of milled white rice. *Cereal Chem., 35,* 111-126.

Marco, M., Bollich, C., Webb, B., Jackson, B., McClung, A., Scott, J., & Hung, H. (1997). *Registration of 'Jasmine 85' Rice* (Peer Review Journal, Last modified: August 20, 2013).

Singh, N., Singh, L., Kaur, L., Singh-Sodhi, N., Singh-Gill, B. (2003). Morphological thermal and rheological properties of starches from different botanical sources. *Elsevier Science Food Chemistry, 81,* 219-231. http://dx.doi.org/10.1016/S0308-8146(02)00416-8

Tester, R. F., & Morrison, W. R. (1990). Swelling and gelatinization of cereal starches. *Cereal Chemistry, 67,* 551-557.

World Bank. (2013). *Growing Africa: Unlocking the potential of Agribusiness.* Washington, DC: World Bank.

15

Identification and Screening of Citrus Vein Phloem Degeneration (CVPD) on Brastagi Citrus Variety Brastepu (*Citrus nobilis* Brastepu) in North Sumatra Indonesia

Isnaini Nurwahyuni[1], Justin A. Napitupulu[1], Rosmayati[1] & Fauziyah Harahap[2]

[1] Doctorate Program, Department of Agriculture, Faculty of Agriculture, North Sumatra University, Medan, North Sumatera, Indonesia

[2] Departement of Biology, FMIPA, State University of Medan, Jl. Willem Iskandar, Psr V, Medan, Sumatera Utara, Indonesia

Correspondence: Isnaini Nurwahyuni, Doctorate Program, Department of Agriculture, Faculty of Agriculture, North Sumatra University, Medan, North Sumatera 20155, Indonesia.

Abstract

Identification and screening of Citrus Vein Phloem Degeneration (CVPD) on Brastagi citrus variety Brastepu (*Citrus nobilis* Brastepu) in North Sumatra Indonesia is explained. The aim of the study is to explore the technique to obtain healthy Brastepu citrus to be used as a source of explants for in vitro propagation as a step in the preservation of threatened local citrus. The studies were conducted through collection of survived Brastepu citrus followed by screening the CVPD in the plants. Various procedures in the field and in the laboratory have been conducted to screen infected CVPD in Brastepu citrus. Visual examination from leaf, fruit, and seed appearance becomes preliminary information for CVPD infection in the citrus plant. Iodine test was used to support results obtained by visual observation. Histochemical test and the PCR analysis have confirmed that 12 samples of Brastepu citrus are free from CVPD and eight citrus are infected by CVPD. Adequate informations have been obtained to confirm healthy citrus tree which are then to be used as sources of explants for *in vitro* propagation in the preservation of *Citrus nobilis* Brastepu.

Keywords: Brastagi citrus, *Citrus nobilis* Brastepu, threatened citrus, screening CVPD, iodine test, histochemistry, PCR

1. Introduction

Brastagi citrus to be known as "Jeruk Brastagi" is very popular local sweet orange many years ago and become one of the well known commodities from Brastagi, North Sumatra, Indonesia. There are some local citrus that are commonly planted and developed in Brastagi namely variety Brastepu (*Citrus nobilis* Brastepu), Boci (*Citrus nobilis* Lour), and Rimokeling (*Citrus nobilis*) (Nurwahyuni & Sinaga, 2014). *Citrus nobilis* Brastepu which was called as Brastepu citrus has superior potential genetic compared with honey citrus and other local citrus, where the fruit is large in size, has sweet taste, and a raped peel is yellow to reddish color. Moreover, Brastagi citrus becomes multifunction plants as sources of fruit and the leaves and peels contains bioactive that are commonly used for Karo traditional medicine to cure some diseases (Simatupang, 2009). However, Brastagi citrus nowadays are grown in the field as unintended plants because they are not planted properly with some reasons, the production quantity and their sensitivity to some diseases such as CVPD. It has been informed by citrus farmer that the CVPD has destroied Brastagi citrus and nowadays it has been replaced by imported citrus. The eruption of Mount Sinabung on 2013-2014, where Brastagi as the impact areas, make the existence of local citrus be decreased in number. If Brastepu citrus is not conserved properly, it is predicted that this variety will disappear in the short time. Therefore, it is need to screen healthy survived trees to be used as sources of explants for in vitro propagation in the conservation of Brastepu citrus. The use of plants pathogen free is believed to be an efficient strategy to control citrus disease, and this strategy is chosen to obtain healthy citrus trees with free from CVPD which is then to be used as explants sources in the propagation of Brastepu citrus.

Some serious diseases that can destroy citrus has been identified such as Citrus Vein Phloem Degeneration

(CVPD), Citrus Sudden Death (CSD), Citrus Variegated Chlorosis (CVC), and Citrus Bacterial Canker (CBC), but the CVPD is the most devastating citrus diseases because there is no cure for the infection trees has been found (Doddapaneni et al., 2008; Vojnov et al., 2010; Halbert & Manjunath, 2004). Infection of CVPD or Huanglongbing (HLB), known as greening diseases (yellow shoot), is a very dangerous disease caused by the infection of bacteria *Candidatus Liberibacter africanus* and *Candidatus L. asiaticus*, an it has infected various type of citrus in many countries including Indonesia. The main target of CVPD infection is phloem tissue with the symptom on small-size and narrow yellow leaves, yellowing on some of the new shoot tip in the green canopy, greening on the leaves, chlorosis, leaf senescence, nutritional deficiency, low production of fruit, and shortening the lifespan of infected trees. Infection of CVPD can be seen from vegetative descriptions mainly in blotchy mottle leaves, small size fruits with the fruit do not color up and remains green when raper, dry mesocarp with bitter, salty or sour taste, and accompany with abnormal and sterile seed. The CVPD was responsible for the decline of citrus tree, lost production of citrus and destroy citrus plants in a short time such in Asia, South-East Asia, south and east Africa, Arabian Peninsula and America (Phahladira et al., 2012; Halbert & Manjunath, 2004). The CVPD was discovered in many countries such as United States, Mexico, South America, Africa, the Middle East, Reunion and Mauritius islands (Ammar et al., 2013; Hall et al., 2012). It is difficult to control the CVPD and infected trees results in low quality and quantity fruit productions (Bassanezi et al., 2006, Bassanezi et al., 2009; do Carmo et al., 2005; Kim & Wang, 2009). In Asia, infection of CVPD has been reported in China, Taiwan, Philippine, and Indonesia, and has destroyed sweet orange (Titrawidjaja, 1984; Bove, 2006).

Many studies have been conducted to identify the infection of CVPD in citrus. Screening methods to identify CVPD infection can be conducted by iodine test eventhough the test is not specific but it can be used to predict plant infection. The polymerase chain reaction (PCR) method is known as a specific method for determination of CVPD. It can be used to identify the presence of *Liberibacter* such as *L. Asiaticus* and *L. Africanus* (Kim & Wang, 2009). Grafting and vector transmission tests and PCR have been used to investigate CVPD symptoms (Ahmad et al., 2011). Identification of infected CVPD with intra-species specific molecular mechanisms associated with Las-induced responses in lemon plants based on the protein accumulation and the concentrations of cationic elements by using MS and ICPS analyses have also been reported (Nwugo et al., 2013). The PCR method has been used to identify CVPD infection in mandarin (*Citrus reticulata*) fruit based on physical and biochemical characteristic (Shokrollah et al., 2011). The PCR method has been claimed to be a rapid detection and identification of CVPD infection in citrus (Fujikawa et al., 2013; Kawabe et al., 2006). The Random Amplified Polymorphic DNA (RAPD) method has also been used to identify polymorphic DNA fragments of an infected citrus by the bacteria (Hocquellet et al., 1999). Semi quantitative RT-PCR method for expression pattern of Nucleotide Binding Site Leucien-Rich Repeats genes has been introduced for CVPD determination (Zamharir et al., 2014). Another method by using 2-DE and mass spectrometry analyses, as well as ICP spectroscopy analysis were employed to elucidate the global protein expression profiles and nutrient concentrations in leaves of Las-infected grapefruit plants at pre-symptomatic or symptomatic stages for CVPD (Nwugo et al., 2013).

The use of antibiotics has been claimed to be promising as potential control strategy for citrus CVPD (Zhang et al., 2014), but until now the removal and destruction of infected trees, and the use of plants resistant are still known to be efficient strategies to stop the spread of citrus disease (Weinert et al., 2004). The aim of the research is to explain a rapid screening technique to determine the presence of infected CVPD in *Citrus nobilis* Brastepu, a threat species of Brastagi citrus in North Sumatra Indonesia, in order to obtain healthy plants to be used as sources of explants for *in vitro* propagation of Brastepu citrus as a step in the preservation of threatened local citrus.

2. Methods

Research methods are consisted of collection of Brastagi citrus variety *Citrus nobilis* Brastepu, identification and screening of CVPD. Collection of Brastagi citrus was conducted to obtain survived local citrus in the villages at Brastagi, Kabupaten Karo North Sumatra, Indonesia. Identification of CVPD was conducted in the field by using visual characterization based on plants performance on their leaves, fruits and seeds, followed the procedures explained earlier (Nurwahyuni & Sinaga, 2014). All laboratory experiments are conducted in Biology Laboratory FMIPA USU, Medan, and Agriculture Biotechnology Laboratory, Gadjah Mada University Yogyakarta. Screening of CVPD on *Citrus nobilis* Brastepu was conducted from its leaves by using iodine test with KI solution of iodine kit. Molecular analysis with histochemical test was performed for molecular tissue by using a microscope. Extraction of the DNA, PCR, and electrophoresis are conducted followed the procedures explained in the references (Fontarnau & Hernandez-Yago, 1982; Weising et al., 2005; Amani et al., 2011). Confirmation test for infection of CVPD is conducted by using PCR. The details of the procedures are explained in previous report (Nurwahyuni & Sinaga, 2014).

3. Result and Discussion

3.1 Characterization of Citrus nobilis Brastepu

Citrus Brastepu (*Citrus nobilis* Brastepu) is a local variety of sweet orange in Brastagi North Sumatra which is known as Brastagi citrus. The characteristic of healthy Brastepu citrus is presented in Figure 1. Mature Brastepu citrus can reach up to 3-4 metres (Figure 1a) and a healthy plant is productive up to 25 years old. The discription of citrus flower is five petals with white color and yellow anther (Figure 1b). Brastepu citrus is a seasoning fruit that only produce fruit for big harvest once a year, and the typical of green young fruit is presented in Figure 1c. The characteristic of citrus fruit is large in size, about 4.1-5.3 cm in diameter, and the taste is sweet. The color of fruit skin when young is green and the peel becomes orange to reddish color when raped (Figure 1d). The Brastepu fruit is with flat basal, the thick of the skin is 0.30-0.50 mm, where in flavedo 0.25-0.40 mm. The skin of Brastepu fruit is easily peeled with orange lith (fruit segment), and contains less seed about 4-8 seeds each fruit (Figure 1e). Brastepu citrus have single leaf, the avarage of which is 7 cm and 2.5 cm in length and wide, repectively. The stipula is clearly seen in the leaf. Peels and the leaves of Brastepu citrus contains bioactive compounds that are commonly used for Karo traditional medicine to cure some diseases (Nurwahyuni et al., 2012; Simatupang, 2009). Despite its superiority in the quantity and quality of Brastepu fruits, such as large in size, sweet taste, and the content of medicinal bioactive compounds in the plants, nowdays Brastepu citrus was not planted properly and has tend to be replaced by honey citrus due to its low resistance to CVPD. It is predicted that Brastepu citrus to be demolished soon if an action has not been done to preserve the plant. With this reason, the study is carried out to obtain healthy Brastepu trees that are free from CVPD to be used as sources of explants for *in vitro* propagation in the preservation of *Citrus nobilis* Brastepu. It is believed that survived healthy Brastepu citrus has resistant ability to CVPD, and *in vitro* propagation becomes a good strategies to produce good quality seedling to preserve the potential genetics of local citrus from diminished and to stop Brastagi citrus from loss of plant diversity.

Figure 1. The description of Citrus Brastepu (*Citrus nobilis* Brastepu): (a) A mature healthy Brastepu citrus grown in Desa Bukit, (b) the flower, (c) Immature green fruit, and (d) Mature fruit, (e) Citrus anatomy and its seed

3.2 Visual Investigation to Screen CVPD on Citrus nobilis Brastepu

The survey has been conducted to search survived *Citrus nobilis* Brastepu at Brastagi, Kabupaten Karo, North Sumatra, Indonesia. There were only 20 survived citrus found from six villages, successively at Desa Tongkoh (2 plants), Desa Bukit (10 plants), Desa Melas (2 plants), Desa Beganding (2 plants), Desa Surbakti (3 plants), and Desa Brastepu (3 plants). The condition of plants are varied from 5-30 years old, most of them are categorized as wild plants. Screening of CVPD on Brastepu citrus has been conducted by visual identification as it is known that phatogen detection can be seen based on the typical of leaf tissue and fruits disorder. Identification of infected plant can be seen from its vegetative symptom with one or some of these descriptions of blotchy mottle, small and narrow leaves, greening fruits with small seed and undeveloped flavedo with bitter or salty tasted due to CVPD infection in the floem. The disease has affected on cell defense, transport, photosynthesis, carbohydrate metabolism, and hormone metabolism (Albrecht & Bowman, 2008; Martinelli et al., 2012). The characteristic of Brastepu citrus from different plants is summarized in Table 1. Most of the plants are look healthy with green wide leaves, good pruning, and big size fruits. Some of Brastepu citrus are suspected to be infected by the CVPD with symptoms of small leaves, narrow shape, mottle, and yellowish as shown in Figure 2. Infected plants show greening leaves with necrosis that make the texture of citrus leaf abnormal in isolated branches. Furthermore, the vascular bundel of the leaf is spongy type with yellow color in the whole leaf, transparence, and also containing of green spot in the vascular bundle (Figure 2a). The greening to yellow process is continued in the leaf starting from vascular bundle up to leaf margin, followed by chlorosis, blotchy, and almost all leaves developed small and narrow as a symptom of CVPD infection (Figure 2b).

Figure 2. The symptom of infected CVPD on Brastepu citrus: (a) Greening to yellow leaves with chlorosis in branches, (b) The development of small, narrow, and *blotchy leaves,* (c) Small size and fade greening fruits flavedo and dry, and (d) Formation of black, long and small seed with sterile or abort

During the collection time, most of Brastepu citrus plants did not produce fruits, only few of them produce fruits. A healthy plant produce many big fruits with normal development, and orange to reddish feel color with orange lith, fresh and sweet taste with normal seeds when fruit raped. Unhealthy citrus tree produce few fruits, most of them abort when they are young. The fruits tend to developed abnormal and have small size, and when they raped they remain green with dry locule with bitter and salty taste and they are asymmetric and lopsided with a bent fruit axis (Figure 2c). The formation of green color for raped fruit can be used as an identification of CVPD infection due to the problem in the synthesis of chlorophyl to become chromophyl (Roy & Goldschmidt, 2008). The development of the seed in an infected plants are also abnormal, the shape is long, black in color, aborted seeds and sterile (Figure 2d). The fruit is asymetric with dry and fade flavedo accompany with small size seed is the indication of infected CVPD diseases (Brlansky & Roger, 2007; Bove & Ayres, 2007). These results may be used to predict healthy and infected citrus that to be confirmed in the next examinations, those are the iodine test, histochemical anatomy, and PCR analysis.

Table 1. The description of Brastepu citrus from visual observation in the field for screening of CVPD infection from their leaves, fruits and seeds

| Plants Code | Location | The description for CVPD simpthom | | |
		Color of Leaf	Vascular and Venae of Leaf	Raped Fruits and The seed
1	Desa Tongkah	Green healthy	Normal	No fruit
2	Desa Tongkah	Green healthy	Normal	No fruit
3	Desa Bukit	Green healthy	Normal	Normal, orange, the seed normal
4	Desa Bukit	Green healthy	Normal	Normal, orange, the seed normal
5	Desa Bukit	Green healthy	Normal	Normal, orange, the seed normal
6	Desa Bukit	Green healthy	Normal	Normal, orange, the seed normal
7	Desa Bukit	Green healthy	Normal	Normal, orange, the seed normal
8	Desa Bukit	Green healthy	Normal	Normal, orange, the seed normal
9	Desa Bukit	Green healthy	Normal	Normal, orange, the seed normal
10	Desa Bukit	Green healthy	Normal	Normal, orange, the seed normal
11	Desa Melas	Leaf defoliation, ±50% yellow	Brown	Small fruits, fade flavedo, the seed is black, sterile, abort
12	Desa Melas	Leaf defoliation, ±50% yellow	Brown	Small fruits, fade flavedo, the seed is black, sterile, abort
13	Desa Beganding	Green healthy	Normal	Normal, orange, the seed normal
14	Desa Beganding	Green healthy	Normal	Normal, orange, the seed normal
15	Desa Surbakti	Greening and yellow, *blotchy mottle* leaves	Normal, some are yellow	Normal, orange to green, the seed is steril
16	Desa Surbakti	Greening, and yellow, *blotchy mottle* leaves	Normal, some are yellow	Normal, orange to green, steril
17	Desa Surbakti	Green healthy	Green spot in the vascular bundle, leaf spongy type	Small size fruit, green in color, the seed is black, sterile, and abort
18	Desa Brastepu	Leaf defoliation, ±75% yellow	Brown	No fruit
19	Desa Brastepu	Leaf defoliation, *blotchy mottle* leaves, ±75% yellow	Brown	Small fruits, fade flavedo, the seed is black, sterile, abort
20	Desa Brastepu	Leaf defoliation, ±75% yellow	Brown	No fruit

3.3 Iodine test for Screening CVPD Infection

Screening of CPVD infection by using iodine test was conducted to support the visual investigation in the field as it is known that iodine is reacted with amylum in the leaf to produced dark blue color. This method is effective to diagnose the presence of amylum results from photosynthesis (Su, 2008). There are 5 leaves are randomly taken from every Brastepu citrus plants, and the test were conducted in pastic bag for each of them to observed and the color changing as shown in Figure 3, and the results are summarized in Table 2. It is seen from the test that there are 12 plants are negative to CVPD where the leaves samples are not given change in color with iodine because accumulation of amylum were not found in healthy tissues. Special attentions have been conducted for suspected unhealthy samples (No. 11, 12, 17, and 18), where the iodine give color change from brown to black color. Another four samples gave positive-negative results to CVPD. They are samples No. 15, 16, 19, and 20. The results in iodine test supported the results obtained by visual investigation. There are 12 Brastepu citrus plants free from CVPD and eigh samples suspicious to be infected by CVPD. Further infestigations have to be conducted to confirm these results by using histochemical anatomy test and the PCR analysis.

Figure 3. Screening of CVPD on 20 *Citrus nobilis* Brastepu by using iodine test. Brown color represent healthy leaves, and blue to black color are suspicious to be infected by CVPD

Table 2. Screening analysis of CVPD infection for sample of *Citrus nobilis* Brastepu that are obtained from various villages in Brastagi, Kabupaten Karo, North Sumatra, Indonesia

No Citrus Plants	Villages Location	Analysis and Confirmation of CVPD infection		
		Iodine test*	Amylum – Anatomy Test**	Analysis PCR
1	Desa Tongkah	5 Negative	Negative	Negative
2	Desa Tongkah	5 Negative	Negative	Negative
3	Desa Bukit	5 Negative	Negative	Negative
4	Desa Bukit	5 Negative	Negative	Negative
5	Desa Bukit	5 Negative	Negative	Negative
6	Desa Bukit	5 Negative	Negative	Negative
7	Desa Bukit	5 Negative	Negative	Negative
8	Desa Bukit	5 Negative	Negative	Negative
9	Desa Bukit	5 Negative	Negative	Negative
10	Desa Bukit	5 Negative	Negative	Negative
11	Desa Melas	5 Positif	Positive	Positive
12	Desa Melas	5 Positif	Positive	Positive
13	Desa Beganding	5 Negative	Negative	Negative
14	Desa Beganding	5 Negative	Negative	Negative
15	Desa Simpang Empat	5 Positive	Positive	Positive
16	Desa Simpang Empat	5 Positive	Positive	Positive
17	Desa Simpang Empat	2 Positive, 3 Negative	Positive	Positive
18	Desa Brastepu	1 Positive, 4 Negative	Positive	Positive
19	Desa Brastepu	1 Positive, 4 Negative	Positive	Positive
20	Desa Brastepu	1 Positive, 4 Negative	Positive	Positive

Note. * There are 5 leaves are randomly selected from one plant; ** There are 3 leaves are randomly selected from one plant.

3.4 Histochemical Anatomy Test for CVPD Infection

Followed by the results obtained by visual investigation and iodine test as stated above, another screening test for CVPD infection has been carried out through histochemical test for molecular anatomy investigation in a preparat that was chosen closed to petiol of the leaf by using microscope. The histochemical anatomy results for the tissue of a healthy and infected Brastepu citrus are presented in Figure 4. The tissues of healthy and infected tissues are clearly shown in the figure where the pholem cell in a healthy leaf is normal where the amylum is distributed well in the cell (Figure 4a). In unhealthy leaf tissue, the accumulations of amylum was found in certain areas of a pholem cell, that were shown with the observation of dark to black spots of amylum on the parenchym tissues (Figure 4b). The preparat of infected leaf consisted of empty room without amylum because the amylum could not distribute well due the miss orientation of the tissue as a results of infection of pholem cell. Histochemical anatomy test have been conducted randomly for three leaves in evey plants and the results are summarized in Table 2. The tissues of 12 Brasitepu citrus plants are healthy, and eight of them are suspected to be infected by the CVPD. Screening results from visual observations, iodine test, and histochemical anatomy test have given the information on the status of Brastepu citrus, and the results are all agree and consistent to show the indication of negative and positive infection of CVPD in Brastepu citrus. The PCR analysis is then to be used for confirmation of CVPD in the citrus trees.

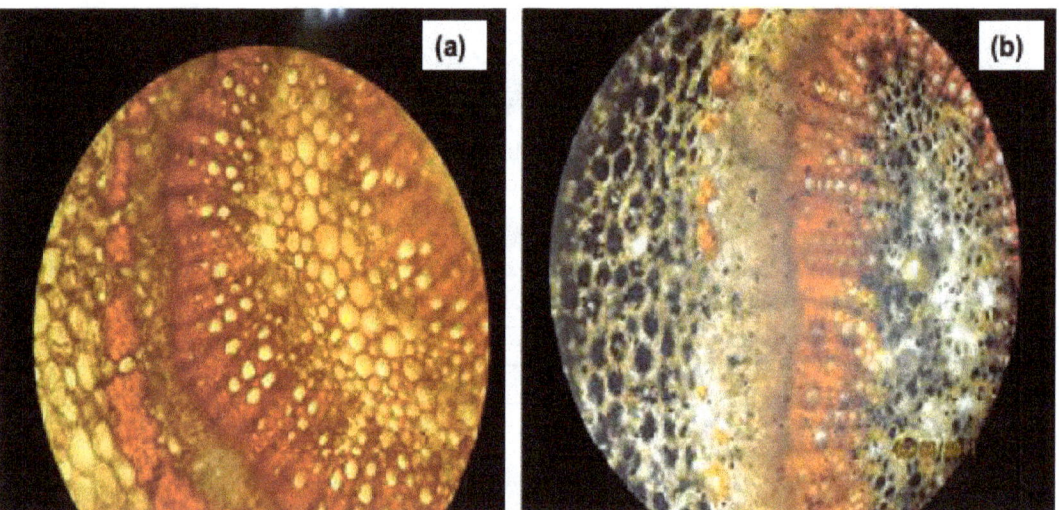

Figure 4. The anatomy of a petiol to show infection of CVPD in citrus leaves at 40 x amplification: (a) Healthy tissue, (b) Unhealthy tissue with accumulation of amylum

3.5 Isolation of DNA for PCR Assays

Isolation of the DNA from Brastepu citrus leaf has been carried out for PCR analysis. It is known that the presence of certain protein in the DNA can be used to confirm the CVPD infection (Akarapisan et al., 2008). Isolation of citrus DNA is conducted by extraction of lysis cell, separation and purification. The electrophoresis results for the DNA Brastepu citrus has proved that the DNA was isolated well, and the quantity of the Brastepu citrus DNA has been measured by using spectrophotometric method which are adequate for PCR analysis and confirmation. To confirm CVPD infection in Brastepu citrus, the DNA of Brastepu citrus is analysed followed the procedures explained with using CVPD protein-specific primers (Padmalatha & Prasad, 2006; Sahasrabudhe & Deodhar, 2010). The PCR results are shown in Figure 5 and the confirmation results are summarized in Table 2. The results have showed that there are 12 citrus plants have confirmed free from CVPD and eight citrus plants are infected by CVPD.

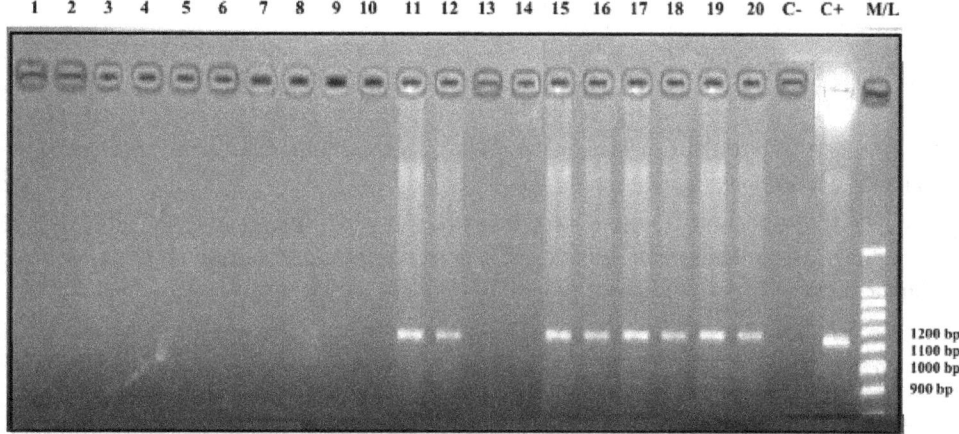

Figure 5. The PCR analysis of citrus DNA for confirmation of CVPD infection. Infected samples are shown in the figure

The PCR test has confirmed that Brastepu citrus number 1, 2, 3, 4, 5, 6, 7, 8, 9, 10, 13, and 14 are free from CVPD, and another Brastepu citrus plants number 11, 12, 15, 16, 17, 18, 19, and 20 are positively infected by the CVPD. Therefore, healthy citrus plants has been selected to be used as the source of explant for *in vitro* propagation, those are the Brastepu trees grown in Desa Bukit Brastagi because the all plants are healthy with plenty of fruits. Potential genetics of healthy Brastepu citrus would be inherited to the comming Brastepu seedling that to be produced by *in vitro* propagation in the preservation of threated *Citrus nobilis* Brastepu in further study.

4. Conclusion

Screening method to investigate the presence of infected CVPD on Brastagi citrus variety Brastepu (*Citrus nobilis* Brastepu) in North Sumatra Indonesia has been conducted. Visual identification from the leaf, tissue, fruits, and seed of Brastepu citrus have given the informations that some plants are infected by CVPD. Screening program by iodine test for leaves tissues and anatomy of petiol were used to support the results from visual observation. The PCR analysis has confirmed that 12 Brastepu plants are free from CVPD and eight Brastepu plants are positive to be infected by the CVPD. The study is assigned to be a good strategy to select healthy citrus plant to be used as source of explant for *in vitro* propagation, that is a step in the preservation of threatened Brastagi citrus variety Brastepu.

References

Ahmad, K., Sijam, K., Hashim, H., Abdu, A., & Rosli, Z. (2011). Assessment of CitrusSusceptibility towards CandidatusLiberibacter Asiaticus-Terengganu Isolate Based on Vector and Graft Transmission Tests. *Journal of Agricultural Science, 3*(3), 159-166. http://dx.doi.org/10.5539/jas.v3n3p159

Akarapisan, A., Piwkhao, K., Chanbang, Y., Naphrom, D., & Santasup, C. (2008). Occurrence of Huanglongbing Disease of Pomelo (*Citrus grandis*) in Northern Thailand. *IRCHLB Proceedings Dec. 2008*. Retrieved from http://www.plantmanagementnetwork.org

Albrecht, U., & Bowman, K. D. (2008). Gene expression in Citrus sinensis (L.) Osbeck following infection with the bacterial pathogen Candidatus Liberibacter asiaticus causing Huanglongbing in Florida. *Plant Science, 175*, 291-306. http://dx.doi.org/10.1016/j.plantsci.2008.05.001

Amani, J., Kazemi, R., Abbasi, A. R., & Salmanian, A. H. (2011). A simple and rapid leaf genomic DNA extraction method for polymerase chain reaction analysis. *Iranian Journal Of Biotechnology, 9*(1), 69-71.

Ammar, E., Hall, D. G., & Shatters, R. G. (2013). Stylet Morphometrics and Citrus Leaf Vein Structure in Relation to Feeding Behavior of the Asian Citrus Psyllid Diaphorina citri, Vector of Citrus Huanglongbing Bacterium. *PLOS ONE, 8*(3), 1-12 (e59914). http://dx.doi.org/10.1371/journal.pone.0059914

Bassanezi, R. B., Montesino, L. H., & Stuchi, E. S. (2009). Effects of huanglongbing on fruit quality of sweet orange cultivar in Brazil. *European Journal of Plant Pathology, 125*, 565-572. http://dx.doi.org/10.1007/s10658-009-9506-3

Bove, J. M., & Ayres, A. J. (2007). Etiology of three recent diseases of citrus in Sao Paulo State: Sudden death, variegated chlorosis and huanglongbing. *Iubmb Life, 59,* 346-354. http://dx.doi.org/10.1080/15216540701299326

Bove, J. M. (2006). Huanglongbing: A destructive, newly-emerging, century old disease of citrus. *Journal of Plant Pathology, 88*(1), 7-37. http://dx.doi.org/10.1080/15216540701299326

Brlansky, R. H., & Roger, M. E. (2007). Citrus huanglongbing: Understanding the vector pathogen interaction for disease management. *APS net Features, 1207.*

Do Carmo, T. D., Luc, D. J., Eveillard, S., Cristina, M. E., de Jesus, J. W. C., Takao, Y. P., ... Bove, J. M. (2005). Citrus huanglongbing in Sao Paulo State, Brazil: PCR detection of the 'Candidatus' Liberibacter species associated with the disease. *Mol Cell Probes, 19*(3), 173-179. http://dx.doi.org/10.1016/j.mcp.2004.11.002

Doddapaneni, H., Liao, H., Lin, H., Bai, X., Zhao, X., Civerolo, E. L., Irey, M., ... Pietersen, G. (2008). Comparative phylogenomics and multi-gene cluster analyses of the Citrus Huanglongbing (HLB)-associated bacterium *Candidatus* Liberibacter. *BMC Research Notes, 1,* 72. http://dx.doi.org/10.1186/1756-0500-1-72

Fontarnau, A., & Hernandez-Yago, J. (1982). Characterization of Mitochondrial DNA in Citrus. *Plant Physiol., 70,* 1678-1682. http://dx.doi.org/10.1104/pp.70.6.1678

Fujikawa, T., Miyata, S. I., & Iwanami, T. (2013). Convenient Detection of the Citrus Greening (Huanglongbing) Bacterium 'CandidatusLiberibacter asiaticus' by Direct PCR from the Midrib Extract. *PLOS ONE, 8*(2), 1-7 (e57011). http://dx.doi.org/10.1371/journal.pone.0057011

Halbert, S. E., & Manjunath, K. L. (2004), Asian Citrus Psyllids (Sternorrhyncha: Psyllidae) and Greening Diseases of Citrus: A Literature Review and Assessment of Risk in Florida. *Florida Entomologist, 87*(3), 330-353. http://dx.doi.org/10.1653/0015-4040(2004)087%5B0330:ACPSPA%5D2.0.CO;2

Hall, D. G., Richardson, M. L., Ammar, E. D., & Halbert, S. E. (2012). Asian citrus psyllid, Diaphorina citri (Hemiptera: Psyllidae), vector of citrus huanglongbing disease. *Entomologia Experimentalis et Applicata, 146,* 207-223. http://dx.doi.org/10.1111/eea.12025

Hocquellet, A., Bove, J. M., & Garnier, M. (1999). Isolation of DNA from the Uncultured "CandidatusLiberobacter"Species Associated with Citrus Huanglongbing by RAPD. *Current Microbiology, 38,* 176-182. http://dx.doi.org/10.1007/PL00006783

Kawabe, K., Truc, N. T. N., Lan, B. T. N., Hong, L. T. T., & Onuk, M. (2006). Quantification of DNA of citrus huanglongbing pathogen in diseased leaves using competitive PCR. *J Gen Plant Pathol, 72,* 355-359. http://dx.doi.org/10.1007/s10327-006-0306-8

Kim, J. S., & Wang, N. (2009). Characteristization of copy numbers of 16 S DNA and 16 S RNA of *Candidatus* Liberibacter asiaticus and the implication in detection in planta using quantitative PCR. *BMC Research Notes, 2,* 37-40. http://dx.doi.org/10.1186/1756-0500-2-37

Martinelli, F., Uratsu, S. L., Albrecht, U., Reagan, R. L., Phu, M. L., Britton, M., ... Dandekar, A. M. (2012). Transcriptome Profiling of Citrus Fruit Response to Huanglongbing Disease. *PLOS ONE, 7*(5), 1-16 (e38039).

Nurwahyuni, I., & Sinaga, R. (2014). *In vitro* Propagation For Bioconservation Of Threatened Brastagi Citrus In North Sumatra Indonesia. *Int J Pharm Bio Sci, 5*(4, B), 863-873.

Nurwahyuni, I., & Sinaga, R. (2014). Bioconservation of Brastagi Citrus (*Citrus nobilis* Brastepu) A Local Citrus of Sumatra Utara through Cutting Propagation. *Proceeding of SEMIRATA BKSPTN-B* (In press). IPB Bogor, Indonesia.

Nurwahyuni, I., & Sinaga, R. (2014). *Strategi Biokonservasi Jeruk Keprok Brastagi (Citrus nobilis Brastepu) Menggunakan Teknik Kultur Jaringan Tanaman Untuk Mengatasi Kepunahan Jeruk Lokal Sumatera Utara.* Research Report FMIPA USU, Medan Indonesia.

Nurwahyuni, I., Napitupulu, J. A., Rosmayati, & Harahap, F. (2012). Pertumbuhan Okulasi Jeruk Keprok Brastepu (*Citrus nobilis Var.* Brasitepu) Menggunakan Jeruk Asam Sebagai Batang Bawah. *Jurnal Penelitian Saintika, 12*(1), 24-35.

Nwugo, C. C., Duan, Y., & Lin, H. (2013). Study on Citrus Response to Huanglongbing Highlights a Down-Regulation of Defense-Related Proteins in Lemon Plants Upon 'Ca. Liberibacter asiaticus' Infection. *PLOS ONE, 8*(6), 1-13 (e67442).

Nwugo, C. C., Lin, H., Duan, Y., & Civerolo, E. L. (2013). The effect of 'CandidatusLiberibacter asiaticus' infection on the proteomic profiles and nutritional status of pre-symptomatic and symptomatic grapefruit (*Citrus paradisi*) plant. *BMC Plant Biology, 13*, 1-24. http://dx.doi.org/10.1186/1471-2229-13-59

Phahladira, M. N. B., Viljoen, R., & Pietersen, G. (2012). Widespread occurrence of "Candidatusliberibacter africanus subspecies capensis"in Calodendrum capensein South Africa. *Eur J Plant Pathol, 134*, 39-47. http://dx.doi.org/10.1007/s10658-012-0020-7

Roy, P. S., & Goldschmidt, E. E. (2008). *Biology of citrus.* Cambridge, UK: Cambridge University Press.

Sahasrabudhe, A., & Deodhar, M. (2010). Standardization of DNA extraction and optimization of RAPD-PCR condition in Garcinia indica. *International Journal of Botany, 6*(3), 293-298. http://dx.doi.org/10.3923/ijb.2010.293.298

Shokrollah, H., Abdullah, T. L., Sijam, K., & Abdullah, S. N. A. (2011). Identification of physical and biochemical characteristic of mandarin (*Citrus reticulata*) fruit infected by Huanglongbing (HLB). *AJCS, 5*(2), 181-186.

Simatupang, S. (2009). Karakterisasi and Pemanfaatan Plasma Nutfah Jeruk in Situ oleh Masyarakat Lokal Sumatra Utara. *Buletin Plasma Nutfah, 15*(2), 70-74.

Su, H. J. (2008). *Production and cultivation of virus free citrus saplings for citrus rehabilitation in Taiwan.* New Delhi, India: APCoAB.

Tirtawidjaja, S., Hadewidjaja, T., & Lasheen, A. M. (1984). Citrus vein phloem degeneration virus, a possible cause of citrus chlorosis in Java. *Proc Am Soc Hort Sci, 86*, 235-243.

Vojnov, A. A., do Amaral, A. M., Dow, J. M., Castagnaro, A. P., & Marano, M. R. (2010). Bacteria causing important diseases of citrus utilise distinct modes of pathogenesis to attack a common host. *Appl Microbiol Biotechnol, 87*, 467-477. http://dx.doi.org/10.1007/s00253-010-2631-2

Weinert, M. P., Jacobson, S. C., Grimshaw, J. F., Bellis, G. A., Stephens, P. M., Gunua, T. G., ... Davis, R. I. (2004). Detection of Huanglongbing (citrus greening disease) in Timor-Leste (East Timor) and in Papua New Guinea. *Australasian Plant Pathology, 33*, 135-136. http://dx.doi.org/10.1071/AP03089

Wesing, K., Nybom, H., Wolff, K., & Kahl, G. (2005). *DNA finger printing in plants: Principles, methods and applications* (2nd ed., p. 444). London: CRC Press. http://dx.doi.org/10.1201/9781420040043

Zamharir, M. G., Hamzeh, K., Alizadeh, A., & Kachoei, S. (2014). Expression Patterns of Nbs-Lrr Family Genes in Infectedfected Citrus by Candidatus Leiberibacter Asiaticus. *International Journal of Agriculture and Crop Sciences, 7*(15), 1509-1513.

Zhang, M., Guo, Y., Powell, C. A., Doud, M. S., Yang, C., & Duan, Y. (2014). Effective Antibiotics against 'CandidatusLiberibacter asiaticus' in HLB-Affected Citrus Plants Identified via the Graft-Based Evaluation. *PLOS ONE, 9*(11), 1-11 (e111032).

16

Relative Efficiency of Zinc-Coated Urea and Soil and Foliar Application of Zinc Sulphate on Yield, Nitrogen, Phosphorus, Potassium, Zinc and Iron Biofortification in Grains and Uptake by Basmati Rice (*Oryza sativa* L.)

Yashbir Singh Shivay[1], Rajendra Prasad[1], Rajiv Kumar Singh[2,3] & Madan Pal[1]

[1] Division of Agronomy, Indian Agricultural Research Institute, New Delhi, India

[2] Krishi Vigyan Kendra, Aligarh of Chander Shekhar Azad University of Agriculture & Technology, Kanpur, India

[3] Directorate of Seed Research, Village Kushmaur, Post Office, NBAIM, District Mau, Uttar Pradesh, India

Correspondence: Yashbir Singh Shivay, Division of Agronomy, Indian Agricultural Research Institute, New Delhi 110 012, India. E-mail: ysshivay@hotmail.com

Abstract

Two on-farm trials conducted one each in Aligarh and Meerut districts of the state of Uttar Pradesh, India on zinc (Zn) deficient soils during the rainy season (July-October) showed that Zn application increased not only Zn concentration and uptake by rice but also increased protein content of rice kernels and concentrations of Fe, N, P and K due to the overall improvement in crop growth. Foliar application of Zn was better from the viewpoint of Zn biofortification of rice kernels; nevertheless much of the foliar applied Zn was retained in husk. Since, foliar application of Zn is made at a late stage of crop growth, hence it was not as effective as soil application in increasing yield attributes, yield and concentration and uptake of Fe, N, P and K in rice. This study brought out that adequate soil application of Zn sulphate followed by its foliar application is the best approach. Zn coated urea applying less than half the amount of Zn as applied through soil + foliar application was very close to it and is quite promising.

Keywords: basmati rice, ferti-fortification, nutrient concentrations and uptake, Zn fertilization

1. Introduction

Zinc (Zn) is now recognised as the fourth major micronutrient deficiency in humans and comes after vitamin A, iron and iodine deficiencies (Bell & Dell, 2008). Zn deficiency leads to diarrhoea and pneumonia in infants and children (Black et al., 2008; Graham, 2008) and is considered for responsible for childhood dwarfism (Hotz & Brown, 2004). Zn plays an important role in production of protein and thus helps in wound healing, blood formation and growth and maintenance of tissue (Bell & Dell, 2008).

Rice is staple food in South and Southeast Asia, where about 90% of it is grown and consumed. Polished rice contains only 13-15 mg Zn kg^{-1} (Welch, 2005) and a rice based diet is thus likely to be deficient in Zn as compared to diet containing grain legumes and animal proteins, which are rich in Zn (Prasad, 2003). Asian, especially Indian soils are low in available Zn (Prasad, 2006; Singh, 2011) and this leads to production of low Zn containing rice. Keeping this in view, programmes such as HarvestPlus and Golden Rice are underway to develop rice varieties capable of producing grains denser in Zn, Fe and other micronutrients (Stein et al., 2007).

Our research at the Indian Agricultural Research Institute (IARI), New Delhi (Shivay et al., 2007; Shivay et al., 2008a, 2008b, 2008c; Shivay et al., 2010; Shivay & Prasad, 2012; Shivay et al., 2013; Shivay & Prasad, 2014) has shown that Zn concentration in rice, wheat, barley and corn grains can be easily increased by adequate Zn fertilization of crops and agronomic biofortification of cereals with Zn and Fe, which are essential plant micronutrients (Prasad & Power, 1997) is a faster and easier way for achieving the goal of obtaining cereal grains denser in Zn. As a follow-up of our research at IARI, two on-farm field trials were therefore conducted to familiarize the farmers about the biofortification of rice grains with Zn and this paper reports the results of these on-farm trials.

2. Materials and Methods

2.1 Description of Study Area

On-farm experiments were conducted at two different places. First on-farm trial was conducted at the Krishi Vigyan Kendra (KVK), Aligarh of Chandra Shekhar Azad University of Agriculture & Technology, Kanpur, Uttar Pradesh, India during rainy season (July-October) of 2008 on a sandy clay loam soil (*typic Ustochrept*). The soil of experimental field had 133.7 kg ha^{-1} alkaline permanganate oxidizable nitrogen (N) (Subbiah & Asija, 1956), 12.3 kg ha^{-1} available phosphorus (P) (Olsen et al., 1954), 295.1 kg ha^{-1} 1 N ammonium acetate exchangeable potassium (K) (Hanway & Heidel, 1952) and 0.36% organic carbon (C) (Walkley & Black, 1934). The pH of soil was 8.5 (1: 2.5 soil and water ratio) (Prasad et al., 2006) and diethylene triamine penta acetic acid (DTPA) extractable Zn and Fe (Lindsay & Norvell, 1978) in soil were 0.33 and 3.7 mg kg^{-1} of soil, respectively.

The second on-farm trial was conducted at Kulimanpur village of Meerut district, Uttar Pradesh, India during rainy season (July-October) of 2008 on a sandy loam soil. The soils of experimental field at second site had 127.3 kg ha^{-1} alkaline permanganate oxidizable nitrogen (N) (Subbiah & Asija, 1956), 14.7 kg ha^{-1} available phosphorus (P) (Olsen et al., 1954), 278.9 kg ha^{-1} 1 N ammonium acetate exchangeable potassium (K) (Hanway & Heidel, 1952) and 0.35% organic carbon (C) (Walkley & Black, 1934). The pH of soil was 8.2 (1:2.5 soil and water ratio) (Prasad et al., 2006) and diethylene triamine penta acetic acid (DTPA) extractable Zn and Fe (Lindsay & Norvell, 1978) in soil were 0.35 and 4.9 mg kg^{-1} of soil, respectively. The critical level of DTPA extractable Zn for rice grown on alluvial soils in the rice-wheat belt of north India varies from 0.38-0.90 mg kg^{-1} soil (Takkar et al., 1997) and thus the response of *Basmati* rice to Zn application was expected on the experimental field.

2.2 Experimental Treatments and Design

The experiment was conducted in a randomized block design with six replications on both sites. The treatments were: check (control, no Zn), 5 kg Zn ha^{-1} to soil as Zn sulphate (soil), 1 kg Zn ha^{-1} as Zn suphate to foliage (foliar), 5 kg Zn ha^{-1} (soil) + 1 kg Zn ha^{-1} (foliar) and 2.83 kg Zn ha^{-1} through Zn-coated urea (soil).

2.3 Application of Treatments and Fertilizers

The experimental field was disk-ploughed twice, puddled three times with a puddler in standing water and levelled. At final puddling 26 kg P ha^{-1} as single superphosphate and 33 kg K ha^{-1} as muriate of potash was broadcasted. Addition of 26 kg P as single super phosphate also supplied 45 kg S ha^{-1} and takes care of S deficiency (if any). Nitrogen at 130 kg N ha^{-1} as prilled urea or Zn-coated urea (ZnCU) was band applied in two equal splits, half 10 days after transplanting (DAT) and the other half at panicle initiation (40 DAT). Thus, Zn in ZnCU was band applied. When applied at the site, ZnCU supplied 2.83 kg Zn ha^{-1} for the 1.0% Zn coatings onto prilled urea. Foliar application of Zn was made at flowering and 0.5% $ZnSO_4 \cdot 7H_2O$ foliar application at the rate of 1,000 liters of solution ha^{-1} (+ 2.5 kg ha^{-1} lime) which supplied 1.0 kg Zn ha^{-1}.

2.4 Rice Transplanting

Two 25-day-old seedlings of rice were transplanted per hill at 20 cm × 10 cm in the first fortnight of July at both the places of study. The 'Pusa Sugandh 5', a derivative of Pusa 3A Karnal *Basmati*, is a semi-dwarf (90–100 cm height), high yielding basmati rice variety released in the year 2004 by the Indian Agricultural Research Institute, New Delhi, for commercial cultivation was transplanted in Meerut district of Uttar Pradesh. However, in Aligarh district 'Pusa *Basmati* 1' was transplanted which is an aromatic (*Basmati*) variety released from Indian Agricultural Research Institute, New Delhi, India during 1989 for its commercial cultivation. It is a cross between 'Pusa 150' and 'Karnal Lokal'. It produces long slender grains with good aroma and excellent cooking qualities (Rani et al., 2009; V. P. Singh & A. K. Singh, 2009; Siddiq et al., 2012). Irrigation channels measuring 1 m wide were placed between the replications to ensure easy and uninterrupted flow of irrigation water where an individual plot was independently irrigated from the irrigation channels. Rice crop was grown as per recommended package of practices and was harvested in the second fortnight of October at both the places.

2.5 Recording of Growth, Yield Attributes and Yields of Rice

Ten hills were randomly selected in each plot for measuring plant height and fertile tillers hill^{-1} 10 days before harvest and the average values were computed. Similarly, 10 panicles were randomly selected from each plot for recording the data on yield attributes (fertile tillers hill^{-1}, panicle length, grains panicle^{-1} and 1,000-grain weight). At harvest, grain, straw and biological yields have been recorded for each plot and finally converted into tonnes ha^{-1}. The rice grain yield was recorded at 14% moisture. The rough rice grains were hulled and rice kernel and rice husk yields were obtained which were used for chemical analysis of mineral nutrients such as Zn, Fe, N, P and K.

2.6 Chemical Analysis of Zn and Fe Concentration in Rice Kernel, Rice Husk and Rice Straw

At harvest, samples of rice kernel, rice husk and rice starw were drawn from each plot of the experiment for the chemical analysis of Zn and Fe concentrations. Zn and Fe in rice kernel, husk and straw samples was analysed on a di-acid ($HClO_4$ + HNO_3 in 3:10 ratio) digest on an Atomic Absorption Spectrophotometer (Prasad et al., 2006). Thereafter, the uptake of the Zn and Fe was calculated by multiplying Zn and Fe concentrations with respective plot yield of rice kernel, rice husk and rice straw yields.

2.7 Chemical Analysis of N, P and K in Rice Kernel, Rice Husk and Rice Straw

At harvest, samples of rice kernel, rice husk and rice starw were drawn from each plot of the experiment for the chemical analysis of N, P and K as per the procedure described by Prasad et al. (2006). Finally recorded data in all the six replications were subjected to statistical analysis and final tabulation were done of the statistically analyzed data.

2.8 Statistical Analysis

All the data obtained from rice for this study were statistically analyzed using the *t*-test as per the procedure given by K. A. Gomez and A. A. Gomez (1984). LSD values at $P = 0.05$ were used to determine the significance of differences between treatment means.

3. Results

A. Aligarh Site

3.1 Yield Attributes

Zn application significantly increased tillers m^{-2} and grains panicle^{-1} in rice, but not the panicle length and 1,000-grain weight (Table 1). Tillers m^{-2} and grains panicle^{-1} were the most with soil + foliar application of Zn sulphate (ZnS), significantly more than soil application of ZnS or Zn-coated urea (ZnCU), which in turn was significantly superior to foliar application of ZnS.

Table 1. Effect of various zinc treatments on yield attributes of aromatic rice [Aligarh, Uttar Pradesh site]

Treatments	Plant height (cm)	Tillers m^{-2}	Panicle length (cm)	Grains panicle^{-1} (Nos.)	1,000-grain weight (g)
Check	104	310	24	85	21.1
5 kg Zn ha^{-1} (soil)	107	326	26	91	22.2
1 kg Zn ha^{-1} (foliar)	105	318	25	88	22.0
5 kg Zn ha^{-1} (soil) + 1 kg Zn ha^{-1} (foliar)	108	342	27	94	22.7
2.83 kg Zn ha^{-1} through Zn-coated urea (soil)	107	328	26	91	22.3
SEm±	2.12	3.51	0.79	1.03	0.38
LSD ($p = 0.05$)	NS	9.95	NS	2.93	NS

3.2 Yields

Zn application significantly increased grain and straw yield as well as harvest index of rice (Table 2). The highest values of grain, kernel, husk, straw and biological yield were obtained with soil + foliar application of ZnS, followed by soil application of ZnCU, which in turn was significantly superior to soil application of ZnS (Table 2). Foliar application of ZnS recorded the lowest values for all yields. Harvest index was also the highest for soil + foliar application of ZnS followed by ZnCU, which in turn was followed by soil or foliar application of ZnS.

Table 2. Effect of various zinc treatments on yields of aromatic rice [Aligarh, Uttar Pradesh site]

Treatments	Rice grain yield (t ha^{-1})	Rice kernel yield (t ha^{-1})	Rice husk yield (t ha^{-1})	Rice straw yield (t ha^{-1})	Biological yield (t ha^{-1})	Harvest index (%)
Check	3.58	2.40	1.18	6.80	10.38	34.5
5 kg Zn ha^{-1} (soil)	3.93	2.63	1.30	7.45	11.38	34.4
1 kg Zn ha^{-1} (foliar)	3.80	2.55	1.25	7.25	11.05	34.4
5 kg Zn ha^{-1} (soil) + 1 kg Zn ha^{-1} (foliar)	4.52	3.03	1.50	8.12	12.63	35.7
2.83 kg Zn ha^{-1} through Zn-coated urea (soil)	4.10	2.75	1.35	7.63	11.73	35.1
SEm±	0.03	0.02	0.01	0.03	0.04	0.21
LSD ($p = 0.05$)	0.08	0.06	0.03	0.09	0.11	0.60

3.3 Zn Concentration and Uptake

Zn application increased Zn concentration in rice kernls, husk and straw and the highest values were obtained for soil + foliar application of ZnS (Table 3). In the case of kernels and straw soil + foliar application of ZnS recorded significantly higher Zn concentration than ZnCU, which in turn was superior to soil or foliar application of ZnS. In the case of husk ZnCU was superior to to foliar application of ZnS, which in turn was superior to soil application of ZnS. Foliar application of ZnS significantly increased Zn concentration in husk but not in kernels.

Zn application increased Zn uptake by rice. In husk and straw as well as in total Zn uptake by rice various treatments were in the following order: soil + foliar application of ZnS > ZnCU > foliar application of Zn > soil application of Zn; soil + foliar application recording the highest Zn uptake. However in the case of rice kernels, soil or foliar application of Zn recorded the same value for Zn uptake.

Table 3. Effect of various zinc treatments on zinc concentrations in rice kernel, rice husk, rice straw and their uptake in aromatic rice [Aligarh, Uttar Pradesh site]

Treatments	Zn concentration in rice kernel (mg kg^{-1} rice kernel)	Zn concentration in rice husk (mg kg^{-1} rice husk)	Zn concentration in rice straw (mg kg^{-1} rice straw)	Zn uptake in rice kernel (g ha^{-1})	Zn uptake in rice husk (g ha^{-1})	Zn uptake in rice straw (g ha^{-1})	Total Zn uptake in rice crop (g ha^{-1})
Check	20.0	125.0	91.0	48.0	147.5	618.8	814.3
5 kg Zn ha^{-1} (soil)	21.3	130.0	100.0	56.0	169.0	745.0	970.0
1 kg Zn ha^{-1} (foliar)	22.0	147.0	102.0	56.1	183.8	739.5	979.4
5 kg Zn ha^{-1} (soil) + 1 kg Zn ha^{-1} (foliar)	25.0	175.0	107.0	75.7	262.5	868.8	1207.0
2.83 kg Zn ha^{-1} through Zn-coated urea (soil)	23.8	170.0	105.0	65.5	229.5	801.2	1096.2
SEm±	0.30	1.46	0.98	1.07	2.41	7.95	9.27
LSD ($p = 0.05$)	0.86	4.13	2.78	3.02	6.82	22.55	26.26

3.4 Iron Concentration and Uptake

Soil application of ZnS or ZnCU significantly increased Fe concentration in rice kernels, husk and straw. In the case of kernels and husk, soil application of ZnS, ZnCU or soil + foliar application of ZnS were at par (Table 4). However, soil + foliar application of ZnS or ZnCU recorded significantly higher concentration in rice straw than soil application of ZnS. Foliar application of ZnS had no effect on Fe concentration by rice.

Zn application increased Fe uptake by rice. As regards grain and straw and total Fe uptake by rice, different treatments were in the following order: soil + foliar application of ZnS > ZnCU > soil aplication of ZnS > foliar application of ZnS. In the case of husk soil application of ZnS and ZnCU were at par. In general, as a contrast to Zn uptake, soil application of Zn recorded the higher Fe uptake than foliar application of ZnS.

Table 4. Effect of various zinc treatments on Fe concentrations in rice kernel, rice husk, rice straw and their uptake in aromatic rice [Aligarh, Uttar Pradesh site]

Treatments	Fe concentration in rice kernel (mg kg^{-1} rice kernel)	Fe concentration in rice husk (mg kg^{-1} rice husk)	Fe concentration in rice straw (mg kg^{-1} rice straw)	Fe uptake in rice kernel (g ha^{-1})	Fe uptake in rice husk (g ha^{-1})	Fe uptake in rice straw (g ha^{-1})	Total Fe uptake in rice crop (g ha^{-1})
Check	8.2	12.3	42.0	19.7	14.5	285.6	319.8
5 kg Zn ha^{-1} (soil)	9.0	13.4	48.0	23.7	17.4	357.6	398.7
1 kg Zn ha^{-1} (foliar)	8.4	12.8	45.0	21.4	16.0	326.3	363.7
5 kg Zn ha^{-1} (soil) + 1 kg Zn ha^{-1} (foliar)	9.3	14.1	55.0	28.2	21.2	446.6	496.0
2.83 kg Zn ha^{-1} through Zn-coated urea (soil)	9.1	13.8	56.0	25.0	18.6	427.3	470.9
SEm±	0.15	0.39	0.58	0.79	0.53	4.55	4.59
LSD ($p = 0.05$)	0.44	1.10	1.66	2.24	1.51	12.88	13.01

3.5 Nitrogen Concentration and Uptake

Zn application significantly increased N concentration and crude protein content in rice kernel and N concentration in straw. In general, soil + foliar application of ZnS or ZnCU recorded significantly higher N concentration in rice kernel and straw than foliar application of ZnS (Table 5). The N concentration in rice kernel was about 2.5 times of that in straw.

As regards N uptake by rice kernel, straw as well as total N uptake by rice crop, different treatments were in the following order: soil + foliar application of ZnS > soil application of ZnCU > soil application of ZnS > foliar application of ZnS; soil + foliar application recording the highest N uptake.

Table 5. Effect of various zinc treatments on N concentrations in rice kernel, rice husk, rice straw and their uptake in aromatic rice and crude protein content in brown rice [Aligarh, Uttar Pradesh site]

Treatments	N concentration in rice kernel (%)	N concentration in rice straw (%)	N uptake in rice kernel (kg ha^{-1})	N uptake in rice straw (kg ha^{-1})	Total N uptake in rice crop (kg ha^{-1})	Crude protein content in rice kernel (%)
Check	1.50	0.62	36.0	42.2	78.2	8.92
5 kg Zn ha^{-1} (soil)	1.60	0.66	42.1	49.2	91.3	9.52
1 kg Zn ha^{-1} (foliar)	1.53	0.64	39.0	46.4	85.4	9.10
5 kg Zn ha^{-1} (soil) + 1 kg Zn ha^{-1} (foliar)	1.62	0.67	49.1	54.4	103.5	9.64
2.83 kg Zn ha^{-1} through Zn-coated urea (soil)	1.61	0.67	44.3	51.1	95.4	9.58
SEm±	0.007	0.005	0.54	0.41	0.61	0.05
LSD ($p = 0.05$)	0.020	0.015	1.54	1.16	1.72	0.14

3.6 Phosphorus Concentration and Uptake

A significant increase in P concentration in rice due to Zn application was recorded only in kernels, where soil application of ZnS recorded significantly higher P concentration than foliar application of ZnS or soil application of ZnS. Least P concentration was recorded with soil + foliar application of ZnS (Table 6).

Uptake of P by kernel was much more than that by husk or straw. In kernels and total P uptake by rice was the highest with soil application of ZnS followed by soil + foliar application of ZnS or ZnCU, which were at par. Least P uptake by kernels and total P uptake by the rice crop was recorded with foliar application of ZnS.

Table 6. Effect of various zinc treatments on P concentrations in rice kernel, rice husk, rice straw and their uptake in aromatic rice [Aligarh, Uttar Pradesh site]

Treatments	P concentration in rice kernel (%)	P concentration in rice husk (%)	P concentration in rice straw (%)	P uptake in rice kernel (kg ha^{-1})	P uptake in rice husk (kg ha^{-1})	P uptake in rice straw (kg ha^{-1})	Total P uptake in rice crop (kg ha^{-1})
Check	0.64	0.03	0.07	15.36	0.35	4.76	20.47
5 kg Zn ha^{-1} (soil)	0.73	0.04	0.08	19.20	0.52	5.96	27.68
1 kg Zn ha^{-1} (foliar)	0.64	0.03	0.08	16.32	0.38	5.80	22.56
5 kg Zn ha^{-1} (soil) + 1 kg Zn ha^{-1} (foliar)	0.61	0.03	0.08	18.48	0.45	6.50	25.43
2.83 kg Zn ha^{-1} through Zn-coated urea (soil)	0.65	0.04	0.08	17.87	0.54	6.10	24.51
SEm±	0.007	0.002	0.005	0.66	0.021	0.26	0.44
LSD ($p = 0.05$)	0.020	0.004	NS	0.75	0.060	0.75	1.24

3.7 Potassium Concentration and Uptake

Zn application significantly increased K concentration in kernels only and soil application of ZnS and soil + foliar application of ZnS were at par significantly superior to soil application of ZnCU or foliar application of Zn (Table 7).

As regards K uptake in kernels, soil + foliar application of ZnS recorded the highest K uptake followed by the soil application of ZnCU and soil or foliar application of ZnS. In husk the treatments were in the following order: soil + foliar application of ZnS > soil application of ZnCU or ZnS > foliar application of ZnS. Straw contributed most to K uptake by rice and different treatments were in the following order: soil + foliar application of ZnS > soil application of ZnCU > soil application of ZnS > foliar application of ZnS.

Table 7. Effect of various zinc treatments on K concentrations in rice kernel, rice husk, rice straw and their uptake in aromatic rice [Aligarh, Uttar Pradesh site]

Treatments	K concentration in rice kernel (%)	K concentration in rice husk (%)	K concentration in rice straw (%)	K uptake in rice kernel (kg ha^{-1})	K uptake in rice husk (kg ha^{-1})	K uptake in rice straw (kg ha^{-1})	Total K uptake in rice crop (kg ha^{-1})
Check	0.18	0.47	3.40	4.32	5.55	231.2	241.1
5 kg Zn ha^{-1} (soil)	0.20	0.49	3.67	4.26	6.37	273.4	285.0
1 kg Zn ha^{-1} (foliar)	0.19	0.47	3.53	4.84	5.88	255.9	266.6
5 kg Zn ha^{-1} (soil) + 1 kg Zn ha^{-1} (foliar)	0.21	0.50	3.70	6.36	7.50	300.4	314.3
2.83 kg Zn ha^{-1} through Zn-coated urea (soil)	0.19	0.49	3.68	5.22	6.62	280.8	292.6
SEm±	0.004	0.007	0.013	0.12	0.11	1.52	2.11
LSD ($p = 0.05$)	0.012	0.021	0.037	0.34	0.31	4.31	5.99

3.8 Zn, N, P and K Concentration in Fag Leaf

Zn application significantly increased Zn, N and K concentration but not the P concentration in rice flag leaf (Table 8). Soil + foliar application of ZnS and soil applicastion of ZnCU were at par and significantly superior to soil or foliar application of ZnS. As regards N concentration in flag leaf, different treatments were in the following order: soil application of ZnCU > soil + foliar applcation of ZnS > soil application of ZnS > foliar application of ZnS.

Table 8. Effect of various zinc treatments on Zn, Fe, N, P and K concentrations in aromatic rice flag leaf [Aligarh, Uttar Pradesh site]

Treatments	Zn concentration in rice flag leaf (mg kg^{-1} rice leaf DM)	N concentration in rice leaf (%)	P concentration in rice flag leaf (%)	K content in rice flag leaf (%)
Check	95.0	1.29	0.35	4.04
5 kg Zn ha^{-1} (soil)	105.0	1.39	0.36	4.31
1 kg Zn ha^{-1} (foliar)	107.0	1.32	0.40	4.15
5 kg Zn ha^{-1} (soil) + 1 kg Zn ha^{-1} (foliar)	112.0	1.41	0.41	4.47
2.83 kg Zn ha^{-1} through Zn-coated urea (soil)	111.0	1.46	0.42	4.58
SEm±	1.01	0.005	0.025	0.06
LSD ($p = 0.05$)	2.86	0.014	0.072	0.16

B. Meerut Site

3.9 Yield Attributes

Zn application significantly increased tillers m^{-2} and grains panicle^{-1} but not the panicle length and 1,000-grain weight (Table 9). Soil application of ZnS or ZnCU or soil + foliar application of ZnS resulted in significantly more effective tillers than foliar application of ZnS. The highest number of grains panicle^{-1} were produced by soil application of ZnS or ZnCU.

Table 9. Effect of various zinc treatments on yield attributes of aromatic rice [Meerut, Uttar Pradesh site]

Treatments	Plant height (cm)	Effective tillers hill^{-1}	Panicle length (cm)	Grains panicle^{-1} (Nos.)	1,000-grain weight (g)
Check	105	7.1	27.0	124	24.0
5 kg Zn ha^{-1} (soil)	108	7.8	27.7	131	24.5
1 kg Zn ha^{-1} (foliar)	107	7.3	27.4	128	24.3
5 kg Zn ha^{-1} (soil) + 1 kg Zn ha^{-1} (foliar)	109	8.0	28.0	125	24.7
2.83 kg Zn ha^{-1} through Zn-coated urea (soil)	109	7.8	27.8	131	24.5
SEm±	2.25	0.10	0.63	1.20	0.40
LSD ($p = 0.05$)	NS	0.27	NS	3.40	NS

3.10 Yields

Zn application significantly increased grain and straw yield as well as harvest index in rice (Table 10). Soil + foliar application of ZnS recorded significantly more grain, kernel, husk and total biological yield than other treatments. As regards straw yield, soil + foliar appplication of ZnS was at par with soil application of ZnCU and both these treatments were significantly superior to soil or foliar application of ZnS. Harvest index was not significantly affected by Zn application.

Table 10. Effect of various zinc treatments on yields of aromatic rice [Meerut, Uttar Pradesh site]

Treatments	Rice grain yield (t ha^{-1})	Rice kernel yield (t ha^{-1})	Rice husk yield (t ha^{-1})	Rice straw yield (t ha^{-1})	Biological yield (t ha^{-1})	Harvest index (%)
Check	5.40	3.62	1.78	8.50	13.90	38.8
5 kg Zn ha^{-1} (soil)	5.70	3.82	1.88	8.90	14.60	39.0
1 kg Zn ha^{-1} (foliar)	5.60	3.75	1.85	8.70	14.30	39.2
5 kg Zn ha^{-1} (soil) + 1 kg Zn ha^{-1} (foliar)	6.00	4.02	1.98	9.20	15.20	39.5
2.83 kg Zn ha^{-1} through Zn-coated urea (soil)	5.70	3.82	1.88	9.00	14.70	38.8
SEm±	0.09	0.07	0.03	0.09	0.14	0.44
LSD ($p = 0.05$)	0.27	0.20	0.09	0.26	0.40	NS

3.11 Zinc Concentration and Uptake

Zn application increased Zn concentration in rice kernel, husk and straw of rice (Table 11). Zn concentration in rice kernels was the highest with soil application of ZnCU, signicantly more than that obtained with soil + foliar application of ZnS, which in turn was superior to foliar application of ZnS. However, foliar application of ZnS recorded significantly more Zn concentration in rice kernels than soil aplication of ZnS. Zn concentration in rice husk was the highest with soil + foliar application of ZnS, significantly superior to foliar application of ZnS, which in turn was superior to ZnCU. Lowest Zn concentration was obtained with soil application of ZnS. As regards Zn concentration in straw, soil + foliar application of ZnS and soil application of ZnCU were at par and significantly superior to foliar application of ZnS, which in turn was superior to soil application of ZnS.

Zn uptake significantly increased Zn uptake by rice kernels, husk and straw in rice and this resulted in a significant increase in total Zn uptake by rice. As regards rice kernels soil + foliar application of ZnS and soil application of ZnCU were at par and significantly superior to soil or foliar application of ZnS. In the case of rice husk, foliar aplication of ZnS was next only to soil + foliar application of ZnS and was followed by soil application of ZnCU. Soil application of ZnS recorded the least Zn uptake by rice.

Table 11. Effect of various zinc treatments on Zn concentrations in rice kernel, rice husk, rice straw and their uptake in aromatic rice [Meerut, Uttar Pradesh site]

Treatments	Zn concentration in rice kernel (mg kg^{-1} rice kernel)	Zn concentration in rice husk (mg kg^{-1} rice husk)	Zn concentration in rice straw (mg kg^{-1} rice straw)	Zn uptake in rice kernel (g ha^{-1})	Zn uptake in rice husk (g ha^{-1})	Zn uptake in rice straw (g ha^{-1})	Total Zn uptake in rice crop (g ha^{-1})
Check	18.0	17.7	110.3	65.2	31.5	937.6	1034.3
5 kg Zn ha^{-1} (soil)	18.7	23.8	117.0	71.4	44.7	1041.3	1157.4
1 kg Zn ha^{-1} (foliar)	20.2	34.7	124.2	75.8	64.2	1080.5	1220.5
5 kg Zn ha^{-1} (soil) + 1 kg Zn ha^{-1} (foliar)	21.2	37.7	126.0	85.2	74.6	1159.2	1319.0
2.83 kg Zn ha^{-1} through Zn-coated urea (soil)	22.2	25.7	124.8	84.8	48.3	1123.2	1256.3
SEm±	0.28	0.42	0.49	2.57	1.29	15.3	15.76
LSD ($p = 0.05$)	0.79	1.20	1.40	7.30	3.65	43.3	44.65

3.12 Iron Concentration and Uptake

Fe concentration in rice kernels, husk and straw was increased by Zn application and for all the components soil + foliar application of ZnS or ZnCU were at par and significantly superior to foliar application of ZnS (Table 12). Foliar application of ZnS had no significant effect on Fe concentration in rice kernels, husk and straw.

All treatments except foliar application of Zn significantly increased Fe uptake by rice kernels, husk and straw and total Fe uptake by rice crop.

Table 12. Effect of various zinc treatments on Fe concentrations in rice kernel, rice husk, rice straw and their uptake in aromatic rice [Meerut, Uttar Pradesh site]

Treatments	Fe concentration in rice kernel (mg kg^{-1} rice kernel)	Fe concentration in rice husk (mg kg^{-1} rice husk)	Fe concentration in rice straw (mg kg^{-1} rice straw)	Fe uptake in rice kernel (g ha^{-1})	Fe uptake in rice husk (g ha^{-1})	Fe uptake in rice straw (g ha^{-1})	Total Fe uptake in rice crop (g ha^{-1})
Check	21.2	56.0	98.2	76.7	99.7	834.7	1011.1
5 kg Zn ha^{-1} (soil)	22.4	58.0	103.6	85.6	109.0	922.0	1116.6
1 kg Zn ha^{-1} (foliar)	21.6	55.0	100.3	81.0	101.8	872.6	1055.4
5 kg Zn ha^{-1} (soil) + 1 kg Zn ha^{-1} (foliar)	22.8	58.0	105.2	91.7	114.8	967.8	1174.3
2.83 kg Zn ha^{-1} through Zn-coated urea (soil)	22.7	58.0	104.3	86.7	109.0	938.7	1134.4
SEm±	0.37	0.50	1.59	2.26	2.23	24.26	29.07
LSD ($p = 0.05$)	1.06	1.42	4.51	6.40	6.31	68.34	82.37

3.13 Nitrogen Concentration and Uptake

Nitrogen concentration and crude protein content in rice kernels was significntly increased due to Zn fertilization and soil + foliar application of ZnS and soil application of ZnCU were at par and significantly superior to soil application of ZnS, which in turn was significantly superior to foliar application of ZnS (Table 13).

Nitrogen uptake by rice kernels and straw and total N uptake by the rice crop was most with soil + foliar application ZnS or ZnCU, which were at par. Foliar application of ZnS recorded the lowest N uptake.

Table 13. Effect of various zinc treatments on N concentrations in rice kernel, rice straw and their uptake in aromatic rice and also crude protein content [Meerut, Uttar Pradesh site]

Treatments	N concentration in rice kernel (%)	N concentration in rice straw (%)	N uptake in rice kernel (kg ha^{-1})	N uptake in rice straw (kg ha^{-1})	Total N uptake in rice crop (kg ha^{-1})	Crude protein content in rice kernel (%)
Check	1.29	0.79	46.7	67.1	113.8	7.7
5 kg Zn ha^{-1} (soil)	1.43	0.89	54.6	79.2	133.8	8.5
1 kg Zn ha^{-1} (foliar)	1.34	0.85	50.2	74.0	124.2	8.0
5 kg Zn ha^{-1} (soil) + 1 kg Zn ha^{-1} (foliar)	1.50	0.92	60.3	84.6	144.9	8.9
2.83 kg Zn ha^{-1} through Zn-coated urea (soil)	1.47	0.85	56.2	76.5	132.7	8.7
SEm±	0.01	0.005	1.55	1.23	2.39	0.08
LSD ($p = 0.05$)	0.03	0.014	4.41	3.48	6.77	0.23

3.14 Phosphorus Concentration and Uptake

In rice kernels soil application of ZnCU recorded the highest P concentration and was followed by soil + foliar application of ZnS, which in turn was significantly superior to soil application of ZnS (Table 14). In the case of husk, however, soil + foliar application of ZnS was superior to soil application of ZnS or ZnCU, which were at par. In straw, soil application of ZnCU recorded the highest P concentration followed by soil application of ZnS, which in turn was significantly superior to soil + foliar application of ZnS.

Phosphorus uptake by rice kernels was increased by soil application of ZnS or ZnCU or soil + foliar application of ZnS, which were at par. Foliar application of ZnS did not increase P uptake by rice crop. Zn application did not increase P uptake by husk. However in rice straw, different treatrments were in the following order: soil application of ZnCU > soil application of ZnS > soil + foliar application of Zn > foliar application of ZnS. Soil application of ZnCU recorded the highest P uptake, significantly higher than soil + foliar application of ZnS or soil application of ZnS Foliar application of ZnS did not increase P uptake by rice.

Table 14. Effect of various zinc treatments on P concentrations in rice kernel, rice husk, rice straw and their uptake in aromatic rice [Meerut, Uttar Pradesh site]

Treatments	P concentration in rice kernel (%)	P concentration in rice husk (%)	P concentration in rice straw (%)	P uptake in rice kernel (kg ha^{-1})	P uptake in rice husk (kg ha^{-1})	P uptake in rice straw (kg ha^{-1})	Total P uptake in rice crop (kg ha^{-1})
Check	0.58	0.063	0.14	21.00	1.12	11.90	34.02
5 kg Zn ha^{-1} (soil)	0.66	0.065	0.19	25.21	1.22	16.91	43.34
1 kg Zn ha^{-1} (foliar)	0.55	0.063	0.15	20.62	1.16	13.05	34.83
5 kg Zn ha^{-1} (soil) + 1 kg Zn ha^{-1} (foliar)	0.63	0.067	0.16	25.33	1.33	14.72	41.38
2.83 kg Zn ha^{-1} through Zn-coated urea (soil)	0.66	0.065	0.21	25.21	1.22	18.9	45.33
SEm±	0.006	0.0004	0.004	0.49	0.086	0.48	1.035
LSD ($p = 0.05$)	0.016	0.0011	0.011	1.38	0.283	1.37	2.931

3.15 Potassium Concentration and Uptake

A significant increase in K concentration in rice kernels was obtained only with soil + foliar application of ZnS or soil application of ZnCU (Table 15). There was no significant increase in K uptake due to Zn application. In the case of rice straw, soil application of ZnS or ZnCU or soil + foliar application of ZnS were at par and significantly superior to foliar application of ZnS.

Foliar application of ZnS did not significantly increase K uptake in rice kernels or husk, but did so in straw and therefore in total K uptake by rice crop. Soil application of ZnS or ZnCU and soil + foliar application of ZnS recorded significantly more K uptake than foliar application of ZnS.

Table 15. Effect of various zinc treatments on K concentrations of rice kernel, rice husk, rice straw and their uptake in aromatic rice [Meerut, Uttar Pradesh site]

Treatments	K concentration in rice kernel (%)	K concentration in rice husk (%)	K concentration in rice straw (%)	K uptake in rice kernel (kg ha^{-1})	K uptake in rice husk (kg ha^{-1})	K uptake in rice straw (kg ha^{-1})	Total K uptake in rice crop (kg ha^{-1})
Check	0.31	0.34	1.78	11.22	6.05	151.30	168.57
5 kg Zn ha^{-1} (soil)	0.33	0.37	1.87	12.61	6.96	166.43	186.00
1 kg Zn ha^{-1} (foliar)	0.32	0.35	1.84	12.00	6.48	160.08	178.56
5 kg Zn ha^{-1} (soil) + 1 kg Zn ha^{-1} (foliar)	0.34	0.38	1.89	13.67	7.52	173.88	195.07
2.83 kg Zn ha^{-1} through Zn-coated urea (soil)	0.34	0.37	1.87	12.99	6.96	168.30	188.25
SEm±	0.007	0.009	0.013	0.39	0.21	2.59	2.92
LSD ($p = 0.05$)	0.021	NS	0.037	1.10	0.60	7.34	8.26

4. Discussion

Foliar application of Zn resulted in higher Zn concentration in rice kernels than soil application of ZnS in the on-farm trial in Meerut but not at the Farmers Science Centre in Aligarh. However in both the on-farm trials Zn concentration in husk was sugnificantly more with foliar than with soil application of Zn. Impa and Johnson-Beebout (2012) reported that biofortification recovery of Zn with foliar application was 8 times of that obtained with soil application. However data reported by Shivay and Prasad (2010) suggested that Zn concentration in rice grain was 47.5 mg kg^{-1}, while that in kernel was only 40.3 mg kg^{-1}. Thus more of foliar applied Zn tended to be retained by the husk. From a study involving multi-country and multi-location trials, Phattrakul et al. (2012) also reported that a major share of foliar applied Zn was retained in husk.

It needs to be considered that in Asian countries food security is a serious problem alongwith low Zn concentration in rice (Prasad et al., 2014). In the present study, highest values of yield attributes, rice grain yield and concentrations and uptake of Zn, Fe, N, P and K were obtained by soil + foliar application of Zn. This is because soil applied Zn on Zn deficient soils results in overall better growth, higher yield attributes and grain and straw yield in rice (Pooniya & Shivay, 2013), which results in increased uptake of all nutrients. This does not happen with foliar application of Zn, which is made at a much latter stage of crop growth. Soil application of Zn but not the foliar application of Zn also resulted in increased crude protein content in rice kernels, which is important from the viewpoint of wide spread protein malnutrition in India and other Asian countries (Prasad, 2003).

For most of the characters studied ZnCU performed better than ZnS and was next only to soil + foliar application of Zn. These results are in conformity of our results reported earlier (Shivay et al., 2007, 2008a; Prasad et al., 2013). The major advantage with ZnCU is saving in the amount of Zn to be applied; only 2.83 kg Zn ha^{-1} was applied with ZnCU as against 6 kg Zn ha^{-1} in the case of soil + foliar application of ZnS. The ZnCU is therefore a promising fertilizer.

Zn application also increased Fe, N, P and K concentration and uptake in rice. Thus, a soil-foliar Zn biofortification programme also results in enrichment of rice kernels in Fe, N, P and K. A close relationship between Zn, Fe, N, P and K in wheat germplasms has been reported by Gomez-Beccera et al. (2010) and Zhao et al. (2009).

In conclusion, the present study brings out that for increased production of rice alongwith its biofortification with micronutrients Zn and Fe, adequate soil application of Zn should be followed by some foliar application of Zn.

Acknowledgements

The authors are grateful to the Director and Head, Division of Agronomy, Indian Agricultural Research Institute, New Delhi 110 012, India for providing necessary facilities to carry out this research work. Rajendra Prasad is grateful to the Indian National Science Academy for awarding him an Honorary Scientist position, which made this research possible.

References

Bell, D. W., & Dell, B. (2008). *Micronutrients for sustainable food, feed, fibre and bioenergy products* (p. 175). International Fertilizer Industry Association, Paris.

Black, R. E., Lindsay, H. A., Bhutta, Z. A., Caulfield, L. E., DeOnnis, M., Ezzat, M., ... Rivera, J. (2008). Maternal and child under nutrition: Global and regional exposures and health consequences. *Lancet, 371*, 243-260. http://dx.doi.org/10.1016/S0140-6736(07)61690-0

Gomez, K. A., & Gomez, A. A. (1984). *Statistical Procedures for Agricultural Research*. New York: John Wiley & Sons.

Gomez-Becerra, H. F., Yazici, A., Ozturk, L., Budak, H., Peleg, Z., Morgounov, A., ... Cakmak, I. (2010). Genetic variation and environmental stability of grain mineral nutrient concentrations in *Triticum dicoccoides* under five environments. *Euphytica, 171*, 39-52. http://dx.doi.org/10.1007/s10681-009-9987-3

Graham, R. D. (2008). Micronutrient deficiencies in crops and their global significance. In B. J. Alloway (Ed.), *Micronutrient deficiencies in global crop production* (pp. 41-61). Springer, Dordrecht. http://dx.doi.org/10.1007/978-1-4020-6860-7_2

Hanway, J. J., & Heidel, H. (1952). Soil analysis methods as used in Iowa State College Soil Testing Laboratory. *Bulletin 57*. Ames, IA: Iowa State College of Agriculture.

Hotz, C., & Brown, K. H. (2004). Assessment of the risk of zinc deficiency in populations and options for its control. International Zinc Nutrition Consultative Group Technical Document No 1. *Food Nutrition Bulletin, 25*, S91-S204.

Impa, S. M., & Johnson-Beebout, S. E. (2012). Mitigating zinc deficiency and achieving high grain Zn in rice through integration of soil chemistry and plant physiology research. *Plant and Soil, 361*, 3-41. http://dx.doi.org/10.1007/s11104-012-1315-3

Lindsay, W. L., & Norvell, W. A. (1978). Development of DTPA soil test for zinc, iron, manganese and copper. *Soil Science Society of America Journal 42*, 421-428. http://dx.doi.org/10.2136/sssaj1978.03615995004200030009x

Olsen, R., Cole, C. V., Watanabe, F. S., & Dean, L. A. (1954). Estimation of available phosphorus in soils by extraction with sodium bicarbonate. *Circular 939 United States Department of Agriculture.* Washington, DC: US Government Printing Office.

Phattarakul, N., Rerkasem, B., Li, L. J., Wu, L. H., Zou, C. Q., Ram, H., ... Cakmak, I. (2012). Biofortification of rice grain with zinc through zinc fertilization in different countries. *Plant and Soil, 361*, 131-141. http://dx.doi.org/10.1007/s11104-012-1211-x

Pooniya, V., & Shivay, Y. S. (2013). Enrichment of *Basmati* rice grain and straw with zinc and nitrogen through ferti-fortification and summer green manuring under Indo-Gangetic plains of India. *Journal of Plant Nutrition, 36*(1), 91-117. http://dx.doi.org/10.1080/01904167.2012.733052

Prasad, R. (2003). Protein-energy malnutrition in India. *Fertiliser News, 48*(4), 13-26.

Prasad, R. (2006). Zinc in soils and in plant, human and animal nutrition. *Indian Journal of Fertilizers, 2*, 103-119.

Prasad, R., & Power, J. F. (1997). *Soil Fertility Management for Sustainable Agriculture* (p. 356). Boca Raton, New York: CRC, Lewis Publishers.

Prasad, R., Shivay, Y. S., & Kumar, D. (2013). Zinc fertilization of cereals for increased production and alleviation of zinc malnutrition in India. *Agricultural Research, 2*(2), 111-118. http://dx.doi.org/10.1007/s40003-013-0064-8

Prasad, R., Shivay, Y. S., & Kumar, D. (2014). Agronomic biofortification of cereal grains with iron and zinc. *Advances in Agronomy, 125*, 55-91. http://dx.doi.org/10.1016/B978-0-12-800137-0.00002-9

Prasad, R., Shivay, Y. S., Kumar, D., & Sharma, S. N. (2006). *Learning by Doing Exercises in Soil Fertility (A Practical Manual for Soil Fertility)* (pp. 68). Division of Agronomy, Indian Agricultural Research Institute, New Delhi.

Rani, N. S., Prasad, G. S. V., & Viraktamath, B. C. (2009). National system for evaluation of *Basmati* rices for yield and quality traits. *Indian Farming, 59*, 7-11.

Shivay, Y. S., & Prasad, R. (2012). Zinc-coated urea improves productivity and quality of *Basmati* rice (*Oryza sativa* L.) under zinc stress condition. *Journal of Plant Nutrition, 35*, 928-951. http://dx.doi.org/10.1080/01904167.2012.663444

Shivay, Y. S., & Prasad, R. (2014). Effect of source and methods of zinc application on corn productivity, nitrogen and zinc concentrations and uptake by high quality protein corn (*Zea mays*). *Egyptian Journal of Biology, 16*(1), 72-78. http://dx.doi.org/10.4314/ejb.v16i1.10

Shivay, Y. S., Kumar, D., & Prasad, R. (2008a). Effect of zinc-enriched urea on productivity, zinc uptake and efficiency of an aromatic rice-wheat cropping system. *Nutrient Cycling in Agroecosystems, 81*(3), 229-243. http://dx.doi.org/10.1007/s10705-007-9159-6

Shivay, Y. S., Kumar, D., & Prasad, R. (2008c). Relative efficiency of zinc sulphate and zinc oxide coated urea in rice-wheat cropping system. *Communication in Soil Science and Plant Analysis, 39*, 1154-1167. http://dx.doi.org/10.1080/00103620801925869

Shivay, Y. S., Kumar, D., Ahlawat, I. P. S., & Prasad, R. (2007). Relative efficiency of zinc oxide and zinc sulfate coated urea for rice. *Indian Journal of Fertilizers, 3*(2), 51-56.

Shivay, Y. S., Prasad, R., & Rahal, A. (2008b). Relative efficieny of zinc oxide and zinc sulphate enriched urea for spring wheat. *Nutrient Cycling in Agroecosystems, 82*, 259-264. http://dx.doi.org/10.1007/s10705-008-9186-y

Shivay, Y. S., Prasad, R., & Rahal, A. (2010). Genotypic variation for productivity, zinc utilization efficiencies, and kernel quality in aromatic rices under low available zinc conditions. *Journal of Plant Nutrition, 33,* 1835-1848. http://dx.doi.org/10.1080/01904167.2010.503832

Shivay, Y. S., Prasad, R., & Pal, M. (2013). Zinc fortification of oats grain and straw through zinc fertilization. *Agricultural Research, 2,* 375-381. http://dx.doi.org/10.1007/s40003-013-0078-2

Siddiq, E. A., Vemireddy, L. R., & Nagraju, J. (2012). Basmati rices: Genetics, breeding and trade. *Agricultural Research, 1,* 25-36. http://dx.doi.org/10.1007/s40003-011-0011-5

Singh, M. V. (2011). Assessing extent of zinc deficiency for soil factors affecting and nutritional scarcity in humans and animals. *Indian Journal of Fertiliser, 7*(10), 36-43.

Singh, V. P., & Singh, A. K. (2009). History of *Basmati* rice research and development in India. *Indian Farming, 59,* 4-6.

Stein, A. J., Nestel, P., Meenakshi, J. V., Qaim, M., Sachdev, H. P. S., & Bhutta, Z. (2007). Plant breeding to control zinc deficiency in India: How cost effective is biofortification? *Public Health Nutrition, 10,* 492-501. http://dx.doi.org/10.1017/S1368980007223857

Subbiah, B. V., & Asija, G. L. (1956). A rapid procedure for the determination of available nitrogen in soils. *Current Science, 25,* 259-260.

Takkar, P. N., Singh, M. V., & Ganeshmurthy, A. N. (1997). A critical review of plant nutrient supply needs, efficiency and policy issues for Indian agriculture for the year 2000: Micronutrients and Trace elements. In J. S. Kanwar & J. C. Katyal (Eds.), *Plant Nutrient, Supply Efficiency and Policy Issues: 2000-2025* (pp. 238-264). New Delhi: National Academy of Agricultural Sciences.

Walkley, A. J., & Black, I. A. (1934). An examination of the Degtjareff method for determination soil organic matter and a proposed modification of the chromic acid titration method. *Soil Science, 37,* 29-38. http://dx.doi.org/10.1097/00010694-193401000-00003

Welch, R. M. (2005). Harvesting health: agricultural linkages for improving human nutrition. In P. Anderson, J. K. Tuladhada, K. B. Karki & S. L. Maskey (Eds.), *Micronutrients in south and Southeast Asia* (pp. 9-10). International Centre for Integrated Mountain Development, Kathmandu.

Zhao, F. J., Su, Y. H., Dunham, S. J., Rakszegi, M., Bedo, Z., McGrath, S. P., & Shewry, P. R. (2009). Variation in mineral micronutrient concentrations in grain of wheat lines of diverse origin. *Journal of Cereal Science, 49,* 290-295. http://dx.doi.org/10.1016/j.jcs

17

Suppression of Hepatitis B Virus Production and Inflammatory Response *in vitro* and *in vivo* by *Mormodica charantia* Compound EMCDO

Chi-I Chang[1], Chiy-Rong Chen[2], Yo-Chia Chen[1], Kuei-Wen Cheng[1], Bo-Wei Lin[1], Yun-Wen Liao[1] & Wen-Ling Shih[1]

[1] Department of Biological Science and Technology, National Pingtung University of Science and Technology, Pingtung, Taiwan

[2] Department of Life Science, National Taitung University, Taitung, Taiwan

Correspondence: Department of Biological Science and Technology, National Pingtung University of Science and Technology, 1, Shuefu Rd., Neipu, Pingtung, 91201, Taiwan.

Abstract

Eight compounds were purified from *Mormodica charanti*, and their chemical structures were determined in this study. Their anti-HBV and anti-inflammation activities were investigated. Compound EMCDO exhibited the most efficient effect in terms of reducing HBV surface antigen, e antigen and viral DNA levels in HBV particles or surface antigen-producing cells 2.2.15 and PLC/PRF/5, respectively. Tumor suppressor p53 played a significant role in EMCDO-mediated anti-HBV effects. Pretreatment with EMCDO prevented 2.2.15 cells-induced tumor formation in a nude mice subcutaneous model. The anti-HBV and anti-tumor activities of EMCDO were better than those of oltipraz, an inhibitor of HBV transcription. EMCDO reduced the proinflammatory cytokines and mediators in LPS-treated RAW264.7 cells in a dose-dependent manner. The LPS-upregulated phosphorylation level of I$\kappa\alpha$ was reduced in the presence of EMCDO. The transcription activity of NF-κB was increased in cells treated with EMCDO. Utilization of a mouse ear edema model further confirmed the activity of EMCDO against TPA-elicited inflammation.

Keywords: *Mormodica charantia,* HBV, nude mice, RAW264.7, inflammation

1. Introduction

Hepatitis B virus (HBV) is the major cause of acute and chronic hepatitis and a serious liver disease (Maddrey, 2000). Although HBV has been vaccine-preventable since 1981, the infection rates have not declined over the past several years, leading to the conclusion that we have allowed gaps in screening, prevention, and treatment to go unchecked (Chemin, 2010). Hepatocellular carcinoma therefore remains a significant problem, owing to the high mortality rate and difficulty in treatment using existing medical drugs and approaches. In view of the limited treatment options and grave prognosis, preventive control is considered the best strategy to lower the current morbidity and mortality associated with liver-related disease (Elgouhari, Abu-Rajab Tamimi, & Carey, 2008). A close connection between inflammation and cancer has been long suspected. It is well-known that inflammation increases the risk and accelerates the development of various types of cancer. A clear example of inflammation-related cancer is hepatocellular carcinoma (HCC) (Berasain et al., 2009a). Thus, HBV chronic infection and hepatitis are two risk factors for hepato-carcinogenesis, suggesting that the strategy for the prevention of liver carcinogenesis should consist of anti-HBV and anti-inflammation approaches, which could reduce the further development of liver cirrhosis and carcinoma (Berasain et al., 2009b).

There is a growing body of *in vitro* evidence demonstrating the inhibitory effects of certain natural products on liver cancer cells *via* various regulatory mechanisms (Aslam et al., 2012). Triterpenes have previously been purified from the bitter melon, and detailed analysis has shown that the compounds consist in the main of cucurbitane-type triterpenoids (Liao et al., 2012). The anti-carcinogenic activity of cucurbitane-type triterpenoids has been clearly demonstrated in two-stage carcinogenesis testing in skin tumors induced by peroxynitrite (ONOO⁻) as an initiator and TPA as a promoter in a specific pathogen-free mouse model (Takasaki et al., 2003).

Additionally, the inhibitory effects of mogroside V and 11-oxo-morgroside V on Epstein-Barr virus early antigen (EBV-EA) activation by TPA in Raji cells have been confirmed (Akihisa et al., 2007), and an anti-HIV inhibitory activity of some cucurbitane-type triterpenoids isolated from the tubers of *Hemsleya endecaphylla* has also been observed (Chen et al., 2008).

Inflammation is a critical and complex immune event. An aberrant or intensified inflammatory response has been described for numerous diseases, including cancer progression (Marusawa & Jenkins, 2014). Several triterpenoids from various plants, such as lupeol, a triterpenoid isolated from Lonchocarpus araripensis Benth., reduce the production of mucus and overall inflammation in the lung (Vasconcelos et al., 2008); the triterpenoid acetyl-11-keto-β-boswellic acid (AKβBA) isolated from the oleogum resin of *Boswellia carterii* functioned as a NF-κB inhibitor and alleviated skin inflammation in a psoriasis mouse model (Wang et al., 2009).

Although many bioactive beneficial compounds have not yet been well-defined, there exists scientific evidence suggesting potential anti-virus, immune modulation and cancer prevention effects. The aim of this study was to evaluate the potential against HBV and acute inflammation of compound EMCDO isolated from *Mormodica charantia*, as well as elucidate the host factors and mechanisms involved. Taken together, the findings of this project will enable us to improve knowledge and understanding of the anti-HBV, anti-inflammation and chemoprevention effects and mechanisms of some potential natural products, and it is hoped that the outcome will also be of use in assisting the development of biological agents for the prevention of HBV replication and further reducing the incidence of liver cancer.

2. Method

2.1 Cell Culture

2.2.15 cells secrete HBV surface antigen (HBsAg), e antigen (HBeAg), nucleocapsids and virions (Acs et al., 1987). The hepatoma cell line PLC/PRF/5 possesses HBV DNA sequence integration at several sites and produces HBsAg into growth medium. RAW264.7 is a mouse leukemic monocyte macrophage cell line used for an anti-inflammation cell-based system. Three cell lines were cultured in DMEM supplemented with 10% FBS containing 100 U/mL of penicillin and 100 μg/mL of streptomycin at 37 °C in a 5% CO_2 humidified incubator.

2.2 Mormodica Charantia Compounds

Fruit of *M. charantia* were purchased from contracted farmers in Kaohsiung, Taiwan. The plant material was identified by Prof. Sheng-Zehn Yang, Curator of the Herbarium, National Pingtung University of Science and Technology. A voucher specimen was deposited in the laboratory of Dr. Chi-I Chang (Pingtung, Taiwan). Dried fruit (30 kg) of *M. charantia* L. wild variant WB24 was mechanically powdered and extracted five times with methanol (60 L) at room temperature (7 days each). The combined MeOH extract was then evaporated under reduced pressure to afford a black residue, which was suspended in H_2O (4 l) and partitioned sequentially using ethyl acetate (EA) and *n*-butanol (*n*-BuOH) (5 × 4 l). The *n*-BuOH fraction (966 g) was chromatographed on a Diaion HP-20 column (150 × 10 cm) and eluted using mixtures of water and methanol of reducing polarity as eluents to yield twenty-five fractions: fr. 1 [9000 ml, water], fr. 2 [6000 ml, water–methanol (98:2)], fr. 3 [6000 ml, water–methanol (95:5)], fr. 4 [6000 ml, water–methanol (93:7)], fr. 5 [6000 ml, water–methanol (90:10)], fr. 6 [6000 ml, water–methanol (88:12)], fr. 7 [6000 ml, water–methanol (85:15)], fr. 8 [6000 ml, water–methanol (83:17)], fr. 9 [6000 ml, water–methanol (80:20)], fr. 10 [(7000 ml, water–methanol (75:25)], fr. 11 [7000 ml, water–methanol (70:30)], fr. 12 [7000 ml, water–methanol (65:35)], fr. 13 [7000 ml, water–methanol (60:40)], fr. 14 [7000 ml, water–methanol (55:45)], fr. 15 [7000 ml, water–methanol (50:50)], fr. 16 [1000 ml, water–methanol (45:55)], fr. 17 [7000 ml, water–methanol (40:60)], fr. 18 [7000 ml, water–methanol (35:65)], fr. 19 [7000 ml, water–methanol (30:70)], fr. 20 [(7000 ml, water–methanol (25:75)], fr. 21 [7000 ml, water–methanol (20:80)], fr. 22 [7000 ml, water–methanol (15:85)], fr. 23 [7000 ml, water–methanol (10:90)], fr. 24 [7000 ml, water–methanol (5:95)], and fr. 25 [10000 ml, methanol]. Fraction 16 (6.1 g) was further subjected to Sephadex LH-20 column chromatography (5 × 50 cm) with gradient elution (water–methanol,1:1 to 0:1) to afford six fractions (800 ml each), frs 16A–16F. Fr. 16B (0.8 g) was subjected to column chromatography over Si gel with elution by CH_2Cl_2–MeOH (1:0 to 9:1) and semipreparative HPLC with elution by CH_2Cl_2–EA (10:1) to yield compound RA2-117 (95 mg). Fr. 16C (4.2 g) was subjected to column chromatography over Si gel with elution by CH_2Cl_2–MeOH (1:0 to 6:1) and semipreparative HPLC with elution by hexane–acetone (6.5:3.5) to yield compounds RA2-8 (21 mg), RA2-20 (27 mg), RA2-52 (12 mg), and CH93 (16 mg). Fr. 16D (2.2 g) was subjected to column chromatography over Si gel with elution by CH_2Cl_2–MeOH (1:0 to 4:1) and semipreparative HPLC with elution by hexane–EA (7:3) to yield compounds EMCDO (87 mg), RA2-11 (206 mg), and RA2-289 (15 mg).

2.3 Reagents, Kits and Plasmids

Oltipraz, an efficient anti-HBV compound (Chi et al., 1998), was purchased from Sigma. Lipopolysaccharide (LPS) is a well-known inflammation inducer on RWA264.7 cells, and the anti-inflammation compound curcumin could be served as positive control in this cell system. LPS, curcumin, as well as trypan blue were also purchased from Sigma. Dominant negative p53 (p53DN) was purchased from Clontech, which expressed mutated p53 protein by a G to A conversion at nucleotide 1017. HBsAg and HBeAg ELISA kits were purchased from General Biological Corporation (Taipei, Taiwan). An α-fetoprotein (AFP) testing ELISA kit was obtained from Panomics. Three cytokines ELISA kits were purchased from BD Biosciences. All antibodies were purchased from Cell Signaling Tecnology.

2.4 Semi-Quantitative Polymerase Chain Reaction

The primer design was derived from HBV S gene: forward primer 5'-CACATCAGGATTCCTAGGACC-3'(nt 166 to 186); reverse primer 5'-GGTGAGTGATTGGAGGTTG-3'(nt 339 to 321) (Abe et al., 1999). 2.2.15 cells were treated with different concentrations of EMCDO or oltipraz for 3 days, and the HBV particles in conditioned medium were harvested and subjected to HBV DNA isolation using a viral DNA purification kit (Invitrogen). PCR cycles consisted of 3 min initial denaturation at 95 °C, then 30s denaturation at 94 °C, 45s annealing at 55 °C and 90s extension at 72 °C. The linearity of the PCR reactions was checked at different cycle numbers. The plateau phase became apparent after 32 cycles; thus, after a serial check of PCR product, we selected 28 cycles to analyze the HBV DNA level in our system (Netea et al., 1996). The 173bps PCR product was analyzed by electrophoresis on 2% agarose gels stained with ethidium bromide. The EC50s of the compounds were calculated according to the literature (Alexander, Browse, Reading, & Benjamin, 1999).

2.5 Western Blotting and Luciferase Assay

Western blotting analysis was performed as described in our previous studies (Chang et al., 2013; Shih et al., 2013). p53-luc plasmid expressed the firefly luciferase gene is controlled by a synthetic promoter that contains direct repeats of the transcription recognition sequences for the p53 (Stratagene). NF-κB-luciferase was also obtained from Stratagene. Cells were seeded and transfected with luciferase reporter plasmid together with pRKβGAL plasmid using Lipofectamine reagent (Invitrogen). Luciferase activity was determined using a Luciferase Assay System (Promega) and normalized for transfection efficiency by measuring the β-galactosidase activity using a β-Galactosidase Enzyme Assay System (Promega).

2.6 IC$_{10}$ and IC$_{50}$ Determination

Cells were attached onto a 96-well plate then exposed to serial dilutions of compound (1-50 µg/ml) for various durations of incubation. Medium containing the compound was changed every 2 days. The cell viability was measured by MTT assay (Stockert, Blazquez-Castro, Canete, Horobin, & Villanueva, 2012), and the cytotoxic response of RAW264.7 was evaluated by trypan blue exclusion (Shih, Kuo, Chuang, Cheng, & Doong, 2000). The 50% and 10% inhibitory concentrations of the compound were calculated using the formula: %IC = $[1-(A_{570test}/A_{570cont})] \times 100$, where % IC = % inhibition of cell proliferation, $A_{570test}$ = absorbance of test sample, $A_{570cont}$ = absorbance of control sample (Cetin & Bullerman, 2005).

2.7 Toxicity in Mice

Two groups of both sexes mice were formed, each containing 6 mice. The first group was used as the control group, which received corn oil solvent, while the second group of mice was administered 500 mg/kg body weight of EMCDO via the intraperitoneal route. During 14 days observation, the general behavior was recorded every day. At the end of the experiment, all mice were sacrificed, the organs examined, and blood collected and subjected to biochemical analysis. During the period, the behaviors displayed, such as nervous excitation, depression, reflexes, muscular and activities weakness, salivation, food ingestion and diarrhea, were recorded. All collected organs were weighed and subjected to histopathological investigation by two independent pathologists. On the 15th day, serum was collected and the biochemical markers of liver and renal function were studied. Liver and renal functions were evaluated by determination of glutamic oxaloacetic transaminase (GOT), glutamic pyruvic transaminase (GPT), creatinine and blood urea nitrogen (BUN) (Wu et al., 2011).

2.8 Tumor Prevention in a Mouse Model

18 6-week-old nude mice were purchased from BioLASCO (Taiwan). The mice were divided into 3 groups randomly, one negative control group treated with corn oil, one group treated with 50 mg/kg EMCDO, and the final group treated with 50 mg/kg oltipraz. The sample was injected every day via the intraperitoneal route for 7 days, then 1×10^6 2.2.15 cells in100µl medium were inoculated via the subcutaneous route. Mice were injected with sample every 2 days for 8 weeks, and the tumor size was measured. The general behavior was also recorded.

At the 56th day after 2.2.15 cell inoculation, blood was collected and serum was isolated, then subjected to liver and renal function evaluation.

2.9 TPA-Induced Ear Edema Mouse Model

8-week-old Balb/c mice were divided into 5 groups, each containing 5 mice. Edema was induced by topical application of 2 μg/20 μl acetone to the inner and outer ear surfaces of each ear. Thirty minutes later, the inner and outer surfaces of both ears were treated with 20 μl of the tested compounds, including acetone vehicle, anti-inflammation drug indomethacin 1.25 μg/20 μl acetone per ear and EMCDO at three different concentrations. The ear thickness was measured using a pocket thickness gauge before and 30 mins after inflammation induction, followed by compound treatment. The ear thickness was then measured at various time points. The anti-inflammation ability was expressed as a edema inhibition percentage: % inhibition = (Tc-Tt)/Tc, where Tc and Tt respectively represented the average thickness with vehicle and compound treatment.

2.10 Statistical Analysis

Data were analyzed by Student's t-test; the data are presented as the mean ± standard deviation (SD). $P < 0.05$ was taken to indicate significant differences.

3. Results

3.1 Chemical Structure, Physical and Spectral Data (IR, MS, and NMR) of the Eight Tested Compounds

Eight cucurbitane-type triterpenes, 5β,19-epoxy-19(R)-methoxycucurbita-6,23(E),25-triene-3β-ol (RA2-8), 5β,19-epoxy-25-methoxycucurbita-6,23(E)-dien-3β-ol (EMCDO), (23E)-7β,25-dimethoxycucurbita-6,23(E)-dien-3β-ol (RA2-11), 5β,19-epoxy-19(S)-methoxycucurbita-6,23(E)-diene-3β,25-diol (RA2-20), (23E)-3β-hydroxy-7β,25-dimethoxycucbita-5,23-dien-19-al (RA2-52), 5β,19-epoxy-19(S),25-dimethoxycucurbita-6,23(E)-diene-3β-diol (RA2-117), (23E)-7β-methoxycucurbita-6,23(E)-diene-3β,25-diol (RA2-289), and (23E)-3β,7β,25-trihydroxycucbita-5,23-dien-19-al (CH93), were isolated from the n-BuOH soluble fraction of the methanol extract of *M. charantia* fruit (Figure 1). The chemical structures of these compounds were identified by comparing their physical and spectral data (IR, MS, and NMR) with the values described in the literature. The IR, MS, and NMR data of RA2-10, RA2-11, RA2-117, and RA2-289 were reported in our previous study (Chang et al. 2008).Here, we only describe the physical and spectral data of the remaining four compounds, RA2-8, RA2-20, RA2-52 and CH93.5β,19-epoxy-19(R)-methoxycucurbita-6,23(E),25-trien-3β-ol (RA2-8) Amorphous white powder; ^{1}H NMR (400 MHz, CDCl$_3$) δ: 0.83 (3H, s, H-30), 0.85 (3H, s, H-18), 0.87 (1H, d, J = 6.4 Hz, H-21), 0.92 (3H, s, H-28), 1.20 (3H, s, H-29), 1.81 (3H, s, H-26), 3.41 (1H, brd, J = 6.9 Hz, H-3), 3.42 (3H, s, 19-OCH$_3$), 3.94 (1H, brd, J = 9.6 Hz, 3-OH), 4.63 (1H, s, H-19), 4.84 (2H, brs, H-27), 5.60 (1H, brd, J = 15.6 Hz, H-23), 5.60 (1H, dd, J = 3.2, 10.0 Hz, H-7), 5.97 (1H, dd, J = 2.4, 10.0 Hz, H-6), 6.10 (1H, d, J = 15.6 Hz, H-24); ^{13}C NMR (100 MHz, CDCl$_3$) δ: 14.7 (C-18), 17.4 (C-1), 18.7 (C-26), 18.7 (C-21), 19.8 (C-30), 20.5 (C-29), 23.2 (C-11), 24.1 (C-28), 27.2 (C-2), 28.0 (C-16), 30.5 (C-12), 33.5 (C-15), 36.6 (C-20), 37.3 (C-4), 39.8 (C-22), 40.5 (C-10), 41.6 (C-8), 45.1 (C-13), 48.0 (C-9), 48.3 (C-14), 50.3 (C-17), 58.3 (19-OCH$_3$), 76.2 (C-3), 86.7(C-5), 112.1 (C-19), 114.1 (C-27), 129.4 (C-23), 131.0 (C-6), 132.8 (C-7), 134.1 (C-24), 142.2 (C-25); EI-MS (70 eV) m/z (rel. int.): 450 [M-H$_2$O]$^+$ (10), 408 (82), 389 (30), 309 (55), 173 (90), 157 (50), 109 (82), 81 (100). 5β,19-epoxy-19(S)-methoxycucurbita-6,23(E)-diene-3β,25-diol (RA2-20) Amorphous white powder, ^{1}H NMR (400 MHz, CDCl$_3$) δ: 0.82 (3H, s, H-30), 0.85 (3H×2, s, H-18, H-28), 0.86 (3H, d, J = 5.4 Hz, H-21), 1.21 (3H, s, H-29), 1.27 (3H×2, s, H-26, 27), 2.22 (1H, brs, H-8), 2.26 (1H, m, H-10), 3.37 (3H, s, 19-OCH$_3$), 3.40 (1H, brs, H-3), 3.70 (1H, d, J = 10.4 Hz, 3-OH), 4.40 (1H, s, H-19), 5.49 (1H, dd, J = 3.2, 10.0 Hz, H-7), 5.56 (2H, m, H-23, 24), 6.07 (1H, dd, J = 2.0, 10.0 Hz, H-6); ^{13}C NMR (100 MHz, CDCl$_3$) δ: 15.0 (C-18), 16.5 (C-1), 18.6 (C-21), 19.9 (C-30), 20.6 (C-29), 21.4 (C-2), 24.4 (C-28), 27.1 (C-11), 27.8 (C-16), 29.8 (C-26), 29.9 (C-27), 30.3 (C-12), 33.4 (C-15), 36.2 (C-20), 37.0 (C-4), 37.8 (C-10), 39.0 (C-22), 45.1 (C-13), 48.0 (C-14), 48.9 (C-9), 49.7 (C-8), 50.1 (C-17), 57.3 (19-OCH$_3$), 70.6 (C-25), 76.1 (C-3), 85.0 (C-5), 114.7 (C-19), 125.2 (C-23), 130.5 (C-7), 133.0 (C-6), 139.5 (C-24); EI-MS (70 eV) m/z (rel. int.): 426 [M-HCO$_2$CH$_3$]$^+$ (10), 408 (14), 309 (15), 281 (9), 239 (11), 187 (21), 172 (39), 157 (31), 109 (50), 91 (100), 84 (100), 69 (59), 55 (78). 3β-hydroxy-7β,25-dimethoxycucurbita-5,23(E)-dien-19-al (RA2-52) Amorphous white powder; ^{1}H NMR (400 MHz, CDCl$_3$) δ: 0.71(3H, s, H-30), 0.87 (1H, d, J = 6.8 Hz, H-21), 0.88 (3H, s, H-18), 1.02 (3H, s, H-28), 1.20 (3H×2, s, H-26, 27), 1.21 (3H, s, H-29), 3.09 (1H, s, 25-OCH$_3$), 3.16 (1H, brs, H-7), 3.20 (1H, s, 7-OCH$_3$), 3.41 (1H, brd, J = 6.4 Hz, 3-OH), 3.56 (1H, brs, H-3), 5.34 (1H, d, J = 15.6 Hz, H-24), 5.48 (1H, ddd, J = 5.6, 8.4, 15.6 Hz, H-23), 5.90 (1H, d, J = 8.4 Hz, H-6), 9.70 (1H, s, H-19); ^{13}C NMR (100 MHz, CDCl$_3$) δ: 14.8 (C-18), 18.0 (C-30), 18.7 (C-21), 20.7 (C-1), 22.2 (C-11), 25.2 (C-29), 25.7 (C-27), 26.0 (C-26), 26.9 (C-28), 27.3 (C-16), 28.3 (C-2), 28.8 (C-12), 34.7 (C-15), 35.9 (C-10), 36.0 (C-20), 39.2 (C-22), 41.4 (C-4), 44.4 (C-8), 45.5 (C-13), 47.3 (C-14), 49.8 (C-9), 49.9 (C-17), 50.2 (25-OCH$_3$), 55.9 (7-OCH$_3$), 74.8 (C-25), 75.1 (C-7), 76.1 (C-3), 121.7 (C-6), 128.3 (C-

23), 136.7 (C-24), 146.2 (C-5), 206.7 (C-19); EI-MS (70 eV) m/z (rel. int.): 500 [M]$^+$ (8), 470 (20), 455 (22), 440 (40), 309 (13), 203 (33), 187 (31), 172 (97), 157 (47), 149 (52), 133 (49), 121 (60), 109 (100), 99 (72), 81 (37), 69 (34), 55 (60). (23E)-3β,7β,25-trihydroxycucbita-5,23-dien-19-al (CH93) Amorphous powder; IR (KBr)v_{max}: 3425, 1715, 1660, 1470, 1380, 1000, 980 cm^{-1}; ^{1}H NMR (400 MHz, CDCl$_3$) : δ 0.72 (3H, s, H-30), 0.86 (3H, s, H-18), 0.88 (3H, d, J = 6.0 Hz, H-21), 1.03 (3H, s, H-29), 1.21 (3H, s, H-26), 1.22 (3H, s, H-27), 1.23 (3H, s, H-28), 3.55 (1H, br s, H-3), 3.95 (1H, br d, J = 5.2 Hz, H-7), 5.56 (2H, m, H-23, H-24), 5.87 (1H, d, J = 5.2 Hz, H-6), 9.70 (1H, br s, H-19); EI-MS (70 eV) m/z 454 [M-H$_2$O]$^+$ (2), 436 (6), 418 (43), 389 (100), 375 (8), 337 (10), 309 (17), 171 (29), 109 (31), 81 (16).

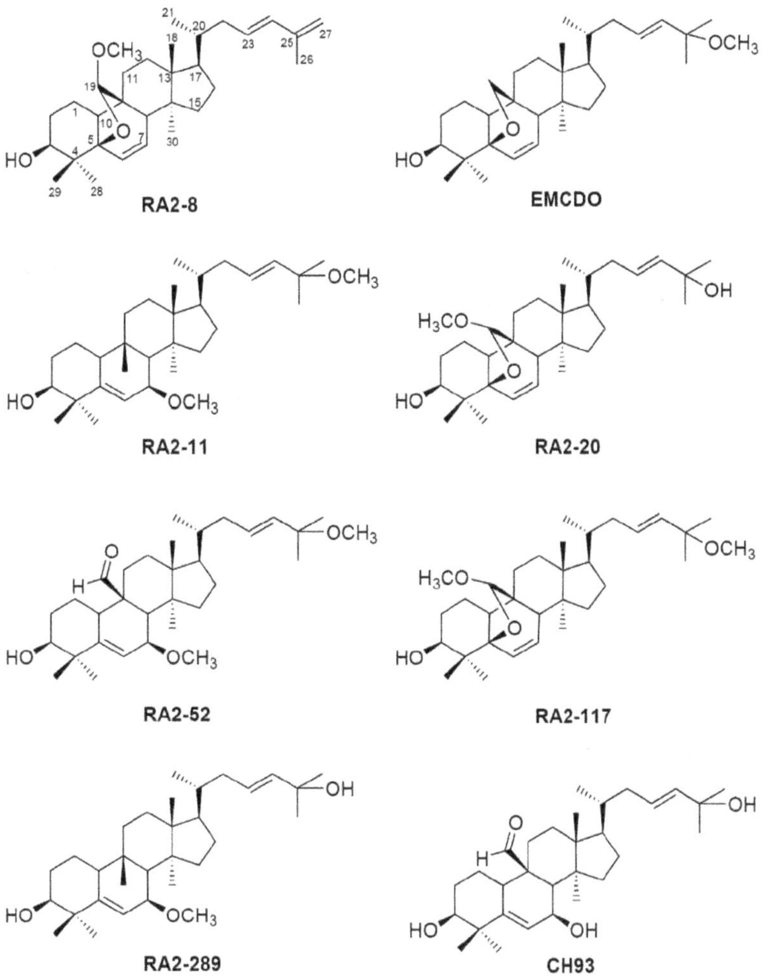

Figure 1. Chemical structures of eight compounds isolated from *M. charantia* fruit. The chemical structures of RA2-8, EMCDO, RA2-11, RA2-20, RA2-52, RA2-117, RA2-289, and CH93 are shown; their formal chemical names are stated in the text

3.2 Cytotoxicities of Compounds towards Two Hepatoma Cell Lines

To determine the non-toxic maximum working concentrations for cell-culture based assay, an MTT assay was utilized to evaluate the cytotoxicities of the isolated *Mormodica charantia* compounds and the anti-HBV positive control oltipraz (Chi et al., 1998). 2.2.15 (Doong, Tsai, Schinazi, Liotta, & Cheng, 1991) and PLC/PRF/5 (Marshall, Coulepis, Pringle, Dimitrakakis, & Gust, 1983) accumulate HBsAg and HBeAg after a longer period of culture; thus, the IC$_{10}$ and IC$_{50}$ of 9 days of treatment for 2.2.15 and 6 days for PLC/PRF/5 were determined. Table 1 illustrates the IC$_{10}$ and IC$_{50}$ concentrations of the tested compounds on two hepatoma cell lines. Based on these results, the working concentrations for further biological assays were: 5 µg/ml of EMCDO, RA2-11,oltipraz on both cell lines; 10 µg/ml of RA2-52 and RA2-117 on both cell lines; 10 µg/ml of RA2-8 on

2.2.15; 10 µg/ml of RA2-20, RA2-289 and CH93 on PLC/PRF/5; 7.5 µg/ml of RA2-8 on PLC/PRF/5; 7.5 µg/ml of RA2-289 and CH93 on 2.2.15; 5 µg/ml of RA2-20 on 2.2.15. Under our treatment conditions, more than 90% cells were viable.

Table 1. IC_{10} and IC_{50} of tested compounds on 2.2.15 and PLC/PRF/5 cell lines

			RA2-8	EMCDO	RA2-11	RA2-20	RA2-52	RA2-117	RA2-289	CH-93	oltipraz
2.2.15	day 3	IC_{10}	21.1 ± 2.1	20.4 ± 2.6	18.4 ± 0.8	12.2 ± 0.9	17.1 ± 1.3	16.8 ± 1.6	12.4 ± 1.0	13.7 ± 1.1	12.4 ± 1.3
		IC_{50}	23.5 ± 2.3	16.8 ± 2.2	21.3 ± 2.0	14.4 ± 1.4	23.8 ± 2.1	24.1 ± 2.3	17.8 ± 1.5	24.8 ± 2.3	18.5 ± 1.5
	day 6	IC_{10}	12.4 ± 1.2	7.8 ± 1.8	9.8 ± 0.3	10.1 ± 0.7	12.4 ± 1.1	13.8 ± 1.1	9.8 ± 0.7	10.1 ± 0.6	9.8 ± 0.7
		IC_{50}	21.4 ± 1.9	12.7 ± 1.3	18.6 ± 1.2	12.7 ± 1.1	19.6 ± 1.3	18.9 ± 1.4	13.1 ± 1.9	13.7 ± 1.2	13.9 ± 0.9
	day 9	IC_{10}	10.9 ± 0.8	5.1 ± 0.2	6.8 ± 0.4	6.4 ± 0.2	10.2 ± 0.9	10.1 ± 0.8	7.6 ± 0.3	7.8 ± 0.8	6.7 ± 0.2
		IC_{50}	19.4 ± 1.8	8.1 ± 0.3	14.2 ± 0.8	9.5 ± 0.7	16.1 ± 1.0	15.2 ± 1.2	10.8 ± 0.6	11.6 ± 1.0	10.2 ± 0.4
PLC/PRF/5	day 2	IC_{10}	14.1 ± 1.2	11.2 ± 1.0	11.9 ± 1.1	16.3 ± 0.9	20.4 ± 2.2	19.4 ± 1.9	14.3 ± 0.8	13.9 ± 0.7	11.1 ± 0.9
		IC_{50}	17.4 ± 1.6	14.8 ± 1.3	14.7 ± 1.2	18.7 ± 0.8	28.4 ± 2.4	21.1 ± 2.0	16.2 ± 1.0	18.7 ± 1.3	13.7 ± 1.1
	day 4	IC_{10}	11.5 ± 0.8	7.1 ± 0.5	8.9 ± 1.0	13.7 ± 0.7	14.5 ± 1.3	13.7 ± 1.3	12.3 ± 0.9	11.9 ± 1.0	8.4 ± 0.8
		IC_{50}	16.2 ± 0.9	8.4 ± 0.4	12.6 ± 1.1	14.9 ± 0.6	19.5 ± 1.2	15.4 ± 1.4	13.4 ± 0.7	14.4 ± 1.2	9.7 ± 0.4
	day 6	IC_{10}	8.7 ± 0.4	5.2 ± 0.3	6.8 ± 0.5	10.6 ± 0.3	11.1 ± 0.8	9.7 ± 0.6	9.8 ± 0.3	10.3 ± 0.6	5.4 ± 0.3
		IC_{50}	13.2 ± 0.6	7.6 ± 0.4	10.4 ± 0.6	13.2 ± 0.7	15.9 ± 1.0	12.1 ± 0.9	11.2 ± 0.7	12.3 ± 1.0	8.8 ± 0.9

3.3 Efficient Inhibition of HBs, HBe Antigen Production and Secreted HBV DNA by EMCDO

First of all, the suppression abilities of the compounds on HBs and HBe antigen production of cultured hepatoma cell lines were measured by ELISA assay. EMCDO revealed a dramatic inhibition effect after 3-9 days of treatment, which was comparable to the well-established HBV inhibitor oltipraz (Figures 2A and 2B). The significant HBsAg suppression activity of EMCDO exhibited an identical pattern on the PLC/PRF/5 cell line (Figure 2C). In addition to measuring the antigen production, the amount of HBV DNA in the culture medium was quantified. 2.2.15 cells were treated with various concentrations of EMCDO or oltipraz for 3 days. The EC_{50} of HBV DNA suppression was 4.58 µg/ml for oltipraz and 1.31 µg/ml for EMCDO; thus, EMCDO exhibited a better anti-HBV ability than oltipraz (Figure 2D).

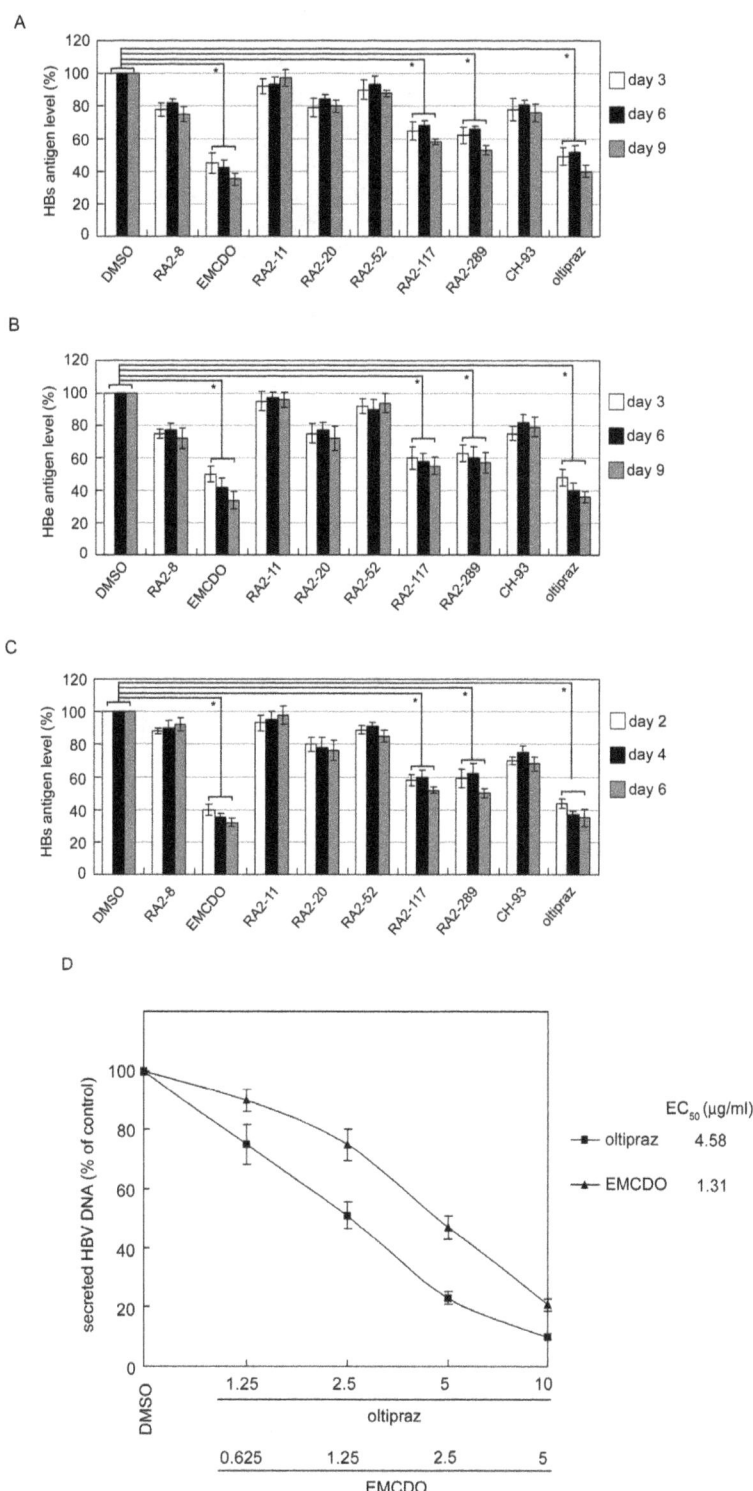

Figure 2. Reduction of HBV antigen and HBV DNA upon compound addition. 2.2.15 cells were treated with the indicated compounds for various time periods. Surface antigen (A) and e antigen HBe (B) levels were determined by ELISA and normalized against DMSO-treated cells. (C) PLC/PRF/5 cells were treated as indicated, and the surface antigen level was determined and normalized against DMSO-treated cells. Significant antigen inhibition is indicated by "*". (D) 2.2.15 cells were treated with EMCDO or oltipraz at various concentrations for 3 days, then the secreted HBV DNA was purified and subjected to semi-quantitative polymerase chain reaction. The EC50 concentration was calculated and is shown

3.4 p53 Involvement in EMCDO-Mediated Anti-HBV Effects

Oltipraz inhibits HBV *via* modulation of p53 (Chi et al., 1998). We then investigated whether EMCDO exerts its anti-HBV function through the activation of p53. The results clearly demonstrated the induction of p53 protein upon EMCDO and oltipraz treatment, and p53 expression showed more significant accumulation in cells treated with EMCDO (Figures 3A and 3B, panel 1). p53 phosphorylation at N-terminal serine residues is closely correlated with the transcriptional activation of its downstream genes (Jenkins, Durell, Mazur, & Appella, 2012). Our results showed greater serine phosphorylation induction and accumulation in EMCDO-treated cells than in oltipraz-treated cells (Figures 3A panels 2-5 and 3B panels 2-3). A p53-luc reporter was transfected into the cells, and activation of the synthetic promoter by the treatment will result in the stimulation of reporter expression. EMCDO and oltipraz induced luciferase activity in a dose-dependent manner, and the luciferase activity was fully inhibited in the cells co-transfected with dominant negative p53 (Figure 3C). Importantly, the p53 activation degree was greater in the EMCDO-treated cells than in the oltipraz-treated cells.

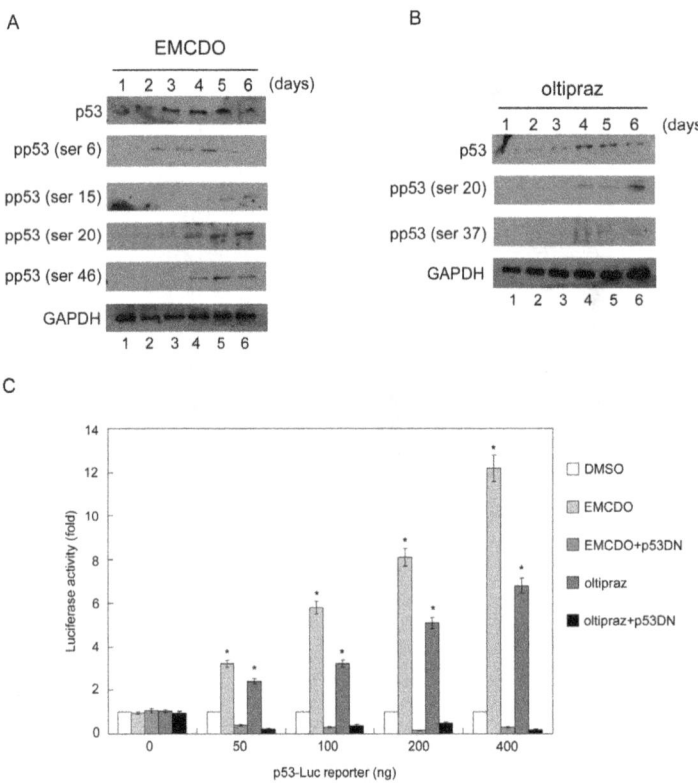

Figure 3. Activation of p53 by EMCDO. Western blotting analysis of (A) EMCDO- and (B) oltipraz-treated 2.2.15 cells. Total p53, phosphor-p53 and internal control GAPDH expression levels were determined by indicated antibodies. (C) Luciferase assay. 2.2.15 cells were treated with DMSO, EMCDO or oltipraz for 2 hrs, then transfected with p53-Luc reporter plasmid combined with pRKβGAL for 48hrs of incubation. p53DN plasmid was cotransfected into the indicated cells. The luciferase activation fold was calculated according to the transfection efficiency and normalized against time "0". Significant activation is indicated by "*"

3.5 Toxicity of EMCDO in Mice

In order to understand the possible application *in vivo*, we then evaluated the safety in a mouse model. Within 14 days observation, intraperitoneal injection of 500 mg/kg EMCDO did not produce any changes as compared with the control group. No mouse deaths suggested that EMCDO is not toxic at a dose of 500 mg/kg. Table 2 shows the mean body weight, organ weight and concentrations of liver and renal function markers. The liver function markers GOT and GPT, and renal function markers BUN and CRE, were all within the normal ranges. The body weight and organ weight of all mice at the end of experiment did not exhibit any significant changes.

Table 2. Biochemical analysis and organ weights of controls and EMCDO-treated mice

	Normal range	Control mice	EMCDO-treated mice
GOT (U/L)	59-247	159.5 ± 68.6	170.5 ± 32.5
GPT (U/L)	28-132	22.5 ± 9.8	26.2 ± 6.31
BUN (mg/dL)	18-29	22.8 ± 3.1	21.4 ± 1.62
CRE (mg/dL)	0.2-0.8	0.45 ± 0.06	0.49 ± 0.01
Heart (g)		0.126 ± 0.001	0.126 ± 0.001
Liver (g)		1.046 ± 0.002	0.965 ± 0.002
Spleen (g)		0.144 ± 0.001	0.161 ± 0.003
Lung (g)		0.160 ± 0.002	0.140 ± 0.004
Kidney (g)		0.301 ± 0.001	0.324 ± 0.002
Stomach (g)		0.264 ± 0.001	0.276 ± 0.003
Body weight (g)		24.5 ± 0.5	24.8 ± 0.4

3.6 Tumor Prevention Ability of EMCDO in a Nude Mouse Model

Subcutaneous inoculation of 2.2.15 cells induces tumor formation (Wu et al., 2011). EMCDO and oltipraz, as well as the negative control corn oil, were injected into nude mice *via* the peritoneal route. Compound injection was performed for 7 days, then 2.2.15 cells were inoculated. The observation period was 8 weeks, and compounds were injected once every 2 days within the 8-week period. At the end of the experiments, serum HBsAg, HBeAg and hepatoma marker AFP were dramatically reduced in EMCDO- and oltipraz-treated mice (Figure 4A). By monitoring the tumor size every 5 days after 2.2.15 inoculation, the average tumor size was observed to be much smaller in EMCDO- and oltipraz-treated mice (Figure 4B). Critically, the tumor suppression ability of EMCDO was a little better than that of oltipraz, especially during the later phase. The results clearly demonstrated the EMCDO is a good tumor prevention agent, the tumor suppression effect could be lasted for more longer time than oltipraz.

A

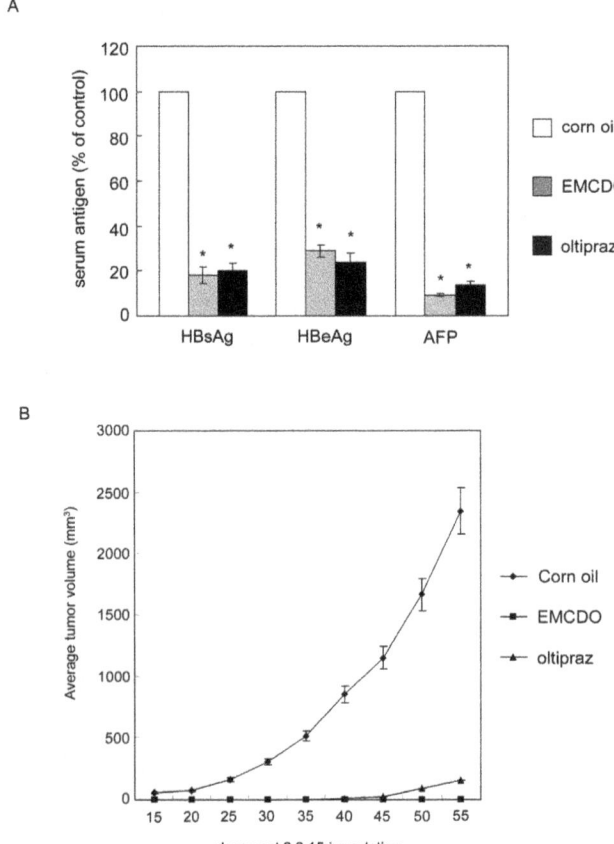

B

Figure 4. Tumor growth inhibition in a mouse model. (A) Serum was collected at the end of the experiment. The levels of HBV surface antigen, e antigen and AFP were determined by ELISA. Corn oil-treated mice served as the control. The antigen reduction percentages in EMCDO- or oltipraz-treated mice are shown in the y-axis. Significant reduction is indicated by "*". (B) Suppression of tumor volume of EMCDO- or oltipraz-injected mice. The x-axis indicates the days post-2.2.15 cells inoculation; the y-axis shows the average tumor volume of 6 mice

3.7 Inhibition of Inflammatory Mediators by EMCDO

Due to the close relationship between inflammation and cancer development, we studied further whether the anti-HBV effect of EMCDO is related to the anti-inflammation mechanism. LPS-treated RAW264.7 is a well-used model to evaluate the anti-inflammatory response (Kim, Hwang, & Han, 2013). The IC_{10} and IC_{50} cytotoxicities were determined in RAW264.7 cells upon EMCDO addition (Figure 5A); thus, the working concentration of the following experiments ensured the more than 90% of cells were viable. In a 24-hr experiment, we used a maximum EMCDO concentration of 5 µg/ml. The suppression dose effect of iNOS by curcumin was monitored (Figure 5B), and the effective concentration was correlated with a previous study (Cheung, Khor, & Kong, 2009). The LPS-stimulated expressions of inflammatory markers inducible nitric oxide synthase (iNOS) and cyclooxygenase-2 (COX-2) were repressed in the presence of EMCDO ranging from 1.25 to 5 µg/ml, and the repression efficiency was better than that of 7.5 µg/ml curcumin, a well-known efficient natural anti-inflammation polyphenol compound (Wang, Sun, Huang, & Zheng, 2013) (Figure 5C). LPS-treated RAW264.7 cells exhibited proinflammatory cytokines production and secretion, including TNF-α, IL-6 and IL-1β (Qi et al., 2012). As expected, the three proinflammatory cytokines were dramatically repressed in the presence of EMCDO in a dose-dependent manner. At a non-cytotoxic concentration of 5 µg/ml EMCDO, the cytokine repression ability was comparable to that of the well-known anti-inflammation compound curcumin (Figure 5D).

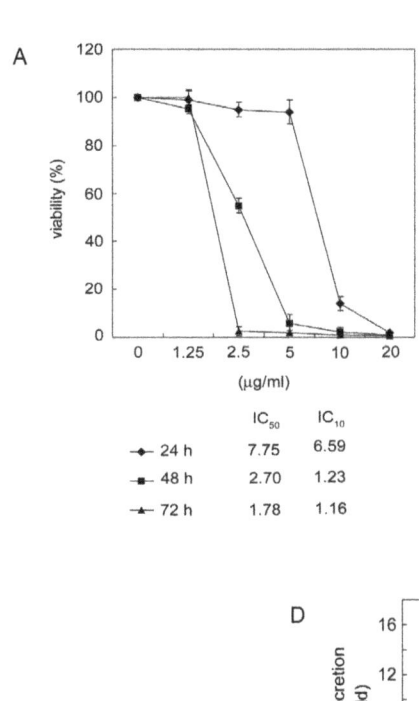

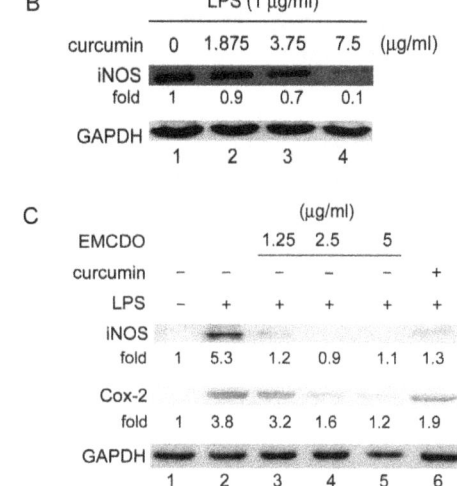

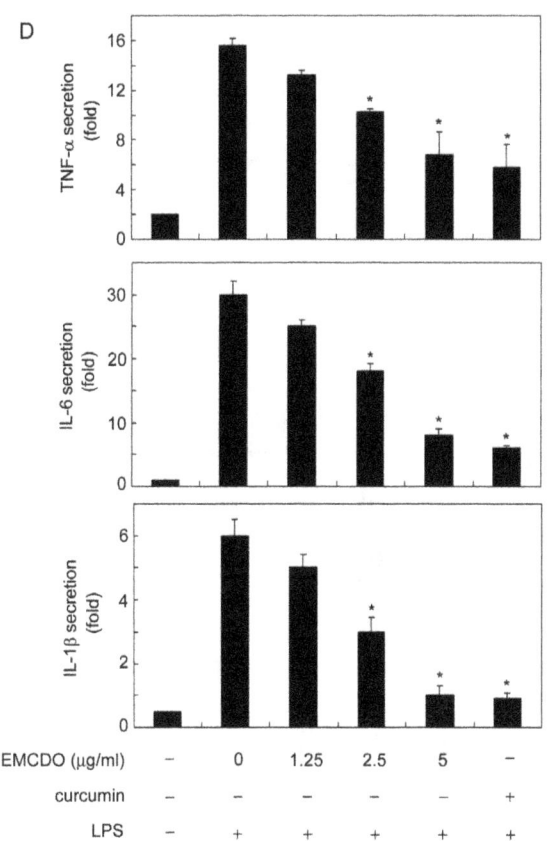

Figure 5. Anti-inflammation ability of EMCDO. (A) Trypan blue exclusion assay. RAW264.7 cells were incubated with various concentrations of EMCDO for 24, 48 and 72 hrs. The viable and dead cells were differentiated by trypan blue dye. The y-axis shows the viable cell percentage normalized against non-treated cells; the IC_{50} and IC_{10} are indicated. (B) Curcumin inhibits LPS-induced iNOS expression. Western blotting analysis of RAW264.7 cells treated with LPS and curcumin for 24 hrs of incubation. 50 µg soluble protein were subjected to SDS-PAGE. The iNOS expression fold was normalized against GAPDH and compared with lane 1. (C) EMCDO suppressed iNOS and COX-2 expression dose-dependently. Western blotting analysis of RAW264.7 cells treated as indicated. The expression level was normalized against GAPDH and compared with lane 1. (D) Measurement of proinflammatory cytokines by ELISA. The y-axis shows the cytokine secretion fold normalized against conditioned media of non-treated RAW264.7 cells. The x-axis indicates the compound treatment of cells

3.8 Activation of NF-κB, Not MAPK Signaling, by EMDCO

Previous studies have indicated that MAPK signaling and NF-κB play roles in mediating anti-inflammation in some cell culture-based systems, including RAW264.7 cells (Yuan, Chen, Sun, Guan, & Xu, 2013). In these assay systems, LPS increases IκBα and MAPKs phosphorylation (Yoon et al., 2010; Meng, Yan, Deng, Gao, & Niu, 2013). To investigate the host cell signaling participation in mediating the EMCDO effects, the phosphorylation levels of JNK, p38, ERK and IκBα were examined by specific phospho-antibodies. The results clearly showed that phosphorylation of JNK, p38 and ERK was not altered upon EMCDO or curcumin treatment (Figure 6A, panel 1-3), while the increased phosphorylation of IκBα by LPS was reduced by 5 μg/ml EMCDO and curcumin (Figure 6A, panel 4). To further confirm the involvement of NF-κB, a luciferase reporter plasmid containing an upstream NF-κB responsive element was transfected into compound-treated cells. Figure 6B illustrates that the up-regulated luciferase activity by LPS was reduced by EMCDO in a concentration-dependent pattern, comparable to the positive control curcumin. Thus, the current findings suggested that NF-κB-related signaling may have a critical regulation in EMCDO's anti-HBV and anti-inflammation effects.

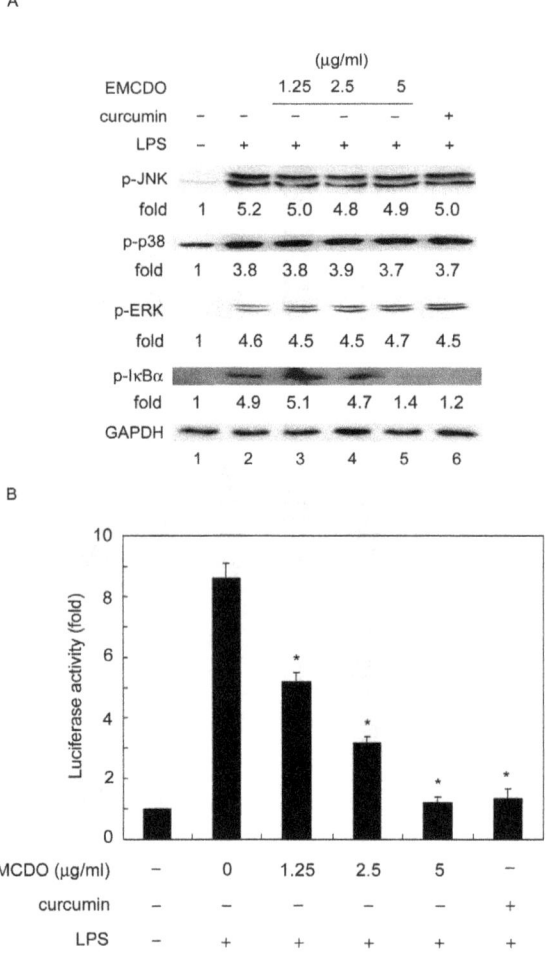

Figure 6. (A) Western blotting analysis. RAW264.7 cells were pre-treated with the indicated compound for 2 hrs, then LPS was added for a further 6 hrs of incubation. Soluble cell lysates were harvested and subjected to SDS-PAGE. The phosphor-protein expression level and internal control GAPDH are shown. The fold induction was determined using software and compared with lane 1. (B) Luciferase assay. Compounds were pre-treated for 2 hrs then co-transfected with NF-κB-Luc and pRKβGAL. After 24 hrs, LPS was added and incubated for an additional 24 hrs. Luciferase and β-galatosidase activities were measured. The data are expressed as the luciferase activation fold by normalization of the transfection efficiency and compared with non-treated control cells. Significant suppression is indicated by "*"

3.9 Anti-Inflammation Ability of EMCDO in vivo

In the case of the ear edema mouse model, EMCDO effectively inhibited the ear thickness in a dose-dependent manner by topical application. Comparison of indomethacin and EMCDO using the same concentration of 1.25 μg/20 μl showed that the edema inhibition ability of EMCDO is better than that of the classic non-steroidal anti-inflammatory drug indomethacin after 6 and 16 hrs of treatment. Importantly, the anti-inflammation efficiency of EMCDO was still significant at 1.25 μg/20 μl after 24 hrs of application. At a lower concentration of EMCDO, the anti-inflammation effect was apparent after 6 hrs of application (Figure 7). Taken together, these results showed that EMCDO is a low-toxicity, highly-effective anti-inflammation compound. The inflammation-relaxing ability of EMCDO is rapid and more potent than that of indomethacin.

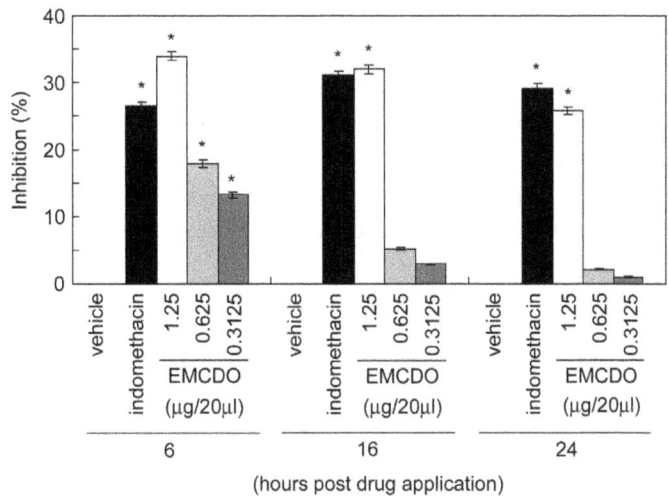

Figure 7. Inhibition of ear edema by EMCDO. Ear thickness was measured before and after TPA application at 30 min. Then, indomethacin and various concentrations of EMCDO were topically applied for different time periods. The y-axis represents the ear thickness reduction percentage. Significant inhibition of ear thickening is indicated by "*"

4. Discussion

The current investigation demonstrated that one *Mormodica charantia* compound, EMCDO, isolated in our laboratory possessed good anti-HBV, anti-tumor and anti-acute inflammation activities. Based upon the HBV-secreting cell line, a mouse macrophage model and *in vivo* assays, it was confirmed that EMCDO is a low-toxicity compound of high biological efficacy.

All the isolated compounds, except for RA2-11 and RA2-52, showed anti-HBV activity. Comparing their structures, the cucurbitane-type triterpenes with a 5β,19-epoxy moiety in the skeleton, RA2-8, EMCDO, RA2-20, and RA2-117, always exhibited a significant anti-HBV effect. In particular, EMCDO without the methoxy substituent at C-19 showed the greatest level of anti-HBV activity. Additionally, RA2-289 possesses a hydroxy group at C-25 in the side chain, and exhibited a higher anti-HBV activity than its methyl ether derivative, RA2-11. Furthermore, the simultaneous methylation at C-7 and C-25 decreased the activity level as compared with the activity results of CH-93 and RA2-52. Thus, for the cucurbitane-type triterpenes, a 5β,19-epoxy moiety in the skeleton, along with the hydroxy free form at C-7 and C-25 in the side chain, may be important for their anti-HBV activity expression.

The underlying molecular mechanisms participating in hepatocellular carcinoma development remain incompletely understood. It is well-known that intrinsic and extrinsic inflammatory signaling pathways converge and cooperate in amplifying the cancer risk and driving cancer cell formation. However, the cancer-associated inflammation response is often persistent and chronic, inducing uncontrolled pathological signals in the tumor microenvironment; thus, cellular proliferation, migration, invasion and angiogenesis will occur (Wai & Kuo, 2012).

HBV and HCV infections result in liver inflammation and further liver cancer development. During acute infection, viral antigens activate immune cells and could induce the infected cells undergoing apoptosis. The result of cell death caused cell regeneration (Bertoletti, Maini, & Ferrari, 2010). Furthermore, current evidence indicates that chronic inflammation will lead to gene mutation and genome instability, and the accumulated mutated gene plays critical roles in liver carcinogenesis (Brechot et al., 2010). For example, HBV X protein and HCV core protein are expressed during infection, and these viral oncoproteins interact with the cellular homeostasis machinery, affect cellular gene expression and regulation, and also participate in liver inflammation (Ray, Steele, Meyer, & Ray, 1997).

Various cellular pathways are activated during a prolonged inflammation process in the liver, such as NF-κB (Assenat, Grebal-Chaloin, Maurel, Vilarem, & Pascussi, 2006) and JAK/STAT3 (Park et al., 2010), as well as MAPK signalings (Ma, Ding, Zhang, & Liu, 2014). EMCDO alleviated the inflammation mediator production and ear inflammation, possibly through the activation of NF-κB in the RAW264.7 system, but no effects on MAPK were observed. Previous studies have indicated a dual function of NF-κB in hepatocarcinogenesis: NF-κB can activate cell-protecting gene expression and then prevent excess cell death, thus limiting the subsequent cell transformation (Luedde et al., 2007). On the other hand, NF-κB may constitute a link between inflammation and cancer (Pikarsky et al., 2004). Additionally, the inflammation response also regulates epigenetic change, including DNA methylation, histone modifications, as well as non-coding RNAs in hepatocytes (Martin & Herceg, 2012).

One important component of the immune system to cope with foreign antigens is cytokines. The liver contains many various cell types that are susceptible to the influences of cytokines. Hepatocytes bear cytokine receptors, such as IL-6, IL-1 and TNF-α. The cytokines released from Th1 cells normally function as proinflammatory cytokines; on the other hand, Th2 cytokines, such as IL-4, IL-8, IL-10, and IL-5, induce anti-inflammatory responses. In fact, many cytokines have pleiotrophic activities, which could function in synergistic or antagonistic manner. Due to the enriched immune cells located in liver, the evidence had been indicated these inflammatory cells secreted cytokines and further cause liver injury characteristic of acute HBV (Rehermann, 2003). In addition, hepatic viruses are able to evade the immune system and persist *via* a variety of mechanisms, the major way being to rapidly mutate under immune pressure. Recently, several studies have described that changes in Th1 and Th2 cytokines expression result from various treatment regimens which correlated in HCC metastasis or recurrence in patients (Budhu & Wang, 2006).

Malignant neoplasia is the result of multiple genetic defects successively accumulating over a period of time in genes mainly controlling proliferation, differentiation and cell death (Stephenson, Abouassaly, & Klein, 2010), yet most current anticancer therapies involve mainly or only the modulation of a single target. As a result, research concentrating on the discovery of multi-target new agents and new treatment strategies is therefore of vital importance (Guruswamy & Rao, 2008). Broadly defined, cancer chemoprevention applies to the inhibition or reversal of oncogenesis, at a variety of time points, to suppress the occurrence of *in situ* or invasive cancers using natural, synthetic, biologic or chemical agents (Desai et al., 2008). Although the principle of chemoprevention has been clearly demonstrated in both animal and clinical trials, none of the existing chemopreventive agents is ideal, because of either a lack of efficacy or toxic side effects (Lam, Macaulay, Leriche, & Gazdar, 2003). Accumulation of *in vitro* evidence demonstrating the inhibitory effects of certain natural products on liver cancer cells (Fang, Hsu, Lin, & Yen, 2010). The anti-carcinogenic activity of cucurbitane-type triterpenoids has been clearly demonstrated in two-stage carcinogenesis testing in skin tumors (Takasaki et al., 2003). The triterpenoid saponin-rich G. oldhamiana root extract (TGOE) selectively inhibited the proliferation of hepatoma SMMC-7721 cells in a dose-dependent manner, while the cytotoxic effects of TGOE on normal hepatocyte L02 cells were much lower. TGOE preferentially induced apoptosis in SMMC-7721 cells due to the regulation of caspase-3 and mitogen-activated protein kinases (MAPKs) (Zhang, Luo, Zhang, & Kong, 2013). In conclusion, the chemical structure determination of 8 isolated *Mormodica charantia* compounds. EMCDO significantly reduced HBV antigens and DNA secretion in hepatoma cell lines, prevented the tumor formation induced by a HBV-producing cell line in nude mice model. In addition, EMCDO downregulated inflammatory mediators in LPS-treated RAW264.7 cells and inhibited edema in the TPA-treated mouse ear. Our data suggested that EMCDO may have potential beneficial effects against HBV and inflammation response, subsequently preventing hepatocellular carcinoma development. To prove the clinical uses of EMCDO, efforts to elucidate the underlying mechanism and more animal tests are needed.

References

Abe, A., Inoue, K., Tanaka, T., Kato, J., Kajiyama, N., Kawaguchi, R., ... Kohara, M. (1999). Quantitation of hepatitis B virus genomic DNA by real-time detection PCR. *J Clin Microbiol, 37*(9), 2899-2903.

Acs, G., Sells, M. A., Purcell, R. H., Price, P., Engle, R., Shapiro, M., & Popper, H. (1987). Hepatitis B virus produced by transfected Hep G2 cells causes hepatitis in chimpanzees. *Proc Natl Acad Sci USA, 84*(13), 4641-4644. http://dx.doi.org/10.1073/pnas.84.13.4641

Akihisa, T., Hayakawa, Y., Tokuda, H., Banno, N., Shimizu, N., Suzuki, T., & Kimura, Y. (2007). Cucurbitane glycosides from the fruits of Siraitia gros venorii and their inhibitory effects on Epstein-Barr virus activation. *J Nat Prod, 70*(5), 783-788. http://dx.doi.org/10.1021/np068074x

Alexander, B., Browse, D. J., Reading, S. J., & Benjamin, I. S. (1999). A simple and accurate mathematical method for calculation of the EC50. *J Pharmacol Toxicol Methods, 41*(2-3), 55-58. http://dx.doi.org/10.1016/S1056-8719(98)00038-0

Aslam, M. N., Bergin, I., Naik, M., Hampton, A., Allen, R., Kunkel, S. L., ... Varani, J. (2012). A multi-mineral natural product inhibits liver tumor formation in C57BL/6 mice. *Biol Trace Elem Res, 147*(1-3), 267-274. http://dx.doi.org/10.1007/s12011-011-9316-2

Assenat, E., Gerbal-chaloin, S., Maurel, P., Vilarem, M. J., & Pascussi, J. M. (2006). Is nuclear factor kappa-B the missing link between inflammation, cancer and alteration in hepatic drug metabolism in patients with cancer? *Eur J Cancer, 42*(6), 785-792. http://dx.doi.org/10.1016/j.ejca.2006.01.005

Berasain, C., Castillo, J., Perugorria, M. J., Latasa, M. U., Prieto, J., & Avila, M. A. (2009a). Inflammation and liver cancer: new molecular links. *Ann N Y Acad Sci, 1155*, 206-221. http://dx.doi.org/10.1111/j.1749-6632.2009.03704.x

Berasain, C., Perugorria, M. J., Latasa, M. U., Castillo, J., Goni, S., Santamaria, M., ... Avila, M. A. (2009b). The epidermal growth factor receptor: a link between inflammation and liver cancer. *Exp Biol Med (Maywood), 234*(7), 713-725. http://dx.doi.org/10.3181/0901-MR-12

Bertoletti, A., Maini, M. K., & Ferrari, C. (2010). The host-pathogen interaction during HBV infection: immunological controversies. *Antivir Ther, 15*(Suppl. 3), 15-24. http://dx.doi.org/10.3851/IMP1620

Brechot, C., Kremsdorf, D., Soussan, P., Pineau, P., Dejean, A., Paterlini-Brechot, P., & Tiollais, P. (2010). Hepatitis B virus (HBV)-related hepatocellular carcinoma (HCC): molecular mechanisms and novel paradigms. *Pathol Biol (Paris), 58*(4), 278-287. http://dx.doi.org/10.1016/j.patbio.2010.05.001

Budhu, A., & Wang, X. W. (2006). The role of cytokines in hepatocellular carcinoma. *J Leukoc Biol, 80*(6), 1197-1213. http://dx.doi.org/10.1189/jlb.0506297

Cetin, Y., & Bullerman, L. B. (2005). Cytotoxicity of Fusarium mycotoxins to mammalian cell cultures as determined by the MTT bioassay. *Food Chem Toxicol, 43*(5), 755-764. http://dx.doi.org/10.1016/j.fct.2005.01.016

Chang, C. D., Lin, P. Y., Liao, M. H., Chang, C. I., Hsu, J. L., Yu, F. L., ... Shih, W. L. (2013). Suppression of apoptosis by pseudorabies virus Us3 protein kinase through the activation of PI3-K/Akt and NF-kappaB pathways. *Res Vet Sci, 95*(2), 764-774. http://dx.doi.org/10.1016/j.rvsc.2013.06.003

Chang, C. I., Chen, C. R., Liao, Y. W., Cheng, H. L., Chen, Y. C., & Chou, C. H. (2008). Cucurbitane-type triterpenoids from the stems of momordica charantia. *J Nat Prod, 71*(8), 1327-1330. http://dx.doi.org/10.1021/np070532u

Chemin, I. (2010). Evaluation of a hepatitis B vaccination program in Taiwan: impact on hepatocellular carcinoma development. *Future Oncol, 6*(1), 21-23. http://dx.doi.org/10.2217/fon.09.158

Chen, J. C., Zhang, G. H., Zhang, Z. Q., Qiu, M. H., Zheng, Y. T., Yang, L. M., & Yu, K. B. (2008). Octanorcucurbitane and cucurbitane triterpenoids from the tubers of Hemsleya endecaphylla with HIV-1 inhibitory activity. *J Nat Prod, 71*(1), 153-155. http://dx.doi.org/10.1021/np0704396

Cheung, K. L., Khor, T. O., & Kong, A. N. (2009). Synergistic effect of combination of phenethyl isothiocyanate and sulforaphane or curcumin and sulforaphane in the inhibition of inflammation. *Pharm Res, 26*(1), 224-231. http://dx.doi.org/10.1007/s11095-008-9734-9

Chi, W. J., Doong, S. L., Lin-Shiau, S. Y., Boone, C. W., Kelloff, G. J., & Lin, J. K. (1998). Oltipraz, a novel inhibitor of hepatitis B virus transcription through elevation of p53 protein. *Carcinogenesis, 19*(12), 2133-2138. http://dx.doi.org/10.1093/carcin/19.12.2133

Desai, A. G., Qazi, G. N., Ganju, R. K., El-Tamer, M., Singh, J., Saxena, A. K., ... Bhat, H. K. (2008). Medicinal plants and cancer chemoprevention. *Curr Drug Metab, 9*(7), 581-591. http://dx.doi.org/10.2174/138920008785821657

Doong, S. L., Tsai, C. H., Schinazi, R. F., Liotta, D. C., & Cheng, Y. C. (1991). Inhibition of the replication of hepatitis B virus in vitro by 2',3'-dideoxy-3'-thiacytidine and related analogues. *Proc Natl Acad Sci USA, 88*(19), 8495-8499. http://dx.doi.org/10.1073/pnas.88.19.8495

Elgouhari, H. M., Abu-Rajab Tamimi, T. I., & Carey, W. D. (2008). Hepatitis B virus infection: understanding its epidemiology, course, and diagnosis. *Cleve Clin J Med, 75*(12), 881-889. http://dx.doi.org/10.3949/ccjm.75a.07019

Fang, S. C., Hsu, C. L., Lin, H. T., & Yen, G. C. (2010). Anticancer effects of flavonoid derivatives isolated from Millettia reticulata Benth in SK-Hep-1 human hepatocellular carcinoma cells. *J Agric Food Chem, 58*(2), 814-820. http://dx.doi.org/10.1021/jf903216r

Guruswamy, S., & Rao, C. V. (2008). Multi-Target Approaches in Colon Cancer Chemoprevention Based on Systems Biology of Tumor Cell-Signaling. *Gene Regul Syst Bio, 2*, 163-176.

Jenkins, L. M., Durell, S. R., Mazur, S. J., & Appella, E. (2012). p53 N-terminal phosphorylation: a defining layer of complex regulation. *Carcinogenesis, 33*(8), 1441-1449. http://dx.doi.org/10.1093/carcin/bgs145

Kim, S. Y., Hwang, J. S., & Han, I. O. (2013). Tunicamycin inhibits Toll-like receptor-activated inflammation in RAW264.7 cells by suppression of NF-kappaB and c-Jun activity via a mechanism that is independent of ER-stress and N-glycosylation. *Eur J Pharmacol, 721*(1-3), 294-300. http://dx.doi.org/10.1016/j.ejphar

Lam, S., MacAulay, C., LeRiche, J. C., & Gazdar, A. F. (2003). Key issues in lung cancer chemoprevention trials of new agents. *Recent Results Cancer Res, 163*, 182-195; Discussion, 264-186. http://dx.doi.org/10.1007/978-3-642-55647-0_17

Liao, Y. W., Chen, C. R., Kuo, Y. H., Hsu, J. L., Shih, W. L., Cheng, H. L., … Chang, C. I. (2012). Cucurbitane-type triterpenoids from the fruit pulp of Momordica charantia. *Nat Prod Commun, 7*(12), 1575-1578.

Luedde, T., Beraza, N., Kotsikoris, V., van Loo, G., Nenci, A., De Vos, R., … Pasparakis, M. (2007). Deletion of NEMO/IKKgamma in liver parenchymal cells causes steatohepatitis and hepatocellular carcinoma. *Cancer Cell, 11*(2), 119-132. http://dx.doi.org/10.1016/j.ccr.2006.12.016

Ma, J. Q., Ding, J., Zhang, L., & Liu, C. M. (2014). Ursolic acid protects mouse liver against CCl-induced oxidative stress and inflammation by the MAPK/NF-kappaB pathway. *Environ Toxicol Pharmacol, 37*(3), 975-983. http://dx.doi.org/10.1016/j.etap.2014.03.011

Maddrey, W. C. (2000). Hepatitis B: an important public health issue. *J Med Virol, 61*(3), 362-366. http://dx.doi.org/10.1002/1096-9071(200007)61:3<362::AID-JMV14>3.0.CO;2-I

Marshall, J., Coulepis, A., Pringle, R., Dimitrakakis, M., & Gust, I. D. (1983). The effect of glucocorticoid hormones on release of HBsAg from PLC/PRF/5 (Alexander) hepatoma cells. *Acta Virol, 27*(5), 429-433.

Martin, M., & Herceg, Z. (2012). From hepatitis to hepatocellular carcinoma: a proposed model for cross-talk between inflammation and epigenetic mechanisms. *Genome Med, 4*(1), 8. http://dx.doi.org/10.1186/gm307

Marusawa, H., & Jenkins, B. J. (2014). Inflammation and gastrointestinal cancer: An overview. *Cancer Lett, 345*(2), 153-156. http://dx.doi.org/10.1016/j.canlet.2013.08.025

Meng, Z., Yan, C., Deng, Q., Gao, D. F., & Niu, X. L. (2013). Curcumin inhibits LPS-induced inflammation in rat vascular smooth muscle cells in vitro via ROS-relative TLR4-MAPK/NF-kappaB pathways. *Acta Pharmacol Sin, 34*(7), 901-911. http://dx.doi.org/10.1038/aps.2013.24

Netea, M. G., Drenth, J. P., De Bont, N., Hijmans, A., Keuter, M., Dharmana, E., … van der Meer, J. W. (1996). A semi-quantitative reverse transcriptase polymerase chain reaction method for measurement of MRNA for TNF-alpha and IL-1 beta in whole blood cultures: its application in typhoid fever and exentric exercise. *Cytokine, 8*(9), 739-744. http://dx.doi.org/10.1006/cyto.1996.0098

Park, E. J., Lee, J. H., Yu, G. Y., He, G., Ali, S. R., Holzer, R. G., … Karin, M. (2010). Dietary and genetic obesity promote liver inflammation and tumorigenesis by enhancing IL-6 and TNF expression. *Cell, 140*(2), 197-208. http://dx.doi.org/10.1016/j.cell.2009.12.052

Pikarsky, E., Porat, R. M., Stein, I., Abramovitch, R., Amit, S., Kasem, S., … Ben-Neriah, Y. (2004). NF-kappaB functions as a tumour promoter in inflammation-associated cancer. *Nature, 431*(7007), 461-466. http://dx.doi.org/10.1038/nature02924

Qi, S., Xin, Y., Guo, Y., Diao, Y., Kou, X., Luo, L., & Yin, Z. (2012). Ampelopsin reduces endotoxic inflammation via repressing ROS-mediated activation of PI3K/Akt/NF-kappaB signaling pathways. *Int Immunopharmacol, 12*(1), 278-287. http://dx.doi.org/10.1016/j.intimp.2011.12.001

Ray, R. B., Steele, R., Meyer, K., & Ray, R. (1997). Transcriptional repression of p53 promoter by hepatitis C virus core protein. *J Biol Chem, 272*(17), 10983-10986. http://dx.doi.org/10.1074/jbc.272.17.10983

Rehermann, B. (2003). Immune responses in hepatitis B virus infection. *Semin Liver Dis, 23*(1), 21-38. http://dx.doi.org/10.1055/s-2003-37586

Shih, W. L., Kuo, M. L., Chuang, S. E., Cheng, A. L., & Doong, S. L. (2000). Hepatitis B virus X protein inhibits transforming growth factor-beta -induced apoptosis through the activation of phosphatidylinositol 3-kinase pathway. *J Biol Chem, 275*(33), 25858-25864. http://dx.doi.org/10.1074/jbc.M003578200

Shih, W. L., Yu, F. L., Chang, C. D., Liao, M. H., Wu, H. Y., & Lin, P. Y. (2013). Suppression of AMF/PGI-mediated tumorigenic activities by ursolic acid in cultured hepatoma cells and in a mouse model. *Mol Carcinog, 52*(10), 800-812. http://dx.doi.org/10.1002/mc.21919

Stephenson, A. J., Abouassaly, R., & Klein, E. A. (2010). Chemoprevention of prostate cancer. *Urol Clin North Am, 37*(1), 11-21. Table of Contents. http://dx.doi.org/10.1016/j.ucl.2009.11.003

Stockert, J. C., Blazquez-Castro, A., Canete, M., Horobin, R. W., & Villanueva, A. (2012). MTT assay for cell viability: Intracellular localization of the formazan product is in lipid droplets. *Acta Histochem, 114*(8), 785-796. http://dx.doi.org/10.1016/j.acthis.2012.01.006

Takasaki, M., Konoshima, T., Murata, Y., Sugiura, M., Nishino, H., Tokuda, H., ... Yamasaki, K. (2003). Anticarcinogenic activity of natural sweeteners, cucurbitane glycosides, from Momordica grosvenori. *Cancer Lett, 198*(1), 37-42. http://dx.doi.org/10.1016/S0304-3835(03)00285-4

Vasconcelos, J. F., Teixeira, M. M., Barbosa-Filho, J. M., Lucio, A. S., Almeida, J. R., de Queiroz, L. P., ... Soares, M. B. (2008). The triterpenoid lupeol attenuates allergic airway inflammation in a murine model. *Int Immunopharmacol, 8*(9), 1216-1221. http://dx.doi.org/10.1016/j.intimp.2008.04.011

Wai, P. Y., & Kuo, P. C. (2012). Intersecting pathways in inflammation and cancer: Hepatocellular carcinoma as a paradigm. *World J Clin Oncol, 3*(2), 15-23. http://dx.doi.org/10.5306/wjco.v3.i2.15

Wang, H., Syrovets, T., Kess, D., Buchele, B., Hainzl, H., Lunov, O., ... Simmet, T. (2009). Targeting NF-kappa B with a natural triterpenoid alleviates skin inflammation in a mouse model of psoriasis. *J Immunol, 183*(7), 4755-4763. http://dx.doi.org/10.4049/jimmunol.0900521

Wang, L. L., Sun, Y., Huang, K., & Zheng, L. (2013). Curcumin, a potential therapeutic candidate for retinal diseases. *Mol Nutr Food Res, 57*(9), 1557-1568. http://dx.doi.org/10.1002/mnfr.201200718

Wu, H. Y., Chang, C. I., Lin, B. W., Yu, F. L., Lin, P. Y., Hsu, J. L., ... Shih, W. L. (2011). Suppression of hepatitis B virus x protein-mediated tumorigenic effects by ursolic Acid. *J Agric Food Chem, 59*(5), 1713-1722. http://dx.doi.org/10.1021/jf1045624

Yoon, W. J., Moon, J. Y., Kang, J. Y., Kim, G. O., Lee, N. H., & Hyun, C. G. (2010). Neolitsea sericea essential oil attenuates LPS-induced inflammation in RAW 264.7 macrophages by suppressing NF-kappaB and MAPK activation. *Nat Prod Commun, 5*(8), 1311-1316.

Yuan, F., Chen, J., Sun, P. P., Guan, S., & Xu, J. (2013). Wedelolactone inhibits LPS-induced pro-inflammation via NF-kappaB pathway in RAW 264.7 cells. *J Biomed Sci, 20*, 84. http://dx.doi.org/10.1186/1423-0127-20-84

Zhang, W., Luo, J. G., Zhang, C., & Kong, L. Y. (2013). Different apoptotic effects of triterpenoid saponin-rich Gypsophila oldhamiana root extract on human hepatoma SMMC-7721 and normal human hepatic L02 cells. *Biol Pharm Bull, 36*(7), 1080-1087. http://dx.doi.org/10.1248/bpb.b12-01069

18

Rooting of Camu-Camu (*Myrciaria dubia*) in Different Propagation Systems and Reproductive Phases

Jhon Paul Mathews Delgado[1], Patrick Mathews Delgado[2], Carlos Abanto Rodriguez[3] & Ricardo Manuel Bardales Lozano[4]

[1] Graduate program in Botany from the National Institute of Amazonian Research-INPA, Manaus, Brazil

[2] Department of Parasitology, Institute of BiolIgly, University of Campinas, São Paulo, Brazil

[3] Peruvian Amazon Research Institute-IIAP, Pucallpa, Ucayali, Peru

[4] Ret Bionorte (Multinsitutional programme of Amazon), Brazil

Correspondence: Jhon Paul Mathews Delgado, National Institute of Amazonian Research-INPA, Manaus, AM, 69067-375, Brazil. E-mail: fedormath@hotmail.com

Abstract

Closed environment and controlled irrigation techniques are conventionally used to root camu-camu stem cuttings. But development of other techniques of camu-camu propagation is required for smallholder growers in Peru, who cultivate camu-camu on *restingas,* old river levee fragments on the margins of white-water rivers. Additionally, the month in which cuttings are collected from camu-camu restingas plantations could influence their rooting capacity, influencing the effectiveness at the propagation bed stage. Here, we test propagation systems (enclosure and outdoor) and types of irrigation (watering daily with without basal reservoir and weekly watering with basal reservoir) and their combined effect on rooting of camu-camu, collected at different phenological stages (flowering and fruiting). An outdoor propagation system and weekly irrigation with basal reservoir was sufficient to root camu-camu, and can be recommended for smallholder growers of camu-camu in Peru. Phenological stage had no significant effect on either rooting percentage or camu-camu root growth. Thus, restingas camu-camu cuttings can be collected during the dry season. There was an overall average of 52% of saplings formed (cuttings with presence of roots and shoots) and 69% of cuttings with roots.

Keywords: cuttings, floodplain, *Myrciaria dubia*, Peruvian, propagation, saplings

1. Introduction

Camu-camu (*Myrciaria dubia* (Kunth) McVaugh, family Myrtaceae) is a shrub or small tree native to the Amazonian floodplain. It has aroused scientific and commercial interest due to the high ascorbic acid content of its fruit pulp, which ranges from 0.8 to 6.1 g per 100 g. The fruit is mainly commercialized in Europe, Japan and the United States, as a source of vitamin C, and is also used in cosmetic and pharmacological products (Yuyama, 2011).

In Peru, camu-camu is usually planted by smallholders on old river levee fragments of white-water rivers called "restingas" (Halme & Bodmer, 2007; Penn, 2006), or called "high-várzea" in Brazil (Wittmann, Junk, & Piedade, 2004; Junk, Onhly, Piedade, & Soares, 2000). Restingas are defined by annual inundations less than 3 m depth, corresponding to a mean flooded period of less than 50 days per year. In contrast, wild camu-camu grows in areas with a mean flood level of 7.5 m; corresponding to a mean flood duration of 230 days per year. In general, restingas, are composed of nutrient-rich Entisols, primarily erosion products from the Andes (Embrapa, 2006; Junk et al., 2011; Moreira & Fageria, 2009). They are used by small farmers who live along the margin of the rivers to cultivate species with short growing periods, such as manioc and maize. Due to the ecological characteristics of the restingas, the Peruvian government promotes camu-camu as a flood-resistant crop, an ideal fruit tree to associate with temporary crops, and as an alternative crop to increase income of wetlands-inhabiting smallholders (Pinedo, 2011; Penn, 2006).

In 1996, Peru began to promote commercial camu-camu cultivation, using seeds collected from wild plants. This initiative resulted in high phenotypic variability within plantations (Penn, 2006). Currently, camu-camu plantation plants are propagated through seeds, making it difficult to stabilize morphological characteristics of commercial interest (Penn, 2006).

As camu-camu is a cross-pollinated plant species, high genetic and phenotypic variability exists within populations (Yuyama & Valente, 2011). However, Yuyama and Valente (2011) indicated that clonal propagation by cutting and clonal selection could enhance genetic gains and provide a means to stabilize morphological characteristics of interest in the shortest possible time.

A greenhouse environment with misting-based irrigation is necessary to achieve 70% to 90% rooting in camu-camu cuttings (Santana, 1998; Yuyama & Valente, 2011). However, the construction of propagation beds demanded by that system can be economically-inviable for camu-camu-growing smallholders in Peru (Pinedo, 2011), who have an average monthly income of 18 to 80 US dollars and for whom there exists a range of additional mitigating factors, whose extent depends on social and environmental circumstances. Such factors include transport costs, drops in product prices by 10% or 20%, and loss of the crops to flooding of up to 25% (Labarta, White, Leguia, Guzman, & Soto, 2007).

The economic realities of camu-camu smallholder growers have driven a search for low-cost techniques for camu-camu propagation and cultivation. In this context, Mathews and Yuyama (2010) report 50%–rooting in camu-camu, when used outdoor propagators and manual irrigation with 20 l of water per square meter per day; dispensing with the need for greenhouses and misting-based irrigation. In Costa Rica, propagators called *propagator non-mist* have been developed. These consist of a wooden frame enclosed in clear polyethylene without irrigation and with a water-tight base and filled with water to a depth of 10 cm (Leakey et al., 1990). In Peru, under this technique, camu-camu shows up to 73% rooting (Abanto et al., 2014, Pinedo, 2011). Although both propagation systems (outdoor and closed) show good results with camu-camu, to assess their suitability as a low-cost propagation method for rural Peruvian smallholders it is necessary to evaluate the in-tandem the effects of daily irrigation and water retention in rooting-beds.

Flooding plays an important role in the physiology (Barrera, Castro, Hernández, Fernández-Trujillo, & Martinez, 2012) and fruit production of camu-camu (Pinedo, 2011). Fruit yield in wild camu-camu varies between years (Peters & Vasquez, 1986/1987), and often occurs only every two years (Yuyama, 2011). On the other hand, because the level of flooding for restingas camu-camu is lower than in wild camu-camu, the harvest period can be longer.

Furthermore, depending on the duration of the dry season, camu-camu usually flowers between August and September, and harvest may occur from December to February (Inga, Pinedo, Delgado, Linares, & Mejia, 2001; Pinedo, 2011). The physiological state of the camu-camu may be different in the flowering and harvest phases, which may influence rooting capacity (Hartmann, Kester, Daviers, & Geneve, 1997). The reproductive phase of camu-camu is during the non-flooded season, and cuttings can be harvested under these conditions. For instance, Mathews and Yuyama (2010) found 55% rooting success in camu-camu cuttings collected in a non-flooded season. But, a year later, in the same plantation - but when flooded - Mathews, Bardales Vasquez and Pinedo (2009) obtained 21% rooting success.

In the current study, the effect of propagation condition and irrigation type on the rooting of camu-camu was evaluated at both flowering and fruiting stages. The study aimed at producing a lower cost technique (without automatic timers and greenhouse system) for camu-camu-growing smallholders.

2. Methods

2.1 Location of the Study and Climatic Data

A rooting study was conducted in the township of Manaquiri, state of Amazonas, Brazil (03°25'41"S, 60°27'34"W), between August 2013 and February 2014. The region has a humid equatorial climate, with a mild short dry season (July-Sep, rainfall of 50-100 mm per month), and the dry-wet transition month (Oct). The wet season extends from May to Nov (200-300 mm monthly); and the mean daily temperature is approx. 27 °C (Fisch, Marenco, & Nobre, 1999).

2.2 Experimental Design and Methods

Treatments were arranged in a split-plot design with three replications. Factors were: propagation system (enclosure and outdoor) in the main plot, and irrigation type (watering daily without basal reservoir and watering weekly with basal reservoir) in the split-plot. Each sample unit had 40 cuttings. These factors were analyzed twice through two similar and independent experiments. In both assays, the plants were sampled randomly. In the first trial, the cuttings were collected from flowering plants (Aug/2013), while in the second trial they were collected from plants bearing fruits (Dec/2013). The cuttings collection period was chosen to coincide with the dry season, which generally occurs between August and December.

Cuttings were harvested from 20 seven year-old plants growing on never-flooded land (terra firme). The cuttings came from basal sprouts (off-shoots). These were cut back to 25 cm lengths 8-10 mm in diameter. Cuttings were collected in the early morning. The basal cut of the cuttings was straight. The base of the cuttings were immersed in indole-3yl-butyric acid solutions (200 ppm) to a depth of about 5 cm for 12 hours after cutting and before being inserted into the rooting medium.

Six wooden on-ground beds (20 cm × 100 cm × 30 cm) were constructed. These were divided into two compartments of equal size. In the bottom of one of the two compartments, a wooden container (15 cm × 95 cm) 5 cm in depth and 15 cm × 95 cm in width acted as post-irrigation reservoir. Finally, three of the beds were enclosed with a frame of transparent polyethylene. The experiment had 60% shading.

Fresh sawdust from a mixture of trees was used as substrate in all treatments.

We used two irrigation treatments: compartments without a basal reservoir were irrigated daily to field capacity at the beginning and middle of the day; weekly, compartments with a basal reservoir were irrigated until the reservoir was filled.

The enclosure and daily irrigation systems attempted to simulate the conventional method of camu-camu rooting, which is normally in greenhouse and under controlled irrigation.

2.3 Data Collection and Analysis

After 90 days, we measured the percentage of formed saplings (cuttings with presence of roots and shoots), the percentage of rooting (cuttings with roots), root number per cutting, and length of the longest root. These measurements were made for two replicates (August/2013 and December/2013). In order to adjust the data to a normal distribution standard, the percentage of formed saplings and the rooting was transformed to square root of arcsine "x" (Gotelli & Ellison, 2011).

$$arcsine \sqrt{x} \tag{1}$$

Treatments were analyzed with analysis of variance (ANOVA). The comparison between treatments was via a Tukey test ($p < 0.05$). The comparison between the two independent assays was performed with a T-test ($p < 0.05$).

3. Results

Propagation system and irrigation type showed no significant interaction with percentage of formed saplings, percentage of rooting, root number or root length (Table 1).

Propagation beds of closed and outdoor system no showed significant effect on rooting. Watering daily without basal reservoir and watering weekly with basal reservoir also showed no significant difference in their effect on rooting (Table 1).

The reproductive phases of camu-camu had no significant effect on rooting. Camu-camu can be collected in the flowering and fruiting phases (Table 1).

The study obtained a general average of 52% of formed samplings and 69% of rooted cuttings, regardless of the factors studied (Table 2).

Table 1. Results of the analysis of variance for propagation system and irrigation type in the rooting of camu-camu, when collected from plants in flowering and fruiting states

Source of variation	Formed saplings (%)	Rooting (%)	Root length (cm)	Root number
Cuttings from plant in flowering (August/2013)				
Propagation system	0.006 [ns]	0.000 [ns]	0.601 [ns]	12.250 [ns]
error (plot)	0.061	0.066	0.291	6.750
Type of Irrigation	0.000 [ns]	0.001 [ns]	0.391 [ns]	6.250 [ns]
Irrigations x Propagation system	0.032 [ns]	0.060 [ns]	3.516 [ns]	1.000 [ns]
error (split-plot)	0.025	0.029	1.193	6.125
Cuttings from plant in fruiting (December/2013)				
Propagation system	0.002 [ns]	0.012 [ns]	0.106 [ns]	7.563 [ns]
error (plot)	0.003	0.005	0.204	5.396
Type of Irrigation	0.009 [ns]	0.139 *	0.526 [ns]	1.563 [ns]
Irrigationx Propagation system	0.004 [ns]	0.000 [ns]	0.276 [ns]	10.563 [ns]
error (split-plot)	0.016	0.018	1.694	4.063

Note. [ns] = not significant; * = Significant ($p < 0.05$). The data are the F values of ANOVA.

Table 2. Average well-developed saplings, rooting, root length and number of roots of camu-camu grown with different propagation systems, irrigation type, and reproductive phases

	Formed saplings (%)	Rooting (%)	Root length (cm)	Root number
Reproductive stage [a]				
Flowering	41 [ns]	57 [ns]	4.3 [ns]	5.8 [ns]
Fruiting	43	61	3.8	4.8
Propagation system [b]				
Closed	45 [ns]	63 [ns]	3.9 [ns]	5.8 [ns]
Outdoor	38	55	4.2	4.8
Type of Irrigation [c]				
daily without basal reservoir	42 [ns]	58 [ns]	4.0 [ns]	4.5 [ns]
weekly with basal reservoir	41	60	4.1	6.1
Mean total [d]	42±14	60±16	4.0±1	5.3±2.4

Note. [ns] = not significant. Comparison tests = a. For T-test ($p < 0.05$); b e c. For ANOVA. d. mean ± standard deviation.

4. Discussion

Both cutting types propagated under closed and outdoor systems exhibited statistically similar rooting frequencies. Closed systems propagation beds of are more commonly used for cuttings which are less resistant to hydric stress, such as those with softwood or herbaceous stems. By contrast, outdoor systems propagation beds are generally used for rooting of hardwood stem cuttings, such as those used in our study (Hartmann, Kester, Daviers, & Geneve, 1997).

In general, the percentage of rooted cuttings achieved in this study (Table 2) is similar to that reported by Mathews and Yuyama (2010) (55%) using a system with outdoor propagation beds and daily irrigation. However, the agronomic advantages of our method include: (i) the use of offshoot-derived cuttings (ii) cutting smaller than 10 mm in diameter. In camu-camu offshoots of this size and origin are abundant in plants in the third and seventh year of planting. This may make them preferable to the method of Mathews and Yuyama (2010) who used larger diameter branches (2.5 to 3.5 cm) derived from 10 year-old trees. In consequence, the type of cutting used in the current study offers the possibility of selecting and propagating plants more quickly than those used by Mathews and Yuyama (2010). In addition, the use of thick cutting requires a drastic pruning of camu-camu main branches.

The percentage of rooting in the present study was lower than that reported by Yuyama and Valente (2011) [52% versus 90% rooting], who also used propagation system enclosure and daily irrigation. However, Yuyama and Valente (2011) did not clarify whether the rooting percentage consisted of cuttings with both roots and shoots. This is important because in camu-camu approximately10% of rooted cuttings fail to develop shoots (Mathews et al., 2009; Santana, 1998).

Using propagation beds in outdoor environments plus daily irrigation, Mathews and Yuyama (2010) obtained higher mean numbers of roots (12±2.1 roots) and greater mean root length (11±1.4 cm) per camu-camu cuttings that in our study (4±1 roots and 5.3±2.4 cm root length). Probably, the use of humus derived from sawdust by these authors had a positive effect on the development of the roots, contrary to our study where fresh sawdust was used. However, camu-camu forms roots regardless of the substrate fertility, and roots improve its development in the nursery (Santana, 1998).

It is likely that the use of off-shoots has facilitated rooting (Yuyama & Valente, 2011), regardless of the reproductive phase in camu-camu. Certainly, basal shoots from young branches root prolifically in several forest species such as *Cordia alliodora* (Ruiz & Pav.) Oken, *Vochysia hondurensis* Sprague, *Albizia guachapale* (Kunth) Dug., *Gmelina arborea* Roxb., and *Eucalyptus deglupta* BL. (Leakey et al., 1990), this latter of Myrtaceae family as *Myrciaria dubia*. Thus, while the cuttings in the study were collected in August and December of the same year, it is likely that camu-camu cuttings can be collected at the beginning and end of the dry season in restingas camu-camu.

5. Conclusion

Outdoor propagation beds with weekly watering and a basal reservoir can be recommended for smallholder camu-camu growers in Peru. This propagation system can replace the closed with controlled irrigation system often used to root camu-camu.

Flowering and fruiting phase not affect the rooting of the camu-camu, and cuttings may be collected from unflooded plantations of restingas camu-camu.

Acknowledgments

The authors would like to thank the graduate program in Botany of the National Institute of Amazonian Research-INPA. This work was financially supported by the CNPq (National Council for Scientific and Technological Development), Brazil. Juan Revilla Cardenas for logistical support. Adrian A. Barnett helped with the English.

References

Abanto, R. A., Chagas, A. E., Sánchez-Choy, J., dos Santos, A. V., Bardales, L. R. M., & Ríos, G. S. (2014). Capacidad de enraizamiento de plantas matrices promisorias de *Myrciaria dubia* (Kunth) Mc Vaugh en cámaras de subirrigación. *Rev. Ceres, 61*(1), 134-140. http://dx.doi.org/10.1590/S0034-737X2014000100018

Barrera, J. A., Castro, S. Y., Hernández, M. S., Fernández-Trujillo, J. P., & Martínez, O. (2012). Maximum Leaf Photosynthetic Light Response for Camu Camu (*Myrciaria dubia* Kunth McVaugh) Plants in Three Growth Stages in the South Colombian Amazonian Region. *Acta Hort., 928*, 193-198.

Empresa Brasileira de Pesquisa Agropecuária. (2006). *Sistema brasileiro de classificação de solos*. Rio de Janeiro: Empresa Brasileira de Pesquisa Agropequaria Solos.

Fisch, G., Marenco, J. A., & Nobre, C. A. (1999). Uma revisão geral sobre o clima da Amazônia. *Acta Amazonica, 28*, 101-126.

Gotelli, N. J., & Ellison, A. M. (2011). *Princípios de estatística em ecologia*. Porto Alegre: ARMED EDITORA S.A.

Halme, K. J., & Bodmer, R. E. (2007). Correspondence between scientific and traditional ecological knowledge: Rain forest classification by the non-indigenous ribereños in Peruvian Amazonia. *Biodivers Conserv, 16*, 1785-1801. http://dx.doi.org/10.1007/s10531-006-9071-4

Hartmann, H. T., Kester, D. E., Daviers, Jr. F. T., & Geneve Robert, L. (1997). *Plant Propagation: Principles and Practice*. New Jersey: Prentice-Hall, Inc.

Inga, H., Pinedo, M., Delgado, C., Linares, C., & Mejia, K. (2001). Fenologia reprodutiva de *Myrciaria dubia*

McVaugh (H.B.K.) camu-camu. *Folia Amazonica, 12*, 1-2.

Junk, W. J., Onhly, J. J., Piedade, M. T. F., & Soares, M. G. M. (2000). *The central Amazon Floodplain: Actual use and options for a sustainable Management.* Leiden: Backhuys. http://dx.doi.org/10.1007/s13157-011-0190-7

Junk, W. J., Piedade, M. T. F., Schöngart, J., Cohn-Haft, M., Adeney, M., & Wittmann, F. (2011). A Classification of major naturally-occurring Amazonian lowland wetlands. *Wetlands, 31*, 623-640.

Labarta, R., White, D., Leguía, E., Guzmán, W., & Soto, J. (2007). La Agricultura en la Amazonia Ribereña del Río Ucayali: Una Zona Productiva pero Poco Rentable? *Acta amazonica, 37*(2), 177-186. http://dx.doi.org/10.1590/S0044-59672007000200002

Leakey, R. R. B., Mesen, J. F., Tchoundjeu, Z., Longman, K. A., Dick, J. McP., ... Muthoka, P. N. (1990). Low-technology techniques for the vegetative Propagation of tropical trees. *Commonw. For. Rev., 69*(3), 247-257.

Mathews, D. J. P., & Yuyama, K. (2010). Cutting lenght of camu-camu with indolebutyric acid for clonal production. *Rev. Bras. Frutic., 32*(2), 522-526.

Mathews, D. J. P., Bardales, R., Vásquez, D. C., & Pinedo, P. M. (2009). Enraizamento de estacas de camu-camu com diferentes diâmetros sob diferentes concentrações do ácido indolbutírico. In A. E. Guamão (Ed.), *I Workshop Agricultura No Trópico Úmido* (pp. 13-15). Manaus, Brazil: Post Graduate Program in Agriculture in the Humid Tropics, National Institute of Amazonian Research.

Moreira, A., & Fageria, N. K. (2009). Soil Chemical Attributes of Amazonas State, Brazil. *Communications in Soil Science and Plant Analysis, 40*, 2912-2925. http://dx.doi.org/10.1080/00103620903175371

Penn, J. W. (2006). The cultivation of camu camu (*Myrciaria dubia*): A tree planting programme in the Peruvian amazon. *Forests, Trees and Livelihoods, 16*, 85-101. http://dx.doi.org/10.1080/14728028.2006.9752547

Peters, C. M., & Vásquez, A. (1986/1987). Estudios ecológicos de camu-camu (Myrciaria dubia). I. Producción de frutos en poblaciones naturales. *Acta Amazonica, 16/17*, 161-174.

Pinedo, P. M. (2011). *Camu-camu (Myrciaria dubia, Myrtaceae). Aportes para su aprovechamiento sostenible en la Amazonía peruana.* Iquitos: Research Institute of the Peruvian Amazon/IIAP

Santana, S. C. (1998). Propagação de camu-camu por estaquia. *Biotecnología Ciência & Desenvolvimento, 29*. http://dx.doi.org/10.1590/S0100-29452011000200001

Wittmann, F., JunK, J. W., & Piedade, M. T. M. (2004). The várzea forests in Amazonia: Flooding and the highly dynamic geomorphology interact with natural forest succession. *Forest Ecology and Management, 196*, 199-212. http://dx.doi.org/10.1016/j.foreco.2004.02.060

Yuyama, K. (2011). A cultura de camu-camu no Brasil. *Rev. Bras. Frutic., 33*, 335-690. http://dx.doi.org/10.1590/S0100-29452011000200001

Yuyama, K., & Valente, J. P. (2011). *Camu-camu (Myrciaria dubia (Kunth) Mac Vaugh).* Curitiba: CRV.

19

Efficacy of Non-Chemical Alternatives on Blue Mold of Apple under Controlled Cold Storage Conditions

Ziad Barakat Al-Rawashdeh[1], Ezz Al-Dein Muhammed Al-Ramamneh[1] & Muwaffaq Ramadan Karajeh[2]

[1] Department of Agricultural Sciences, Al-Shoubak College, Al-Balqa' Applied University, Al-Shoubak, Jordan

[2] Department of Plant Protection & IPM, Faculty of Agriculture, Mu'tah University, Karak, Jordan

Correspondence: Ziad Barakat Al-Rawashdeh, Department of Agricultural Sciences, Al-Shoubak College, Al-Balqa Applied University, Al-Shoubak, Jordan.

Abstract

Fruits of three apple cultivars; Golden Delicious, Granny Smith and Fuji were inoculated with blue mold, *Penicillium expansum*, and kept under cold storage conditions for 75 days after treatment with non-chemical alternatives and the fungicide Topsin as a standard. This aimed at providing new postharvest methods for the control of blue mold incidence and severity in apple fruits. Granny Smith showed lower sensitivity to blue mold disease than Golden Delicious but Fuji was the most sensitive to the disease in relation to fruit firmness and total soluble solid content (TSS). Dipping apple fruits in yeast (*Saccharomyces cerevisiae*) solution at a concentration of 1 g L^{-1} or in hot water at 50 °C for 1 or 5 min., or exposing them to microwave for 10 s resulted in an effective control of blue mold disease under controlled cold storage conditions. Calcium nitrate at 1 or 8 g L^{-1} did not result in an effective control of blue mold but increased the storability of apple fruits. Therefore, a combination of two or more of the alternatives may provide a long-lasting effective control of post-harvest blue mold affecting apple fruits.

Keywords: hot water, *Malus domestica*, microwave, post-harvest disease, yeast

1. Introduction

Pome fruits including apple, pear and quince are cultivated in different regions in Jordan. Domestic apple (*Malus domestica* Borkh.) is the main pome fruits for commercial production and found in about 5% of the total area cultivated with fruit trees in Jordan (Statistical Year Book, 2006).

Best quality of apple fruits is the demand for all consumers (Zeebroeck et al., 2007). Avoiding or delaying of post-harvest disorders and decay could be achieved with pre-harvest measures e.g. the improvement of tree nutritional status (Moore et al., 2006). Physiological disorders including superficial scald, bitter bit and physiological spot were reported to reduce the quality of marketable yield of apple fruits (Moore et al., 2006). Apple fruits are also vulnerable to a wide range of post-harvest pathogens (Penrose et al., 1989). Blue mold caused by (*Penicillium expansum* Link) is a major pathogen that leads to obvious yield loss during post-harvest handling and storage of apples and pears (Schirra et al., 2009; Al-Rawashdeh & Karajeh, 2014; Penrose et al., 1989). Combination of some post-harvest treatments with cold storage was efficient in reducing the incidence and severity of apple blue mold (Al-Rawashdeh, 2013). The efficacy of post-harvest fungicides for the control of blue mold was investigated in several apple cultivars; Iprodione, imazalil, prochloraz were effective for the control of blue mold in cv. Granny Smith under suitable storage conditions (Penrose et al., 1989). Particularly, Carbendazime was effective in controlling blue mould in cultivars; Golden Delicious and Royal Gala (Al-Rawashdeh & Karajeh, 2014). Due to the increasing global awareness of environmental hazards of pesticides and the development of resistant strains of the pathogen to fungicides frequently in use, there is a strong debate about the use of fungicides to control post-harvest diseases (Spotts & Cervantes, 1986; Delp, 1988; Holmes & Eckert, 1999). This led to a strict regulation imposed by many food national and international agencies and high residues in fruits are not anymore acceptable beyond fungicide tolerance limit. Chemical alternatives that are relatively safer to the environment and consumers were recently used to control disease in field and during storage. These treatments include the use of edible wax coatings, hot water, mineral salts, bread yeast (*Saccharomyces cerevisiae*), growth regulators and exposure to microwave heat (Schirra et al., 2009; Han,

1990; Al-Rawashdeh, 2013; Hernández-Montiel et al., 2012; Anbukkarasi, 2013; Al-Rawashdeh & Karajeh, 2014).

Fruits of apple cultivars used in this study were collected from the surrounding of Al-Shouback city. Al-Shouback city, located at 1300 m above sea level in the southern part of Jordan provides a suitable place for apple plantation. Commercial varieties which have 1000-1600 hours chilling requirement (Westwood, 1993) thrive in this environment.

Therefore, study was set up to evaluate the effectiveness of treating apple fruits with yeast, hot water, microwave heat exposure, and calcium nitrate on the incidence and severity of blue mold in Golden Delicious, Granny Smith and Fuji cultivars. The fungicide Topsin was used as a standard to compare with the non-chemical treatments.

2. Materials and Methods

2.1 Fruit Collection

Fruits of apple cultivars; Golden Delicious, Granny Smith and Fuji were collected at commercial maturity from a local orchard in Ma'an district, south of Jordan. Fruits were cold-stored at 2 °C until used in later experiments.

2.2 Fruits Inoculation and Post-Harvest Handling

Spores of *P. expansum* were isolated from Apple fruits that were visualized by having the symptoms of blue mold as described by Al-Rawashdeh and Karajeh (2014). Collected Fruits were initially rinsed with distilled water for a few minutes, then wounded with sterilized sand and finally inoculated through dipping into a spore suspension of *P. expansum* at a concentration of 38,000 spores/ml. Fruits were then dipped into either the fungicide Topsin at (2 or 0.5 g L^{-1}), yeast (*S. cerevisiae*) at (1 or 10 g L^{-1}), calcium nitrate at (1 or 8 g L^{-1}), hot water at 50 °C for (5 or 1 min), or tap water (negative control). Fruits were also placed in a microwave oven (2450 Mhz, R-480J, Thailand) for 10 or 30 s. Spores-inoculated fruits that received no further treatments were considered as (positive control). Three replicates were used for each treatment and each replicate consisted of three fruits placed in a ventilated, 20x30 cm^2 plastic container, and the experiment was replicated twice. Fruits were assessed for disease severity and incidence after 75 days of cold storage at 2 °C and 90 % relative humidity. Disease severity represents the percentage of fruits showing disease lesions and was rated using a scale of 0-5 (0: no lesions, 1: 1-10, 2: 11-25, 3: 26-50, 4: 51-75, 5: 76-100%). Disease incidence was scored based on the number of rot spots that developed on each fruit.

2.3 Evaluation of Total Soluble Solids and Firmness

Fruit firmness and its content of total soluble solids were determined after harvest and at the end of the experiment (after 75 days of cold storage). Fruit firmness was assessed using penetrometer. The content of total soluble solids was measured with a refracto-meter using homogenized juice obtained from sampled fruits.

2.4 Statistical Analyses

Data were analyzed by Generalized Linear Model (GLM) procedure using SPSS (Software version 11.5; SPSS Inc., Chicago, USA). Mean separation of treatments was performed using least significance difference (LSD) at the 0.05 probability level.

3. Results

3.1 Firmness and Total Soluble Solids

At harvest time, there were significant differences among the three apple cultivars in their fruit firmness parameter which was obviously higher for Granny Smith and Fuji than that for Golden Delicious (Table 1).

Table 1. Maturity indices as measured by fruit firmness and content of total soluble solids in apple cultivars; Golden Delicious, Granny Smith and Fuji at harvest and after treatment at the end of the experiment. The mean of all treatments was considered for each cultivar at the end of the experiment

Cultivar	Firmness (lb/cm^2)			Total Soluble Solids (TSS) (%)		
	At harvest	After treatment	% decrease in firmness[1]	At harvest	After treatment	% change in TSS[2]
Golden Delicious	10.4[3] b[4]A[5]	8.9 bB	14.4	16.1 aA	16.9 bA	+ 4.9
Granny Smith	12.4 aA	11.0 aB	11.3	11.8 bB	13.2 aA	+ 11.8
Fuji	12.7 aA	9.6 abB	24.4	15.7 aA	15.2 abA	- 3.2

Note. [1] Percentage decrease in firmness of studied cultivars calculated as [(firmness after treatment − firmness at harvest) / firmness at harvest × 100];

[2] Percentage change in TSS of studied cultivars calculated as [(TSS after treatment − TSS at harvest) / TSS at harvest × 100];

[3] Means of three replicates (containers) were used for each treatment and each replicate consisted of three fruits per a container;

[4] Means within columns followed with the same small letters are not significantly different using LSD test at 0.05 probability level;

[5] Means within rows followed with the same capital letters are not significantly different using LSD test at 0.05 probability level.

Their fruit firmness was significantly reduced after treatment (Table 1) mainly by all treatments used except calcium nitrate treatment at 1 and 8 g/l (Table 2). Total soluble solid content of apple fruit was significantly lower for Granny Smith than the other two cultivars (Table 1) and significantly increased in fruits that were subjected to hot water, Topsin and calcium nitrate treatments (Table 2).

Table 2. Maturity indices as measured by fruit firmness and content of total soluble solids, averaged across the three cultivars, after inoculation with the fungus *P. expansum* and subjecting to different treatments

Treatments	Firmness (lb/cm^2)	Total soluble solids (%)
0.5 g L^{-1} Topsin	9.0* c**	16.5 a
2 g L^{-1} Topsin	10.1 bc	15.5 ab
1 g L^{-1} Yeast	10.0 c	14.3 bc
10 g L^{-1} Yeast	9.0 c	13.7 bc
1min. Hot water	10.0 c	16.0 ab
5min. Hot water	10.0 c	16.3 a
10s Microwave	10.3 abc	14.3 bc
30s Microwave	9.3 c	12.3 c
1 g L^{-1} Calcium nitrate	11.5 ab	15.0 ab
8 g L^{-1} Calcium nitrate	11.7 a	15.0 ab
Inoculated control	10.3 abc	15.3 ab

Note. * Means of three replicates (containers) were used for each treatment and each replicate consisted of three fruits per a container.

** Means within columns followed by the same letters are not significantly different using LSD test at 0.05 probability level.

3.2 Blue Mold Incidence and Severity

Generally, the disease incidence of the blue mold, as the number of decay spots per a fruit, was significantly higher for cv. Fuji then for Golden Delicious and Granny Smith as indicated by the average of means for the different treatments and control (Table 3). Among treatments, 30s microwave exposure, and 1 or 8 g L^{-1} calcium nitrate dipping gave significantly higher blue mold incidence than the control while disease incidence was significantly lower after treating mold-inoculated apple fruits of the three cultivars by 0.5 and 2 g L^{-1} Topsin, 1 g

L^{-1} yeast, 1 and 5 min. hot water, and 10s microwave treatments (Table 3).

Table 3. Effect of post-harvest treatments on disease incidence of blue mold in apple cultivars Golden Delicious, Granny Smith and Fuji after 8 weeks of cold storage. Disease incidence was scored based on the number of rot spots that developed on each fruit

Treatment	Cultivar			Means
	Golden Delicious	Granny Smith	Fuji	
0.5 g L^{-1} Topsin	1.3* bc**	1.6 bcd	1.3 c	1.4 d
2 g L^{-1} Topsin	2.3 abc	3.0 b	2.0 c	2.4 c
1 g L^{-1} Yeast	2.3 abc	2.0 bcd	3.3 bc	2.6 c
10 g L^{-1} Yeast	2.3 abc	1.3 bcd	6.0 a	3.2 b
1min. Hot water	3.3 abc	0.6 cd	2.3 bc	2.1 cd
5min. Hot water	2.3 abc	1.6 bcd	2.3 bc	2.1 c
10s Microwave	1.7 bc	0.3 d	2.0 c	1.3 d
30s Microwave	4.7 a	1.3 bcd	6.0 a	4.0 a
1 g L^{-1} Calcium nitrate	4.7 a	2.6 bc	5.0 ab	4.1 a
8 g L^{-1} Calcium nitrate	3.7 ab	5.3 a	2.0 c	3.9 a
Inoculated control	1.7 bc	3.3 b	5.0 ab	3.3 b

Note. * Means of three replicates (containers) were used for each treatment and each replicate consisted of three fruits per a container;

** Means within columns followed by the same letters are not significantly different using LSD test at 0.05 probability level.

Generally, the severity % of blue mold was the highest for cv. Fuji followed by Golden Delicious and the least in Granny Smith as it is indicated by the average of means for the different treatments and control (Table 4). Lower mold severity % was accompanied with all treatments except 30s microwave exposure, and dipping in 1 or 8 g L^{-1} calcium nitrate that did not significantly differ in their severity % from that of the control. The use of 1 or 10 g L^{-1} yeast, 1 or 5 min. hot water, 10s microwave have resulted in a reduction in mold severity % similar to the use of Topsin fungicide at 0.5 and 2 g L^{-1} on the three apple cultivars (Table 4).

Table 4. Effect of post-harvest treatments on disease severity% of blue mold in apple cultivars; Golden Delicious, Granny Smith and Fuji

Treatment	Cultivar			Means
	Golden Delicious	Granny Smith	Fuji	
0.5 g L^{-1} Topsin	13.3* c**	16.7 bc	36.7 bc	22.2 c
2 g L^{-1} Topsin	20.0 bc	36.7 abc	25.0 c	27.2 bc
1 g L^{-1} Yeast	26.7 bc	16.7 bc	48.3 bc	30.6 b
10 g L^{-1} Yeast	30.0 bc	18.3 bc	68.3 ab	38.9 b
1min. Hot water	51.7 ab	6.7 c	56.7 abc	38.3 b
5min. Hot water	21.7 bc	20.0 bc	31.7 bc	24.4 bc
10s Microwave	30.0 bc	6.7 c	30.0 bc	22.2 c
30s Microwave	65.0 a	8.3 c	88.3 a	53.9 a
1 g L^{-1} Calcium nitrate	80.0 a	53.3 ab	81.7 a	71.7 a
8 g L^{-1} Calcium nitrate	65.0 a	66.7 a	23.3 c	60.6 a
Inoculated control	53.3 ab	70.0 a	63.3 ab	62.2 a

Note. * Means of three replicates (containers) were used for each treatment and each replicate consisted of three fruits per a container;

** Means within columns followed by the same letters are not significantly different using LSD test at 0.05 probability level.

4. Discussion

Among the tested apple cultivars, Granny Smith had exhibited lower sensitivity to blue mold disease than Golden Delicious whereas Fuji was the most sensitive to the disease. The sensitivity to the blue mold was negatively correlated with fruit firmness. The loss in fruit firmness in Fuji was approximately twice that reported for Golden Delicious and Granny Smith in response to the various treatments under cold storage conditions (Table 4). Fruit firmness was a cultivar-dependent factor (Nour et al., 2010; Jan et al., 2012) and obviously increased after treating the fruits with 1 and 8 g L^{-1} calcium nitrate. Total soluble solid content of apple fruit was relatively lower for Granny Smith than Golden Delicious and Fuji cultivars and significantly increased when the fruit was subjected to hot water, Topsin and calcium nitrate treatments.

The control of post-harvest diseases is generally based on the application of chemical fungicides. With the increasing demand of consumer for chemical residue-free products and the development of resistant pathogenic strains, the producers have extended their interest for finding effective alternative solutions (Janisiewicz & Korsten, 2002; Kwasiborski et al., 2014). Similar to Topsin, dipping apple fruit in yeast solution at concentration of 1 g L^{-1} or in hot water at 50 °C for 1 or 5 min., or exposing them to microwave for 10s were effective in controlling the blue mold. Some treatments e.g. 5 min. hot water and 30 s exposure to microwave can decrease fruit firmness thus increasing the susceptibility of fruit to new infection of *Penicillium expansum* and this situation was cultivar dependant. Water dipping of pear fruit cv. Coscia at 50 °C for 1–2 min resulted in 27–31% less blue mould decay, while the 4-min dip yielded 90% less decay than in control fruit (Schirra et al., 2009). Gholamnejad et al. (2009) showed that all tested yeast isolates inhibited growth of *P. expansum* used for *in vitro* assay and that isolate 69 of *S. cervisiae* was the most effective in reducing the decay of apple fruits indicating its potential use as a valuable biological agent for the control of apple blue mold. Mexican lime (*Citrus aurantiifolia*) fruits under cold storage were effectively protected from blue mold by the yeast, *Debaryomyces hansenii* (Hernández-Montiel et al., 2012).

The use of calcium nitrate at 1 g and 8 L^{-1} concentration did not result in sufficient control of the blue mold despite its effects in increasing the storability of apple fruit through increasing its firmness and total soluble solid content. Thus, a combination of calcium nitrate with one or more of the effective treatments may be useful for reducing the onset and severity of the blue mold as well as for extending the storability of apple fruit (Al-Rawashdeh, 2013). The use of sodium bicarbonate was a useful approach to improve the efficacy of yeast antagonists used for post-harvest disease control (Droby et al., 2003). Pre- and/or post-harvest treatments and cold storage are needed to delay senescence through extending the storability of apple fruit and to control post-harvest decay (Schirra et al., 2009).

5. Conclusion

Studied apple cultivars recorded variation in their susceptibility to blue mold with Fuji being the most sensitive to the disease. Treatment of apple cultivars with Topsin (0.5 or 2 g L^{-1}), yeast solution (1 g L^{-1}), 50 °C hot water (1 or 5 min), or microwave exposure (10 s.) were effective means for reducing the severity of blue mold disease.

References

Al-Rawashdeh, Z. (2013). Ability of mineral salts and some fungicides to suppress apple powdery mildew caused by the fungus *Podosphaera leucotricha*. *Asian Journal of Plant Pathology, 7*, 54-59. http://dx.doi.org/10.3923/ajppaj.2013.54.59

Al-Rawashdeh, Z. B., & Karajah, M. R. (2014). Post-harvest control of apple blue mold under cold storage conditions. *American Journal of Agricultural and Biological Sciences, 9*, 167-173. http://dx.doi.org/10.3844/ajabssp.2014.167.173

Anbukkarasi, V., Paramaguru, P., Pugalendhi, L., Ragupathi, N., & Jeyakumar, P. (2013). Studies on pre and post-harvest treatments for extending shelf life in onion – A review. *Agricultural Reviews, 34*, 256-268. http://dx.doi.org/10.5958/J.0976-0741.34.4.011

Delp, C. J. (1988). *Fungicide resistance in North America.* American Phytopathological Society Press, St. Paul, MN, USA.

Droby, S., Wisniewski, M. E., El-Ghaouth, A., & Wilson, C. L. (2003). Influence of Food additives on the control of postharvest rots of apple and peach and efficacy of the yeast-based biocontrol product Aspire. *Postharvest Biology and Technology, 27*, 127-135. Retrieved from http://www.ars.usda.gov/SP2UserFiles/person/6167/Foodadditives and biological control.pdf

Gholamnejad, J., Etebarian, H. R., Roustaee, A., & Sahebani, N. A. (2009). Biological control of apples blue mold by isolates of *Saccharomyces cerevisiae*. *Journal of Plant Protection Research, 49*, 270-275. http://dx.doi.org/10.2478/v10045-009-0042-0

Han, J. S. (1990). Use of antitranspirant epidermal coatings for plant protection in China. *Plant Disease, 74*, 263-266. http://dx.doi.org/10.1094/PD-74-0263

Hernández-Montiel, L. G., Holguín-Peňa, R. J., Larralde-Corona, C. P., Zulueta-Rodríguez, R., Rueda-Puente E., & Moreno-Legorreta, M. (2012). Effect of inoculums size of yeast *Debaryomyces hansenii* to control *Penicillium italicum* on Mexican lime (*Citrus aurantiifolia*) during storage. *Journal of Food, 10*, 235-242. http://dx.doi.org/dx.doi.org/10.1080/19476337.2011.633350

Holmes, G. J., & Eckert, J. W. (1999). Sensitivity of Penicillium digitatum and P. italicum to postharvest citrus fungicides in California. *Phytopathology, 89*, 716-721. http://dx.doi.org/10.1094/PHYTO.1999.89.9.716

Jan, I., Rab, A., Sajid, M., & Ali, A. (2012). Response of apple cultivars to different storage durations. *Sarhad Journal of Agriculture, 28*, 219-225. Retrieved from http://www.aup.edu.pk/sj_pdf/RESPONSE%20OF%20APPLE%20CULTIVARS%20TO%20DIFFERENT%20STORAGE%20-184-11-Horticulture.pdf

Janisiewicz, W. J., & Korsten, L. (2002). Biological control of postharvest diseases of fruits. *Annual Review of Phytopathology, 40*, 411-441. http://dx.doi.org/10.1146/annurev.phyto.40.120401.130158

Kwasiborski, A., Bajji, M., Renaut, J., Delaplace, P., & Jijakli, M. H. (2014). Identification of Metabolic Pathways Expressed by *Pichia anomala* Kh6 in the Presence of the Pathogen *Botrytis cinerea* on Apple: New Possible Targets for Biocontrol Improvement. *PLoS ONE, 9*(3). http://dx.doi.org/10.1371/journal.pone.0091434

Moore, U., Karp, K., Põldma, P., Asafova, L., & Starast, M. (2006). Post-harvest disorders and mineral composition of apple fruits as affected by pre-harvest calcium treatments. *Acta Agriculturae Scandinavica Section B- Soil and Plant Science, 56*, 179-185. http://dx.doi.org/10.1080/09064710500303175

Nour, V., Trandafir, I., & Ionica, M. E. (2010). Compositional characteristics of fruits of several apple (*Malus domestica*) cultivar. *Notulae Botanicae Horti Agrobotanici Cluj-Napoca, 38*, 228-233. Retrieved from http://www.notulaebotanicae.ro/index.php/nbha/article/view/4762

Penrose, L. J., Koffmann, W., & Ridings, H. I. (1989). Factors affecting the efficacy of post-harvest fungicide application for the control of blue mold (*Penicillium expansum*) in stored apple. *Plant Pathology, 38*, 421-426. http://dx.doi.org/10.1111/j.1365-3059.1989.tb02162.x

Schirra, M., D'Aquino, S., Migheli, Q., Pirisi, F. M., & Angioni, A. (2009). Influence of post-harvest treatments with fludioxonil and soy lecithin co-application in controlling blue and grey mould and fludioxonil residues in Coscia pears. *Food Additives and Contaminants, 26*, 68-72. http://dx.doi.org/10.1080/02652030802348080

Spotts, R. A., & Cervantes, L. A. (1986). Populations, pathogenicity and benomyl resistance of Botrytis spp., Penicillium spp. and Mucor piriformis in packinghouses. *Plant Disease, 70*, 106-108. Retrieved from http://www.apsnet.org/publications/plantdisease/backissues/Documents/1986Articles/PlantDisease70n02_106.pdf

Statistical Year Book. (2006). Department of Public Statistics, The Hashemite Kingdome of Jordan.

Westwood, M. N. (1993). *Temperate Zone Pomology: Physiology and Culture* (p. 523). Oregon, USA: Timber Press, Inc.

Zeebroeck, M. V., Linden, V. V., Ramon, H., Baerdemaeker, J. D., Nicolai, B. M., & Tijskens, E. (2007). Impact damage of apples during transport and handling. *Postharvest Biology and Technology, 45*, 157-167. http://dx.doi.org/10.1016/j.postharvbio.2007.01.015

20

Immediate and Transgenerational Regulation of Plant Stress Response through DNA Methylation

A. R. Khan[1], S. M. Shah[1] & M. Irshad[1]

[1] Department of Environmental Sciences, COMSATS Institute of Information Technology, Abbottabad, Pakistan

Correspondence: M. Irshad, Department of Environmental Sciences, COMSATS Institute of Information Technology, Abbottabad, Pakistan. E-mail: mirshad@ciit.net.pk

Abstract

Epigenetics refers to the heritable changes in gene activity without altering the DNA sequence. DNA methylation along with other epigenetic mechanisms is involved in the chromatin remodeling. This remodeling, especially in plants, plays an important role in the activation or silencing of specific genes as well as other genomic regions in response to the developmental and environmental clues. Environmental clues, biotic and abiotic stresses trigger the shift in the site specific as well as genome wide DNA methylation patterns which influences the plant response to these situations through gene regulation. Therefore, it is of prime importance to analyze variation in the DNA methylation pattern under stress conditions. This review summarizes the topic of DNA methylation by providing the basic/conceptual knowledge and some cases of DNA methylation shift due to stresses.

Keywords: epigenetics, biotic and abiotic stresses, locus specific DNA methylation, genome-wide DNA methylation

1. Epigenetic Modification in Plants

Chromatin is a complex structure of nucleoproteins, which packs DNA in a highly organized way to fit it into the nucleus of a eukaryotic cell. The nucleosome is the fundamental unit of chromatin which is composed of 147 base pairs of DNA, wrapped around a core of eight histone protein molecules (Grunstein, 1997; Kornberg, 1977). Since the tight packing of DNA in the nucleosome can cause problem of accessibility of DNA by transcription factors and RNA polymerases, the static nature of this complex might not be so desirable. Interestingly, the nucleosome packaging is dynamic in nature and can be subjected to alteration depending upon the environmental or developmental clues (Narlikar et al., 2002), ultimately leading to the regulation of the processes such as transcription, recombination, DNA repair etc. This manipulation of dynamic chromatin provides an additional layer of information, resulting in the modifications of cell/tissue activities through gene regulation causing the shift in various processes like from seed germination to organogenesis, root and shot growth, flowering, embryo formation, as well as response to abiotic and biotic stresses (Roudier et al., 2011). Various mechanisms play an important role in regulation of the chromatin context to control gene expression. Among these mechanisms, epigenetic modifications have caught the interest of many researchers in recent decades.

Epigenetic modifications involves the heritable changes in the gene activities that are mitotically and/or meiotically transmissible without changing the DNA sequence (Holliday, 1994). It is important to mention at this point that the reversible developmental modifications that are involved in the molecular responses (phenotypic plasticity etc.) to environmental changes are not included into epigenetics. They can rather be called as non-heritable chromatin modifications. As supported by many examples, the heritable chromatin modifications (epigenetic modifications) can be classified into mitotically transmissible modifications (reset in the next generation) and meioticallytransgenerational chromatin modifications (inherited/transmitted to the generations) (Cubas et al., 1999; Khan et al., 2013; Lauria et al., 2004; Manning et al., 2006; Zemach et al., 2010a).

Various epigenetic mechanisms like DNA methylation and post-translational covalent histone modifications (e.g. acetylation, ubiquitination, methylation, phosphorylation), non-coding RNAs along with chromatin remodeling enzymes are involved in the chromatin modification (Berger, 2007; Kouzarides, 2007; Rapp & Wendel, 2005). Out of these, DNA methylation is the best understood and (based on the previous information) considered as the most stable form of epigenetic modifications. Since DNA methylation is the main focus of this review, other epigenetic modifications will not be discussed.

2. DNA Methylation

A chemical modification in the genomic DNA, which is caused when a methyl group is attached at a specific nucleotide base, is known as DNA methylation. The nucleotide base, which couldtake part in DNA methylation, is either cytosine or adenine but most studied, is methylation at cytosine. When the methyl group (-CH3) is attached at 5 carbon of cytosine then it is termed as 5-methycytosine and is denoted as 5 mC. It is historically ancient and with few exceptions, it has been reported in all major groups of eukaryotes (plant, animals and fungi) (Chan et al., 2005; Goll & Bestor, 2005; Henderson & Jacobsen, 2007; Klose & Bird, 2006; Law & Jacobsen, 2010).

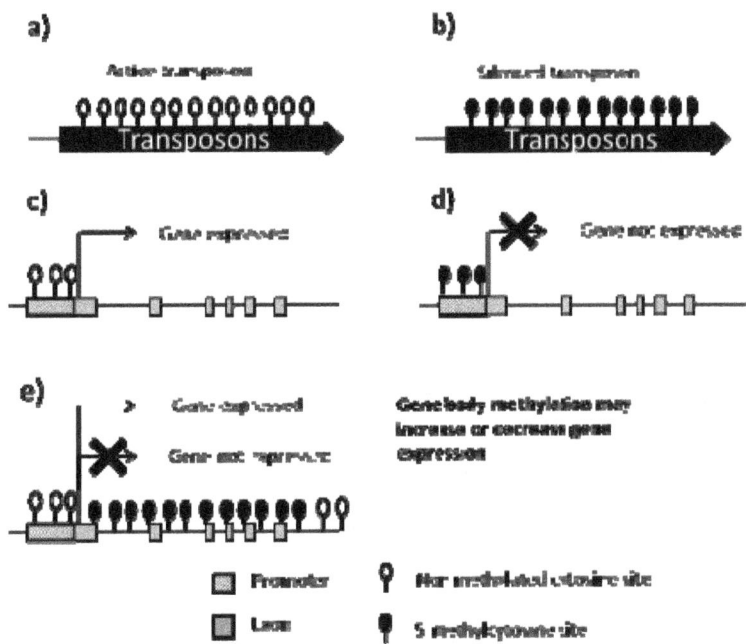

Figure 1. Influence of DNA methylation on gene transposon regulation. a) Transposon in active state due to absence/removal of DNA methylation. b) Transposon in the silence state due to the presence of DNA methylation. c) The gene is expressed in the absence of DNA methylation at the promoter d) DNA methylation at the promoter caused gene silencing. e) Function of gene body methylation is unclear, i.e. depending upon the case study it enhances transcription or it reduces transcription

In plant, different methyltransferase enzymes are responsible for maintenance of DNA methylation. CG methylation is principally maintains by DNA METHYLTRANSFERASE 1 (MET1). MET1 is homologue of mammalian DNA METHYLTRANSFERASE 1 (DNMT1). CHG methylation is maintained by a plant specific CHROMOMETHYASE 3 (CMT3) (Chan et al., 2005). Asymmetric nature of CHH sites needs a complex phenomenon for the maintenance. Plant-specific RNA-dependent DNA Methylation (RdDM) pathway alongwith the *de novo* methyltransferase DOMAINS REARRANGED METHYLTRANSFERASE 2 (DRM2) help to reacquired *de novo* methylation after each replication (Law & Jacobsen, 2010).

5-methylcytosine is mainly characterized in three contexts: CG, CHG and CHH (where H can be replaced by any base other than G) (Law & Jacobsen, 2010). Plants and animals show different patterns of DNA methylation. In plants, it can occur at all three contexts i.e. at CG, CHG and CHH sites, where as in animals it is mostly reported at CG site (Feng et al., 2010; Zemach et al., 2010b). The pattern of DNA methylation in plants is mainly understood through the studies in model plant *Arabidopsis*. The occurrence of 5-methylcitosines is not even when compared on the basis of three contexts. The genome-wide DNA methylation studies on *Arabidopsis* revealed that DNA methylation levels at CG, CHG and CHH are 24%, 6.7% and 1.7% respectively. In eukaryotes, DNA methylation occurs in transposable elements as well as in the genes. The extent of DNA methylation depends upon the region of the genome where it occurs. The studies have shown that DNA methylation heavily occurs in transposons and repetitive elements (in all the contexts: CG, CHG and CHH), and is associated with silencing of these regions (Figure 1), thus playing important role in genome stability (Goll & Bestor, 2005; Henderson & Jacobsen, 2007).

Although, DNA methylation is also associated with gene regulation but the occurrence as well as regulation pattern of DNA methylation in genes depends on the regions within the gene where it occurs (Figure 1). It has negative correlation with the gene expression, if the promoter region is methylated (Zhang et al., 2006). DNA methylation may also occurred within the body of the genes but absent at the start and ends of the gene and provides a bell shaped pattern (known as gene body methylation) (Li et al., 2012; Zemach et al., 2010b). Although this gene body methylation is evolutionarily conserved but the function of DNA methylation within a gene is not yet fully understood (Takuno & Gaut, 2013; Zhang et al., 2010). The review of various studies in this field suggests that it can be positively as well as negatively associated with gene expression. The understanding of various techniques to study these DNA methylation patterns is important.

3. DNA Methylation and Plant Stress Response

DNA methylation, like other epigenetic modifications, is more dynamic than DNA sequence mutations thus they could play essential role in an organism's first response towards changing environment. Therefore studying the variation in the DNA methylation pattern due to variation in environmental condition to regulate the gene expression has become an area of great interest. Grouped according to different stresses, we provide various examples of such studies (especially in crop plants as they are economically and socially important for human survival) (Table 1).

3.1 Salinity

In wheat, shift in the DNA methylationdue to salinity stress, in salinity tolerant cultivar and its salinity sensitive progenitor, has been recently reported (Wang et al., 2014). Salt stress induced shift in DNA methylation in both the promoter and coding regions of some of the 24 selected genes, but only the former were associated with changes in transcript (Wang et al., 2014). Similarly, in another study where two cultivars with different level of salt tolerances were evaluated, dose dependent genome-wide DNA hypomethylation was observed in salt stress conditions (Zhong et al., 2009).

In maize, shift in DNA methylation pattern in response to osmotic and salt stress was studied. It was reported that the osmotic stress-induced methylation of retrotransposons. In addition, salt stress induced DNA methylation which caused down regulation of *zmPP2C* gene expression, which is a negative regulator of the stress response, as well as salinity caused a demethylation that had up regulated the *zmGST* gene expression, which is a positive effecter of the stress response (Tan, 2010).

A set of diverse rice genotypes was analyzed at genome-wide level under salt stress conditionswith the help of MSAP technique. The results revealed a differential methylation pattern in salt stress related gene which lead to the shift of expression of these genes. This differential pattern was also observed in retrotransposons and chromatin modifier genes indicating the involvement of epigenetic markers in stress response (Karan et al., 2012).

Soya bean is an important member of oil seed group. Alteration in DNA methylation profile under salinity stress condition has been observed in soya bean. A significant correlation was observed between shift in DNA methylation pattern under salt stress and gene expression of four genes in this species. The expression profile of one *MYB* (*Myeloblastosis*), one *b-ZIP* (*basic leucine zipper*) and two *AP2/DREB* genes showed differential expression associated with DNA hypomethylation in promoter or coding region under salt stress conditions (Song et al., 2012).

In mangrove plants, a comparative study was carried out. The plant from two different habitats i.e. salt marsh neighborhood and riverside habitat were compared on morphological basis (Lira-Medeiros et al., 2010). The result revealed that the plant growing in riverside were much taller and thicker than those grown in salt marsh neighborhood. Hypermethylation was observed in river side plant on analysis of genome wide DNA methylation with the help of MSAP technique. These results suggested that environmental adaptation will lead to epigenetic variation (Lira-Medeiros et al., 2010). Dyachenko et al. (2006) reported that *Mesembryanthemum crystallinum* plants, when subjected to high saline condition, hypermethylation at CHG methylation in nuclear genome was observed.

3.2 Drought

DNA methylation at genome-wide level was studied under drought conditions in rice. This study compared the variation of DNA methylation pattern under drought and subsequent recovery from it. The results showed that most DNA methylation modifications reversed after the recovery but some were maintained. This indicated that plant may have a mechanism of some sort to remember the previous condition under which it was subjected during its life cycle. This study also demonstrated that environmental stress leads to induced epigenetic variation which play an important role for adaptation of rice or other plant under diverse conditions (Wang et al., 2011).

Table 1. DNA methylation modifications involved in biotic and abiotic stress

Sr. No.	Plant species	Stress	Genomic region	Mode of action	References
1	Wheat	Salinity	24 genes	Stress induced shift in DNA methylation	(Wang et al., 2014)
		Salinity	Genome-wide	Hypomethylation	(Zhong et al., 2009)
		Cold treatment	*VRN-A1*	Site specific Hypermethylation	(Khan et al., 2013)
		Cold treatment	Genome-wide	Demethylation	(Sherman & Talbert, 2002)
2	Maize	Cold stress	*ZmMI1*	Root-specific hypomethylation	(Steward et al., 2002)
		Cold stress	Genome-wide	Global methylation shift. Mainly demethylation of fully methyated fragments	(Shan et al., 2013)
		Osmotic	*Transposon region*	Hypermethylation	(Tan, 2010)
		Salt Stress	Root zmPP2C	Hypermethylation	(Tan, 2010)
		Salt Stress	Leaf zmGST	Hypomethylation	(Tan, 2010)
3	Rice	Salinity	Genome-wide	Differential methylation of salt stress-related genes, retrotransposons and chromatin modifier genes	(Karan et al., 2012)
		Drought	Genome-wide	Genotype dependent differential methyation	(Wang et al., 2011)
4	Tomato	Drought	*Asr2*	CHH hypomethylation in regulatory region	(González et al., 2013)
			Asr1	CG hypermethylation and CHH hypomethylation	(González et al., 2011)
5	Soybean	Salinity	*Glyma11g02400* (Promoter)	-518 to -274 cytosines were demethylated following exposure to salinity stress for 1–24 h	(Song et al., 2012)
			Glyma16g27950 (Promoter)	Hypomethylation at transcription start codon (-24 to -233)	(Song et al., 2012)
6	Mangrove	Salinity	Genome-wide	Global hypomethylation	(Lira-Medeiros et al., 2010)
7	*Mesembryanthem um crystallinum*	Salinity	Genome-wide	CHG hypermethylation	(Dyachenko et al., 2006)
8	Tobacco	Aluminium, Salt and cold	*NtGPDL*	Hypomethylation	(Choi & Sano, 2007)
		Tobacco Mosaic Virus	*NtAlix1*	Hypermethylation	(Wada et al., 2004)

Some work regarding DNA methylation shift in response to stresses has been done in Tomato. Tomato plants are analyzed for DNA methylation pattern under drought condition and it was observed that *Abscisic Acid Stress Ripening1 (Asr1)* gene showed CHH hypomethylationunder more water stress conditions and due to this hypomethylation, an increase in *Asr1* gene expression was observed (González et al., 2011). Along with *Asr1*, DNA methylation at *Abscisic Acid Stress Ripening1 (Asr2)* gene was also reported in its regulatory region at all three contexts of DNA methylation (CG, CHG CHH). Interestingly the gene body methylation was only observed for one context (CG). The site-specific removal of methyl marks from CHH sites in the regulatory region was observed under drought stress. The *Asr2* response is heritable and observed generation after generation. The *Asr2* is thought to have evolutionary importance(González et al., 2013).

3.3 Cold Stress and Cold Treatment

In response to cold treatment (vernalization), site specific DNA hypermethylation has been recently reported (Khan et al., 2013). In this study, the DNA methylation profile of *VRN-A1* gene was studied in winter wheat and differential pattern of methylation at non-CG sites was reported. The CG sites remained unaffected by the treatment while CHG and CHH sites within specific region of intron-1 of *VRN-A1* gene showed an increased due to the cold treatment. At genome-wide level, Sherman and Talbert (2002) compared Near-Isogenic Lines (NILs) for winter and spring wheat alongwith vernalized (cold treatment) and non-vernalized condition for DNA methylation variations. They reported that winter wheat was more methylated as compared to spring wheat and the vernalization (cold treatment) induced demethylation compared to non-vernalized plants (Sherman & Talbert, 2002).

There are some studies in maize also showing a shift in DNA methylation due to cold stress. In one study, DNA methylation pattern of maize was studies in cold stress at seedling stage. A fragment was identified which was only expressed under cold stress and was named as *ZmMI1*. The results indicate that DNA methylation leads to differential gene expression by altering the chromatin structure and was under control of environmental stress (Steward et al., 2002). Genome-wide DNA methylation under cold stress was also investigatedby using Methylation-sensitive amplificationpolymorphism (MSAP) technique. The result was global shift in DNA methylation due to demethylation of fully methylated fragments (Shan et al., 2013).

3.4 Biotic Stress

Apart from abiotic stress, a changing pattern on DNA methylation was also observed under biotic stress. The tobacco plant was investigated for pathogenic response under the infection of tobacco mosaic virus (TMV). This study revealed DNA hypermethylation at *NtAlix1 (Nicotiana ALG-2 Interacting protein X 1)* gene after 24 hrs. of inoculation and a close relation was observed between DNA methylation shift and activation of stress response genes (Wada et al., 2004).

3.5 Transgenerational Stress Response

Because heritability determines the potential of evolutionary changes of a trait, it is essential to determine the degree of heritability of epigenetic modifications, their impact on given ecologically important traits (Falconer & Mackay, 1996; Fisher, 1930), and their role in individual adaptation to changing environment (Hoffmann & Sgrò, 2011; Visser, 2008). Although the epigenetic modifications were initially thought to be reversed in next generation, new investigations have revealed the ability of epigenetic modifications caused by stresses to be transgenerationally transmissible. Some of the examples are reviewed here (Table 2):

Table 2. Transgenerational inheritance of DNA methylation shift due to biotic and abiotic stress

Sr. No.	Plant species	Stress	Genomic region	Mode of action	References
1	Rice	Heavy metal stress	TE & protein coding genes	Hypomethylation at CHG sites	(Ou et al., 2012)
2	*Arabidopsis*	Combined stress	*Genome-wide*	Hypomethylation	(Boyko et al., 2010)
3	Tobacco	Tobacco Mosaic Virus	*Genome-wide & disease resistance gene-like loci*	Hypermethylation	(Boyko et al., 2007)

Recently transgenerational inheritance of modifications in DNA methylation due to heavy metal stress has been reported in rice. Ou et al. (2012) observed that heavy metal stress caused hypomethylation at CHG sites. These modifications were heritable over three studied generations and this heritability induced tolerance in these generations.In an interesting study, where *Arabidopsis* plants were exposed to various stresses, like, cold, heat, UVC salt, and flood, the untreated progeny of these plants showed higher tolerance to stress, higher homologous recombination frequency and genomic hypomethylation (Alex Boyko et al., 2010).

In tobacco, it has been shown that the progeny of plants infected with tobacco mosaic virus (TMV) exhibited a high frequency of rearrangements at disease resistance gene-like loci, global genome hypermethylation, and locus-specific hypomethylation (Alexander Boyko et al., 2007). The above mentioned examples (Tables 1-2) are a good source to understand the importance of epigenetic mechanisms in plant under environmental variations and plant adaptation in these conditions. In this respect, the reported data sets of various plant methylomes could provide the basis for the selection of differential epigenetic regions as probable targets for better understanding the molecular pathways involved in them and use them for the genetic manipulation for crop improvement.

4. Summary

DNA methylation is an epigenetic marker which is involved in the gene regulation as well as genome stability through transponson silencing. The modification in the DNA methylation can occur in response to environmental variations (biotic and abiotic stresses). The examples of the methylation shift, due to stresses which ultimately leads to plant response through gene regulations, provided (in this review) will help in understanding an overall pattern in plant stress responses. Overall this review will help new researchers of plant epigenetics to get an overview of the DNA methylation in terms of basic concepts and their role in plant stress response.

References

Berger, S. L. (2007). The complex language of chromatin regulation during transcription. *Nature, 447*, 407-412. http://dx.doi.org/10.1038/nature05915

Boyko, A., Blevins, T., Yao, Y., Golubov, A., Bilichak, A., Ilnytskyy, Y., ... Kovalchuk, I. (2010). Transgenerational Adaptation of Arabidopsis to Stress Requires DNA Methylation and the Function of Dicer-Like Proteins. *PLoS ONE, 5*, e9514. http://dx.doi.org/10.1371/journal.pone.0009514

Boyko, A., Kathiria, P., Zemp, F. J., Yao, Y., Pogribny, I., & Kovalchuk, I. (2007). Transgenerational changes in the genome stability and methylation in pathogen-infected plants: (virus-induced plant genome instability). *Nucleic Acids Research, 35*, 1714-1725. http://dx.doi.org/10.1093/nar/gkm029

Chan, S. W. L., Henderson, I. R., & Jacobsen, S. E. (2005). Gardening the genome: DNA methylation in Arabidopsis thaliana. *Nature Reviews. Genetics, 6*, 351-360. http://dx.doi.org/10.1038/nrg1601

Choi, C.-S., & Sano, H. (2007). Abiotic-stress induces demethylation and transcriptional activation of a gene encoding a glycerophosphodiesterase-like protein in tobacco plants. *Molecular Genetics and Genomics, 277*, 589-600. http://dx.doi.org/10.1007/s00438-007-0209-1

Cubas, P., Vincent, C., & Coen, E. (1999). An epigenetic mutation responsible for natural variation in floral symmetry. *Nature, 401*, 157-161. http://dx.doi.org/10.1038/43657

Dyachenko, O. V., Zakharchenko, N. S., Shevchuk, T. V., Bohnert, H. J., Cushman, J. C., & Buryanov, Y. I. (2006). Effect of hypermethylation of CCWGG sequences in DNA of *Mesembryanthemum crystallinum* plants on their adaptation to salt stress. *Biokhimiia, 71*, 461-465. http://dx.doi.org/10.1134/S000629790604016X

Falconer, D. S., & Mackay, T. F. C. (1996). *Introduction to Quantitative Genetics* (4th ed.). Benjamin Cummings.

Feng, S., Cokus, S. J., Zhang, X., Chen, P.-Y., Bostick, M., Goll, M. G., ... Jacobsen, S. E. (2010). From the Cover: Conservation and divergence of methylation patterning in plants and animals. *Proceedings of the National Academy of Sciences, 107*, 8689-8694. http://dx.doi.org/10.1073/pnas.1002720107

Fisher, R. A. (1930). *The Genetical Theory of Natural Selection*. At The Clarendon Press. Retrieved from http://archive.org/details/geneticaltheoryo031631mbp

Goll, M. G., & Bestor, T. H. (2005). Eukaryotic Cytosine Methyltransferases. *Annual Review of Biochemistry, 74*, 481-514. http://dx.doi.org/10.1146/annurev.biochem.74.010904.153721

González, R. M., Ricardi, M. M., & Iusem, N. D. (2011). Atypical epigenetic mark in an atypical location: cytosine methylation at asymmetric (CNN) sites within the body of a non-repetitive tomato gene. *BMC Plant Biology, 11*, 94. http://dx.doi.org/10.1186/1471-2229-11-94

González, R. M., Ricardi, M. M., & Iusem, N. D. (2013). Epigenetic marks in an adaptive water stress-responsive gene in tomato roots under normal and drought conditions. *Epigenetics: Official Journal of the DNA Methylation Society, 8*, 864-872. http://dx.doi.org/10.4161/epi.25524

Grunstein, M. (1997). Histone acetylation in chromatin structure and transcription. *Nature, 389*, 349-352. http://dx.doi.org/10.1038/38664

Henderson, I. R., & Jacobsen, S. E. (2007). Epigenetic inheritance in plants. *Nature, 447*, 418-424. http://dx.doi.org/10.1038/nature05917

Hoffmann, A. A., & Sgrò, C. M. (2011). Climate change and evolutionary adaptation. *Nature, 470*, 479-485. http://dx.doi.org/10.1038/nature09670

Holliday, R. (1994). Epigenetics: An overview. *Developmental Genetics, 15*, 453-457. http://dx.doi.org/10.1002/dvg.1020150602

Karan, R., DeLeon, T., Biradar, H., & Subudhi, P. K. (2012). Salt Stress Induced Variation in DNA Methylation Pattern and Its Influence on Gene Expression in Contrasting Rice Genotypes. *PLoS ONE, 7*, e40203. http://dx.doi.org/10.1371/journal.pone.0040203

Khan, A. R., Enjalbert, J., Marsollier, A.-C., Rousselet, A., Goldringer, I., & Vitte, C. (2013). Vernalization treatment induces site-specific DNA hypermethylation at the VERNALIZATION-A1 (VRN-A1) locus in hexaploid winter wheat. *BMC Plant Biology, 13*, 209. http://dx.doi.org/10.1186/1471-2229-13-209

Klose, R., & Bird, A. (2006). Genomic DNA methylation: The mark and its mediators. *Trends in Biochemical Sciences, 31*, 89-97. http://dx.doi.org/10.1016/j.tibs.2005.12.008

Kornberg, R. D. (1977). Structure of Chromatin. *Annual Review of Biochemistry, 46*, 931-954. http://dx.doi.org/10.1146/annurev.bi.46.070177.004435

Kouzarides, T. (2007). Chromatin modifications and their function. *Cell, 128*, 693-705. http://dx.doi.org/10.1016/j.cell.2007.02.005

Lauria, M., Rupe, M., Guo, M., Kranz, E., Pirona, R., Viotti, A., & Lund, G. (2004). Extensive maternal DNA hypomethylation in the endosperm of *Zea mays*. *The Plant Cell, 16*, 510-522. http://dx.doi.org/10.1105/tpc.017780

Law, J. A., & Jacobsen, S. E. (2010). Establishing, maintaining and modifying DNA methylation patterns in plants and animals. *Nature Reviews Genetics, 11*, 204-220. http://dx.doi.org/10.1038/nrg2719

Lira-Medeiros, C. F., Parisod, C., Fernandes, R. A., Mata, C. S., Cardoso, M. A., & Ferreira, P. C. G. (2010). Epigenetic Variation in Mangrove Plants Occurring in Contrasting Natural Environment. *PLoS ONE, 5*, e10326. http://dx.doi.org/10.1371/journal.pone.0010326

Li, X., Zhu, J., Hu, F., Ge, S., Ye, M., Xiang, H., ... Wang, W. (2012). Single-base resolution maps of cultivated and wild rice methylomes and regulatory roles of DNA methylation in plant gene expression. *BMC Genomics, 13*, 300. http://dx.doi.org/10.1186/1471-2164-13-300

Manning, K., Tör, M., Poole, M., Hong, Y., Thompson, A. J., King, G. J., Giovannoni, J. J., & Seymour, G. B. (2006). A naturally occurring epigenetic mutation in a gene encoding an SBP-box transcription factor inhibits tomato fruit ripening. *Nature Genetics, 38*, 948-952. http://dx.doi.org/10.1038/ng1841

Narlikar, G. J., Fan, H.-Y., & Kingston, R. E. (2002). Cooperation between Complexes that Regulate Chromatin Structure and Transcription. *Cell, 108*, 475-487. http://dx.doi.org/10.1016/S0092-8674(02)00654-2

Ou, X., Zhang, Y., Xu, C., Lin, X., Zang, Q., Zhuang, T., ... Liu, B. (2012). Transgenerational Inheritance of Modified DNA Methylation Patterns and Enhanced Tolerance Induced by Heavy Metal Stress in Rice (*Oryza sativa* L.). *PLoS ONE, 7*, e41143. http://dx.doi.org/10.1371/journal.pone.0041143

Rapp, R., & Wendel, J. (2005). Epigenetics and plant evolution. *NEW PHYTOLOGIST, 168*, 81-91. http://dx.doi.org/10.1111/j.1469-8137.2005.01491.x

Roudier, F., Ahmed, I., Bérard, C., Sarazin, A., Mary-Huard, T., Cortijo, S., ... Colot, V. (2011). Integrative epigenomic mapping defines four main chromatin states in Arabidopsis. *The EMBO Journal, 30*, 1928-1938. http://dx.doi.org/10.1038/emboj.2011.103

Shan, X., Wang, X., Yang, G., Wu, Y., Su, S., Li, S., ... Yuan, Y. (2013). Analysis of the DNA methylation of maize (*Zea mays* L.) in response to cold stress based on methylation-sensitive amplified polymorphisms. *Journal of Plant Biology, 56*, 32-38. http://dx.doi.org/10.1007/s12374-012-0251-3

Sherman, J., & Talbert, L. (2002). Vernalization-induced changes of the DNA methylation pattern in winter wheat. *Genome, 45*, 253-260. http://dx.doi.org/10.1039/G01-147

Song, Y., Ji, D., Li, S., Wang, P., Li, Q., & Xiang, F. (2012). The Dynamic Changes of DNA Methylation and Histone Modifications of Salt Responsive Transcription Factor Genes in Soybean. *PLoS ONE, 7*, e41274. http://dx.doi.org/10.1371/journal.pone.0041274

Steward, N., Ito, M., Yamaguchi, Y., Koizumi, N., & Sano, H. (2002). Periodic DNA Methylation in Maize Nucleosomes and Demethylation by Environmental Stress. *Journal of Biological Chemistry, 277*, 37741-37746. http://dx.doi.org/10.1074/jbc.M204050200

Takuno, S., & Gaut, B. S. (2013). Gene body methylation is conserved between plant orthologs and is of evolutionary consequence. *Proceedings of the National Academy of Sciences of the United States of America, 110*, 1797-1802. http://dx.doi.org/10.1073/pnas.1215380110

Tan, M. (2010). Analysis of DNA methylation of maize in response to osmotic and salt stress based on methylation-sensitive amplified polymorphism. *Plant Physiology and Biochemistry, 48*, 21-26. http://dx.doi.org/10.1016/j.plaphy.2009.10.005

Visser, M. E. (2008). Keeping up with a warming world; assessing the rate of adaptation to climate change. *Proceedings of the Royal Society B: Biological Sciences, 275*, 649-659. http://dx.doi.org/10.1098/rspb.2007.0997

Wada, Y., Miyamoto, K., Kusano, T., & Sano, H. (2004). Association between up-regulation of stress-responsive genes and hypomethylation of genomic DNA in tobacco plants. *Molecular Genetics and Genomics, 271*, 658-666. http://dx.doi.org/10.1007/s00438-004-1018-4

Wang, M., Qin, L., Xie, C., Li, W., Yuan, J., Kong, L., ... Liu, S. (2014). Induced and Constitutive DNA Methylation in a Salinity Tolerant Wheat Introgression Line. *Plant and Cell Physiology*, pcu059. http://dx.doi.org/10.1093/pcp/pcu059

Wang, W.-S., Pan, Y.-J., Zhao, X.-Q., Dwivedi, D., Zhu, L.-H., Ali, J., ... Li, Z.-K. (2011). Drought-induced site-specific DNA methylation and its association with drought tolerance in rice (*Oryza sativa* L.). *Journal of Experimental Botany, 62*, 1951-1960. http://dx.doi.org/10.1093/jxb/erq391

Zemach, A., Kim, M. Y., Silva, P., Rodrigues, J. A., Dotson, B., Brooks, M. D., & Zilberman, D. (2010a). Local DNA hypomethylation activates genes in rice endosperm. *Proceedings of the National Academy of Sciences*, 201009695. http://dx.doi.org/10.1073/pnas.1009695107

Zemach, A., McDaniel, I. E., Silva, P., & Zilberman, D. (2010b). Genome-Wide Evolutionary Analysis of Eukaryotic DNA Methylation. *Science, 328*, 916-919. http://dx.doi.org/10.1126/science.1186366

Zhang, M., Kimatu, J. N., Xu, K., & Liu, B. (2010). DNA cytosine methylation in plant development. *Journal of Genetics and Genomics, 37*, 1-12. http://dx.doi.org/10.1016/S1673-8527(09)60020-5

Zhang, X., Yazaki, J., Sundaresan, A., Cokus, S., Chan, S. W.-L., Chen, H., ... Ecker, J. R. (2006). Genome-wide High-Resolution Mapping and Functional Analysis of DNA Methylation in Arabidopsis. *Cell, 126*, 1189-1201. http://dx.doi.org/10.1016/j.cell.2006.08.003

Zhong, L., Xu, Y., & Wang, J. (2009). DNA-methylation changes induced by salt stress in wheat Triticumaestivum. *African Journal of Biotechnology, 8*. Retrieved from http://www.ajol.info/index.php/ajb/article/view/66122

21

Evaluation of Growth Performance, Haematological and Serum Biochemical Response of Broiler Chickens to Aqueous Extract of Ginger and Garlic

Vivian U. Oleforuh-Okoleh[1,2], Harriet M. Ndofor-Foleng[3], Solomon O. Olorunleke[1] & Joesph O. Uguru[1]

[1] Department of Animal Science, Ebonyi State University, Abakaliki, Ebonyi State, Nigeria

[2] Department of Animal Science, Rivers State University of Science and Technology, Nkpolu-Oroworukwo, Port Harcourt, Rivers State, Nigeria

[3] Department of Animal Science, University of Nigeria, Nsukka, Enugu State, Nigeria

Correspondence: Vivian U. Oleforuh-Okoleh, Department of Animal Science, Rivers State University of Science and Technology, Nkpolu-Oroworukwo, PMB 5080, Port Harcourt, Rivers State, Nigeria.

Abstract

An experiment which lasted for 56 days was carried out to investigate the growth performance, haematological and serum biochemical response of broiler chickens to aqueous extract of ginger and garlic. Eighty day-old Marshal Strain broiler chickens were used for the experiment. The birds were randomly allotted into four treatment groups consisting of four replicates with five birds per replicate. The aqueous extract was obtained by infusing 14 g of each test ingredient in 1 litre of hot boiled water for 12 hours and 50 ml of the filtrate/litre of drinking water given to birds' *ad-libitum*. T1 (control), T2, T3, and T4 contained 0, and 50 mls of ginger, garlic and a 1:1 ratio mixture of ginger and garlic in drinking water respectively. T2 gave the best performance (p < 0.05) in all growth performance traits – final body weight, weekly weight gain, weekly feed intake and feed conversion ratio and T1 the least. Significant (p < 0.01) increases were observed in haemoglobin concentration, packed cell volume, white blood cell, and red blood cell of the ginger and garlic treated birds. The serum biochemical parameters measured were significantly (p < 0.05) different, with T2 and T3 showing a better response. Cholesterol decreased significantly (p > 0.05) while there was significant increase in the total protein, albumin, and globulin of the treated birds (p < 0.01). Administration of ginger and garlic to broiler chickens increased their performance, boosted their immunity as well as improved their general well-being. It is, thus, recommended in broiler chicken production.

Keywords: aqueous extract, body weight, chicken, cholesterol, feed intake, haematology, total protein

1. Introduction

In Nigeria, the effect of inadequate animal protein intake is felt more by a large proportion of the population especially in the rural areas. Poultry meat is a good source of animal protein and can contribute immensely in boosting the consumption level of animal protein. The prohibitive increase in the cost of input especially that of feed is among the constraints in commercial broiler production (Madubuike & Ekenyem, 2001). Ensuring more net return and minimizing high expenditure for feed are the main challenges, for which many research strategies have been trying to address through the inclusion of feed supplements and feed additives in the diets of broiler chicken.

A major feed additive that has been extensively used is antibiotics. Antibiotics use in livestock is the use of antibiotics for any purpose in the husbandry of livestock, which include not only the treatment or prophylaxis of infection but also the use of sub-therapeutic doses in animal feed to promote growth and improve feed efficiency in contemporary intensive animal farming (Ogle, 2013). Incidentally, their use in animal feed has shown several side effects such as resistance towards the drug and evidence of resistant strains that become zoonotic (Wegener et al., 1999). The emergence of antibiotic resistance by pathogenic bacteria has led to international restriction on the use of antibiotics in animal feeds. Consequently, the poultry industry is under great pressure to minimize their use in animal feed and seek alternatives. These alternatives can be found in the use of herbs and spices materials as supplements. According to Manesh et al. (2012) natural alternatives to antibiotics, such as herbs and medicinal plants, have attracted attention due to their wide range of potential beneficial effects.

Natural medicinal products originating from herbs and spices have been used as feed additives for farm animals (Guo, 2003). The efficacy and importance of a particular feedstuff/feed ingredient in poultry production is evaluated from its effect on the production performance/traits of the birds. Furthermore, valuable information can be obtained from the study of the haematological parameters. This stems from the fact that the blood serves as an important index of physiological, pathological and nutritional status of an animal. Information obtained from haematological assay, apart from being useful for diagnostic and management purposes could equally be incorporated into breeding programmes (Elagib & Ahmed, 2011). Two herbal plants which are nutritionally adequate and locally available in Nigeria that can be harnessed as feed additives are ginger (Ademola et al., 2009) and garlic (Gbenga et al., 2009). Ginger is a rhizomatous herbaceous plant, whose rhizome is used medicinally. Ginger contains several compounds and enzymes including gingerdiol, gingerol, gingerdione and shogaols (Rivlin, 2001; Zhao et al., 2011). These compounds have been reported to have antimicrobial, antioxidative and pharmacological effects (Al-Amin et al., 2006; Tapsell et al., 2006; Ali et al, 2008). Garlic is best known as a spice and herbal medicine for treatment and prevention of an array of diseases (Adibmoradi et al., 2006). The key active ingredient in garlic is a powerful plant chemical called allicin which rapidly decomposes to several organosulphur compounds with bioactivities (Chang & Cheong, 2008). Several reports are available advocating the roles of these herbs in improving growth performance, meat quality, anti-cholesteremic effects and as well as immuno-modulating effects on broiler chickens (Gardzielewska et al., 2003; Aji et al., 2011; Ashayerizadeh et al., 2009; Hanieh et al., 2010; Ayasan, 2011).

The present study was carried out to evaluate the growth performance, haematological and serum biochemical indices of broiler chicks fed ginger and garlic aqueous extract.

2. Materials and Methods

2.1 Study Location

The present study was carried out at the Poultry Unit of the Teaching, and Research Farm of the Department of Animal Science, Ebonyi State University Abakaliki. Abakaliki is located within the south eastern guinea savannah ecological zone between latitude $8°30'$ and $9°40'$ North and longitude $5°40'$ and $6°45'$ East (Nwakpu, 2005). The experiment lasted for 49 days.

2.2 Test Ingredients

The fresh ginger and garlic used were purchased from Abakaliki market. The garlic and ginger were peeled, cut into chips and sundried for a period of six weeks. The dried garlic and ginger chips were ground into smooth powder and stored separately in an air tight container. Each aqueous extract was prepared by adding one liter of boiled hot water to 14 g of either ground ginger or garlic or a mixture of the two (at 7 g each) in separate non-metallic containers. The mixtures were allowed to infuse and cool at room temperature overnight for twelve hours. The next morning, the extract was obtained by filtering the infusion using a filter paper, and then administered to the chicks in their drinking water at 50 ml/liter of water. The aqueous extract was made available to the birds for *ad libitum*. Fresh infusion was prepared daily. The procedure used for preparing the aqueous extract is in line with Leila (1977).

2.3 Experimental Animals and Treatments

A total of eighty day-old Marshal broiler chicks used for the study were purchased from Obasanjo Farms Nigeria Limited, Ota in Ogun State. The chicks were kept for seven days to acclimatize; within this period, they were fed commercial broiler starter diet only and given plain drinking water. On the 8th day, the 80 chicks, having an average body weight of 121.66 ± 5.00 g were randomly allotted to four experimental treatments in a completely randomly design (CRD). Each treatment was replicated four times with 5 chicks per replicate. The birds were housed in a wooden three-tier battery cage, each replicate cell measuring $30 \times 44 \times 21$ inches.

Four experimental treatments identified as T_1, T_2, T_3, and T_4 were studied. Birds on T_1 (control treatment) received basal diet (commercial starter diet -22% CP, 2900 kcal/kg ME- fed for the first 28 days and finisher diet fed from the 28 to 49 day (18% CP, 2900 kca/kg ME), and water without ginger or garlic. Those on T_2 received the basal diet and were given ginger extract at 50 ml/litre of water; birds on T_3 received the basal diet and were given garlic extract at 50 ml/litre of water, and those T_4 were fed the basal diet and received extract of both ginger and garlic at 25 ml each/litre of water. Proper management, necessary vaccinations and good environmental condition were maintained throughout the period of study.

2.4 Data Collection

2.4.1 Growth Performance

Data were collected on growth performance traits (such as daily feed intake, weekly body weight, final body weight and feed conversion ratio).

2.4.2 Haematological and Serum Biochemical Assay

On the 56th day of study, blood samples were randomly collected from four birds/treatment. The blood samples were collected via the wing veins using sterile needles and syringes. The blood samples for haematological parameters were collected into well-labeled and sterilized bottles containing ethylene diamine tetra acetic acid (EDTA), as anti-coagulant. The samples were investigated for the following haematological parameters – packed cell volume (PCV), red blood cell count (RBC), white blood cell (WBC), haemoglobin and platelets (Lamb, 1991). Blood samples for biochemical indices were collected into another sample bottles without the anticoagulant. Plasma samples were analyzed for cholesterol, total protein, albumin, globulin and urea. The serum biochemical indices were done using the clinical routine procedures outlined by Olorede et al. (1996)

2.5 Statistics and Data analysis

The data collected were statistically analyzed by analysis of variance using Repeated Measures in General Linear Model in the statistical package SPSS (2009) using the following statistical model:

$$X_{ij} = \mu + H_i + E_{ij} \tag{1}$$

Where,

X_{ij} = any observation made in the experiment; μ = the population mean; H_i = effect of type of herb (I = ginger, garlic); E_{ij} = residual error.

3. Results and Discussion

3.1 Growth Performance

The summary of the effect of the different herbs on growth performance traits of broiler chickens is presented in Table 1. There were significant differences (P < 0.05) between treatments in performance traits. Birds on ginger infusion had a better performance for all traits studied, there was a 29.07, 16.60 and 14.98% significant (P < 0.05) increase in final body weight relative to T1, T3 and T4 respectively. The present findings affirm the work of Herawati (2010) on Hubbard broiler strains. The authors observed a significant increase in final body weight, higher feed intake and better feed conversion ratio of birds fed 2% supplemented red ginger in their diet). Similar result was obtained by Al-Moramadhi (2010) when broiler chicks were given ginger orally at 100 mg/kg body weight for six weeks. Minh et al. (2010) and Onu (2010) also reported that supplementation of dried ginger to broiler diets resulted in improved performance. Ademola et al. (2009) also observed that ginger increased body weight when included in the diet up to 2% level in the diet. The better performance observed in T2 could be attributed to some medicinal properties contained in ginger. For instance, a protein digesting enzyme (zingibain) found in ginger is believed to improve digestion as well as kill parasites and their eggs. Furthermore, properties in ginger tends to enhance antibacterial and anti inflammatory factors (Mohammed & Yusuf, 2011). Tekeli (2007) stated that due to the active ingredients in these herbs, there is the formation of more stable intestinal flora and improved feed conversion efficiency in consequence of a better digestion.

Table 1. Growth performance of broiler chicks given aqueous infusion of ginger and garlic

Parameters	T1 0 ml	T2 50 ml ginger	T3 50 ml galic	T4 50 ml ginger & garlic	SEM	P-Value
Final body weight (g)	1930.80[c]	2684.40[a]	2238.20[b]	2282.20[b]	24.07	0.00
Weekly weight gain(g)	274.99[c]	383.49[a]	319.74[b]	326.03[b]	3.41	0.00
Weekly feed intake(g)	538.73[b]	567.08[a]	535.30[b]	534.48[b]	4.19	0.00
Feed Conversion Ratio	1.96[a]	1.48[c]	1.67[b]	1.64[b]	0.125	0.00

Note. [abc]Means on the same row followed by different superscripts are significantly different (p < 0.05); SEM – Standard Error of Mean.

Though in the present study, birds on T2 performed better than those on T3 and T4, these two treatments were significantly (p < 0.05) better than the control. Mahmood et al. (2009) reported that garlic had positive effect on the growth rate of broiler chicks. Meraj (1998) noted that the presence of antibiotic substances in garlic is responsible for the improvement of weight gain. In consonance, Rehman et al. (2012) reported that mean feed conversion ratio was significantly influenced by water based infusion of garlic and *Withania somnifera*. The reports of Ashayerizadeh et al. (2009) and Mohebbifar and Torki (2011) showed that inclusion of garlic powder in broiler feed did not change body weight, feed intake and efficiency/feed conversion ratio. Lawson et al. (1992) observed that allicin is unstable and poorly absorbed from the digestive tract. Present findings also support the findings of Javed et al. (2009) which showed a positive effect of aqueous extract of plant mixture (*Zingiber officinale, Carum apticum, Withania somnifera, Trigonella Foenum-Graecum, Silybum marianum, Allium sativum* and *Berberis lyceum*) on the performance of broiler chicks in term of weight gain and Feed Conversion Ratio.

3.2 Haematological and Serum Biochemical Assay

The results of the haematological and serum biochemical response of broiler chickens administered aqueous extract of test ingredients are presented in Table 2.

Table 2. Haematological and serum biochemical response of broiler chickens administered aqueous extract of test ingredients

Parameters	T1 0 ml	T2 50 ml ginger	T3 50 ml galic	T4 50 ml ginger & garlic	SEM	P-Value
PCV (%)	24.89[b]	28.22[ab]	29.14[a]	25.5[b]	1.96	0.010
HB(g/dl)	8.00[b]	9.30[a]	9.73[a]	8.38[b]	0.38	0.001
RBC (10^{12}/mm^3)	2.15[c]	2.62[ab]	2.80[a]	2.46[b]	0.16	0.001
WBC(10^9/mm^3)	6.30[c]	7.10[ab]	7.49[a]	6.80[b]	0.22	0.001
Platelets	166.00	140.00	150.00	170.00	15.42	0.173
Cholesterol(mg/dl)	245.43[a]	225.77[a]	200.00[b]	230.37[a]	14.59	0.048
Protein(g/l)	4.50[c]	5.85[b]	6.54[a]	5.38[b]	0.29	0.000
Albumin (g/l)	2.56[c]	3.44[b]	4.02[a]	3.32[b]	0.27	0.000
Globulin(mg/dl)	1.78[c]	2.40[a]	2.53[a]	2.04[b]	0.18	0.004
Urea(mg/dl)	42.00	40.00	40.36	41.10	1.67	0.503

Note. [abc]Means on the same row followed by different superscripts are significantly different (p < 0.05); SEM – Standard Error of Mean.

There was significant increase (P < 0.05) in the PCV, Hb, RBC, and WBC of birds on the ginger and garlic infusion than those on control treatment. Mitruka et al. (1977) stated that the number of erythrocytes (RBC) in chicken is influenced by the conditions of the animal. The increase in PCV, Hb, and RBC contents of the blood of birds fed the test ingredients is an indication of improved oxygen carrying capacity of the cells which translated to a better availability of nutrients to the birds consequently affecting their well-being. Sole administration of ginger and garlic numerical reduced the platelets in the blood. Muhammed and Lakshmi (2007) opined that inhibiting the transformation of arachidonic acid to thromboxane and decreasing the sensitivity of platelets to aggregating agents may be possible with the administration of ginger in fatty diets. This implies that ginger could be potentially useful in improving blood circulation on account of its inhibitory effects on platelet aggregation (Muhammed & Lakshmi, 2007). Similar inhibitory effect was observed in garlic by Lawson et al. (1992).

The result on serum biochemical indices indicates that inclusion of ginger and garlic in the water of broiler chickens successfully reduced the cholesterol in the serum. The present findings also reveal that the ginger extract caused reduction in the levels of serum cholesterol, though the mixture of the two (ginger and garlic) did not show any variation from the control where the cholesterol statistically increased. Saeid et al. (2010) observed that aqeous extract of ginger significantly reduced the level of cholesterol in the blood of broilers. Bhandari et al. (1998) and Akhani et al. (2004) also reported that ginger treatment significantly decreased serum cholesterol. This affirms the findings of Mansoub (2011) who reported reductions in total cholesterol when broilers were supplemented with 1 g/kg garlic. The results of present study is, also, in agreement with Stanacev et al. (2011) who reported that garlic manifested hypocholesterolemic effects on chickens through inhibition of the most important enzymes that participate in the synthesis of cholesterol and lipids. Konjufca et al. (1997) reported that garlic reduced plasma cholesterol by decreasing the activity of 3-hydroxy-3-methlyglutaryl reductase. Allicin has been proposed as the active compound in garlic responsible for health promotion and hypocholesterolaemic benefits (Lawson, 1998). The result of the present study affirms the findings of Al-Moramadhi (2010) on the effect of ginger root infusion on haematological parameters in broiler chickens. The reduction in cholesterol level can be traced to the presence of gingerols and shagols components in ginger which inhibits lipid peroxidation (Verma et al., 2004; Ashani & Verma, 2009).

4. Conclusion

A mixture of the two herbs was not as beneficial as sole treatment (treatment either with ginger or garlic) in all parameters studied. Although, birds on aqueous extract of ginger, had better performance in terms of final body weight and feed conversion ratio, administration of aqueous extract of garlic in the broilers drinking water improved the haematological and serum parameters studied. The result of the present study, therefore, suggests that the use of ginger and garlic as aqueous extract in the diets of broiler chickens improved their performance as well as their health status. Generally, administration of aqueous extract of ginger and garlic is recommended in broiler production for improved nutritional and physiological traits in broiler chickens.

References

Ademola, S. G., Farinu, G. O., & Babatunde, G. M. (2009). Serum lipid growth and hematological parameters of broilers fed garlic, ginger and their mixtures. *World Journal of Agricultural Science, 5*(1), 9-104.

Adibmoradi, M., Navidshad, B., Seifdavati, J., & Royan, M. (2006). Effect of dietary garlic meal on histological structure of small intestine in broiler chickens. *Journal of Poultry Sciences, 43*, 378-383. http://dx.doi.org/10.2141/jpsa.43.378

Aji, S. B., Ignatuius, K., Ado, A. Y., Nuhu, J. B., & Abdulkarim, A. (2011). Effect of feeding onion (*Allium cepa*) and garlic (*Allium sativum*) on some performance characteristics of broiler chickens. *Research Journal of Poultry Science, 4*, 22-27. http://dx.doi.org/10.3923/rjpscience.2011.22.27

Akhani, S. P., Vishwakarma, S. L., & Goyal, R. K. (2004). Anti-diabetic activity of *Zingiber officinale* in Streptozotocin-induced type I diabetic rats. *J. Pharmacy Pharmacol., 56*, 101-105. http://dx.doi.org/10.1211/0022357022403

Al-Amin, Z. M., Thomson, M., Al-Qattan, K. K., Peltonen-Shalaby, R., & Ali, M. (2006). Anti-diabetic and hypolipideamic properties of ginger (Zingiber officinale) in streptozotocin-induced diabetic rats. *Br. Journ. Nutr., 96*, 660-666. http://dx.doi.org/10.1079/BJN20061849

Ali, B. H., Blunden, G., Tanira, M. O., & Nemmar, A. (2008). Some phytochemical, pharmacological and toxicological properties if ginger (*Zingiber officinale* Roscoe): A review of recent research. *Food Chemistry and Toxicology, 46*, 409-420. http://dx.doi.org/10.1016/j.fct.2007.09.085

Al-Moramadhi, S. A. H. (2010). The effect of Zingiber officinali roots infusion on some physiological parameters in broiler chickens. *Kufa Journal for Veterinary Medical Sciences, 1*(2), 67-76.

Ashani, V. M., & Verma, R. J. (2009). Ameliorative effects of ginger extract on paraben-induced lipid peroxidation in the liver of mice. *Acta Pol Pharm., 66*(3), 225-228.

Ashayerizadeh, O., Daster, B., & Shargh, M. S. (2009). Use of garlic *(Allium sativum)*, black cumin seeds and wild mint *(Mentha longifolia)* in broiler chicken diets. *Journal of Animal and Veterinary Advance, 8*, 1860-1863.

Ayaşan, T. (2011). Black Cumin and Usage of Poultry Nutrition. *National Poultry Congressi 14-16 September 2011* (pp. 1228-1236). University of Cukurova, Agricultural of Faculty, Animal Science Division, Adana, Turkey.

Bhandari, U., & Grover, J. K. (1998). Effect of ethanolic extract of ginger on hyperglycemic rats. *Int. J. Diabetes, 6*, 95-96.

Chang, K. J., & Cheong, S. H. (2008). Volatile organosulfur and nutrient compounds from garlic by cultivating areas and processing methods. *Fed. Am. Soc. Exp. Bio. J., 22*, 1108.2.

Elagib, H. A. A., & Ahmed, A. D. A. (2011). Comparative study on haematological values of blood of indigenous chickens in Sudan. *Asian Journal of Poultry Science, 5*(1), 41-45.

http://dx.doi.org/10.3923/ajpsaj.2011.41.45

Gardzielewska, J., Pudyszak, K., Majewska, T., Jakubowska, M., & Promianowski, J. (2003). Effect of plant-supplemented feeding on fresh and frozen storage quality of broiler chicken meat. *Animal Husbandry Series of Electronic J. Polish Agric-Univ., 6*(2). Retrieved from http://www.ejpau-media.pl/series/volumes/issues/animal/art-12.html

Gbenga, O. E., Adebisi, O. E., Fajemisin, A. N., & Adetunji, A. V. (2009). Response of broiler chickens in terms of performance and meat quality to garlic Allium sativum supplementation. *African Journal of Agric. Res., 4*, 511-517.

Guo, F. C. (2003). *Mushroom and herb polysaccharides as alternative for antimicrobial growth promoters in poultry* (Unpublised doctoral dissertation). Wageningen University, Netherlands.

Hanieh, H., Narabara, K., Piao, M., Gerile, C., Abe, A., & Kondo, Y. (2010). Modulatory effects of two levels of dietary alliums on immune response and certain immunological variables, following immunization, in white leghorn chicken. *Animal Science Journal, 81*, 673-680. http://dx.doi.org/10.1111/j.1740-0929.2010.00798.x

Herawati, O. (2010). The effect of red ginger as phytobiotic on body weight gain, feed conversion and internal organs condition of broiler. *Int. J. Poult. Sci., 9*(10), 963-967. http://dx.doi.org/10.3923/ijps.2010.963.967

Javed, M., Durrani, F., Hafeez, A., Khan, R. U., & Ahmad, I. (2009). Effect of aqueous extract of plant mixture on carcass quality of broiler chicks. *ARPN JABS, 4*(1), 37-40.

Konjufca, V. H., Pesti, G. M., & Bakalli, R. I. (1997). Modulation of cholesterol levels in broiler meat by dietary garlic and copper. *Poultry Science, 76*, 1264-1271. http://dx.doi.org/10.1093/ps/76.9.1264

Lamb, G. N. (1991). *Manual of veterinary laboratory technique* (pp. 98-99). CIBA-GEIGY, Kenya.

Lawson, L. D. (1998). *Garlic: A review of its medicinal effects and indicated active compounds.* In L. D. Lawson & R. Bauer (Eds.), *Phytomedicines of Europe: chemistry and biological activity* (Vol. 91, pp. 176-209). Washington, ACS symposium Series, USA.

Lawson, L. D., Ransom, D. K., & Hughes, B. G. (1992). Inhibition of whole blood platelet-aggregation by compounds in garlic clove extracts and commercial garlic products. *Thromb. Res., 65*, 141-156. http://dx.doi.org/10.1016/0049-3848(92)90234-2

Leila, S. M. (1977). A manual on some Philippine medicinal plants (preparation of drug materials). *Bot. Soc., U.P., 20*, 78-82.

Madubuike, F. N., & Ekeyem, B. U. (2001). *Non ruminant Livestock Production in the Tropics* (p. 185). Gust-chuks Grapics, Owerri, Nigeria.

Mahmood, S., Hassan, M. M., Alam, M., &Ahmad, F. (2006). Comparative efficacy of *Nigella sativa* and *Allium sativum* as growth promoters in broilers. *Inter. Jour. of Agric. and Biol., 11*, 775-778.

Manseh, M. K., Kazemi, S., & Asfari, M. (2012). Influence of poly germander (trucrium polium) and watercress *(Nasturtium officinale)* extract on performance, carcass quality and blood metabolites of males broilers. *Research Options in Animal & Veterinary Sciences, 2*, 66-68.

Mansoub, N. H. (2011). Comparative effects of using garlic as probiotic on performance and serum composition of broiler chickens. *Annals of biological Research, 2*, 486-490.

Meraj, I. C. A. (1998). Effect of garlic and neem leaves supplementation on the performance of broiler chickens. M. Sc. Thesis, Department of Poultry Science, University of Agriculture, Faisalabad, Pakistan.

Minh, D. V., Huyen, L. V., Theun, P., Tuan, T. Q., Nga, N. T., & Khiem, N. Q. (2010). *Effect of supplementation of ginger (Zingiber officinale) and garlic (Allium sativum) extracts (phyto- antibiotics on digestibility and performance of broiler chicken.* MFKARN conference on livestock production, climate change and resource depletion. Retrieved from http://www.mekarn.org/workshops/pakse/abstracts/minh_nias.htm

Mitruka, B. M., Rawnsley, H. M., & Vadehra, B. V. (1977). *Clinical biochemical and haematological reference values in normal experimental animals.* Masson Publishing USA Inc. 272.

Mohammed, A. A., & Yusuf, M. (2011). Evaluation of ginger *(Zingiber officinale)* as a feed additive in broiler diets. *LRRD, 23*(9), Article # 202. Retrieved from http://www.lrrd.org/lrrd23/9/moha23202.htm

Mohebbifar, A., & Torki, M. (2011). Growth performance and humoral response of broiler chicks fed diet containing graded levels of ground date pits with a mixture of dried garlic and thyme. *Global Veterinaria, 6*,

389-398.

Muhammed, M., & Lakshmi, P. (2007). *Ginger (Zingibe officinale): Product Write-Up*. Retrieved from http://www.sabinsa.com/products/standardized-phytoextracts/ginger/ginger.pdf

Nwakpu, C. (2005). *Practical guide to lowland rice production in Nigeria*. SNAAP Press Ltd. Enugu.

Ogle, M. (2013). Riots, Rage, Resistance: A Brief History of How Antibiotics Arrived on the Farm. *Scientific American*. Retrieved November 5, 2014.

Olorede, B. R., Onifade, A. A., Okpara, A. O., & Banatunde, G. M. (1996). Growth, nutrient retention, haematoology and serum chemistry of broiler chickens fed sheabutter cake or palm kernel cake in the humid tropics. *J. of Applied Animal Research, 10*, 173-180. http://dx.doi.org/10.1080/09712119.1996.9706146

Onu, P. N. (2010). Evaluation of two herbal spices as feed additives for finisher broilers. *Biotechnology in Animal Husbandry, 26*, 383-392. http://dx.doi.org/10.2298/BAH1006383O

Rehman, Z. U., Khan, S., Chand, N., Tanweer, A. J., Sultan, A., Akhtar, A., & Tauqeer, A. M. (2012). Effect of water based mixture infusion of *allium sativum* and *withania somnifera* on performance of broiler chicks. *Pakistan Journal of Science, 64*(3), 180-183.

Rivlin, R. S. (2001). Historical perspective on the use of garlic. *J. Nutr., 131*(35), 957-954.

Saeid, J. M., Mohamed, A. B., & AL-Baddy, M. A. (2010). Effect of aqueous extract of ginger (*zingiber officinale*) on blood biochemistry parameters of broiler. *Int. J. Poult. Sci., 9*(10), 944-947. http://dx.doi.org/10.3923/ijps.2010.944.947

SPSS. (2009). *Computer software SPSS Inc*. Headquarters, Wacker Drive, Chicago, Illinois-U.S.A.

Stanacev, V., Glamoci, D., Miloševic, N., Puvacac, N., Stanacev, V., & Plavša N. (2011). Effect of garlic (*allium sativum l.*) in fattening chicks Nutrition. *African Journal of Agricultural Research, 6*(4), 943-948.

Tapsell, L. C., Hemphill, I., Cobiac, L., Patch, C. S., Sullivan, D. R., Fenech, M., … Inge, K. E. (2006). Health benefits of herbs and spices: The past, the present, the future. *Med. J. Aust., 185*, 4-24. http://dx.doi.org/10.3923/crpsaj.2011.12.23

Tekeli, A., Kutlu, H. R., & Celik, L. (2011). Effect of *Z. officinale* and propolis extracts on the performance, carcass and some blood parameters of broiler chicks. *Current Research in Poultry Science, 1*(1), 12-23.

Verma, S. K., Singh, M., Jain, P., & Bordia, S. (2004). Protective effect of ginger, *(Zingiber officinale Rosc)*, on experimental atherosclerosis in rabbits. *Indian J. Exp. Biol., 42*, 736-738.

Wegener, H. C., Aarestrup, F. M., Benner-Smidt, P., & Bager, F. (1999). Transfer of antibiotic resistant bacteria from animal to man. *Acta. Vet. Scand. Suppl., 92*, 51-57.

Zhao, X., Yang, Z. B., Yang, W. R., Wang, Y., Jiang, S. Z., & Zhang, G. G. (2011). Effects of ginger roots (*Zingiber officinale*) on laying performance and antioxidant status of laying hens and on dietary oxidation stability. *Poultry Science, 90*, 1720-1727. http://dx.doi.org/10.3382/ps.2010-01280

Permissions

List of Contributors

Abiodun O. Joda
Department of Crop Production, Olabisi Onabanjo University, Ago-Iwoye, Nigeria

Francis K. Ewete
Department of Crop Protection and Environmental Biology, University of Ibadan, Ibadan, Nigeria

B. N. Singh
Formerly of the West African Rice Development Association, WARDA, Ibadan, Nigeria

Olufemi O. R. Pitan
Department of Crop Protection, Federal University of Agriculture, Abeokuta, Nigeria

Charles L. Webber III
USDA, Agriculture Research Service, Sugarcane Research Unit, Houma, LA, USA

Paul M. White Jr.
USDA, Agriculture Research Service, Sugarcane Research Unit, Houma, LA, USA

Dwight L. Myers
East Central University, Chemistry Department, Ada, OK, USA

Merritt J. Taylor
Oklahoma State University, Division of Agriculture Sciences and Natural Resources, Department of Agricultural Economics, Durant, OK, USA

James W. Shrefler
Oklahoma State University, Division of Agriculture Sciences and Natural Resources, Cooperative Extension Service, Durant, OK, USA

Ganesh C. Bora
Agricultural and Biosystems Engineering, North Dakota State University, Fargo, ND, USA

Donqing Lin
Agricultural and Biosystems Engineering, North Dakota State University, Fargo, ND, USA

Pritha Bhattacharya
Agricultural and Biosystems Engineering, North Dakota State University, Fargo, ND, USA

Sukhwinder Kaur Bali
Natural Resources Management, North Dakota State University, Fargo, ND, USA

Rohit Pathak
Agricultural and Biosystems Engineering, North Dakota State University, Fargo, ND, USA

M. N. Muli
Department of Land Resource Management and Agricultural Technology, University of Nairobi, Nairobi, Kenya

R. N. Onwonga
Department of Land Resource Management and Agricultural Technology, University of Nairobi, Nairobi, Kenya

G. N. Karuku
Department of Land Resource Management and Agricultural Technology, University of Nairobi, Nairobi, Kenya

V. M. Kathumo
Department of Land Resource Management and Agricultural Technology, University of Nairobi, Nairobi, Kenya

M. O. Nandukule
Department of Land Resource Management and Agricultural Technology, University of Nairobi, Nairobi, Kenya

Quirine M. Ketterings
Department of Animal Science, Cornell University, Ithaca, NY, USA

Shona Ort
Department of Animal Science, Cornell University, Ithaca, NY, USA

Sheryl N. Swink
Department of Animal Science, Cornell University, Ithaca, NY, USA

Greg Godwin
Department of Animal Science, Cornell University, Ithaca, NY, USA

Thomas Kilcer
Department of Animal Science, Cornell University, Ithaca, NY, USA

Jeff Miller
Cornell Cooperative Extension, Oneida County, Oriskany, NY, USA

William Verbeten
Cornell Cooperative Extension, Northwest New York Dairy, Livestock and Field Crops Team, Lockport, NY, USA

Muhammad Zahir Ahsan
Plant Breeding Section, Central Cotton research Institute Sakrand, Sindh, Pakistan

Muhammad Saffar Majidano
Plant Breeding Section, Central Cotton research Institute Sakrand, Sindh, Pakistan

Hidayatullah Bhutto
Plant Breeding Section, Central Cotton research Institute Sakrand, Sindh, Pakistan

Abdul Wahab Soomro
Plant Breeding Section, Central Cotton research Institute Sakrand, Sindh, Pakistan

Faiz Hussain Panhwar
Plant Breeding Section, Central Cotton research Institute Sakrand, Sindh, Pakistan

Abdul Razzaque Channa
Plant Breeding Section, Central Cotton research Institute Sakrand, Sindh, Pakistan

Karim Buksh Sial
Agronomy Section, Central Cotton research Institute Sakrand, Sindh, Pakistan

Robert W. Blake
Department of Animal Science and director of the Center for Latin American and Caribbean Studies (2009-2014), Michigan State University, East Lansing, Michigan, USA Professor Emeritus, Department of Animal Science, Cornell University, Ithaca, New York, USA

Elvira E. Sanchez-Blake
Department of Romance and Classical Studies, Michigan State University, East Lansing, Michigan, USA

Debra A. Castillo
Department of Comparative Literature, Stephen H. Weiss Presidential Fellow, and Emerson Hinchcliff Professor of Hispanic Studies, Cornell University, Ithaca, New York, USA

Odette V. Delfin-Ortega
Institute of Economic Research and Enterprise from Universidad Michoacana de San Nicolás de Hidalgo Morelia, Michoacán, México

Joel Bonales Valencia
Institute of Economic Research and Enterprise from Universidad Michoacana de San Nicolás de Hidalgo Morelia, Michoacán, México

Harbans L. Bhardwaj
Agricultural Research Station, Virginia State University, Petersburg, Virginia, USA

Anwar A. Hamama
Agricultural Research Station, Virginia State University, Petersburg, Virginia, USA

Mark E. Kraemer
Agricultural Research Station, Virginia State University, Petersburg, Virginia, USA

D. Ray Langham
Sesaco Corporation, San Antonio, Texas, USA

Vidas Damanauskas
Institute of Power and Transport Machinery Engineering, Aleksandras Stulginskis University, Lithuania

Algirdas Janulevičius
Institute of Power and Transport Machinery Engineering, Aleksandras Stulginskis University, Lithuania

Gediminas Pupinis
Institute of Power and Transport Machinery Engineering, Aleksandras Stulginskis University, Lithuania

Hongxia Liu
Institute of Chemical Ecology, Shanxi Agricultural University, Shanxi, China

Zhixiong Liu
Institute of Chemical Ecology, Shanxi Agricultural University, Shanxi, China

Haixia Zheng
Institute of Chemical Ecology, Shanxi Agricultural University, Shanxi, China

Meihong Yang
Institute of Chemical Ecology, Shanxi Agricultural University, Shanxi, China

Jinlong Liu
Institute of Chemical Ecology, Shanxi Agricultural University, Shanxi, China

Jintong Zhang
Institute of Chemical Ecology, Shanxi Agricultural University, Shanxi, China

Marilyn Theuma
Institute of Earth Systems, Division of Rural Sciences and Food Systems, University of Malta, Msida, MSD, Malta

Claudette Gambin
Permanent Crops - Agriculture Directorate, Department for Rural Affairs and Aquaculture, Agricultural Research and Development Centre, Għammieri, Malta

Everaldo Attard
Institute of Earth Systems, Division of Rural Sciences and Food Systems, University of Malta, Msida, MSD, Malta

O. T. Akintayo
Kwame Nkrumah University of Science and Technology, Kumasi, Ghana

B. K. Maalekuu
Kwame Nkrumah University of Science and Technology, Kumasi, Ghana

J. K. Saajah
Ghana Cocoa Board/Newmont Mining Cooperation, Ghana

Isnaini Nurwahyuni
Doctorate Program, Department of Agriculture, Faculty of Agriculture, North Sumatra University, Medan, North Sumatera, Indonesia

Justin A. Napitupulu
Doctorate Program, Department of Agriculture, Faculty of Agriculture, North Sumatra University, Medan, North Sumatera, Indonesia

Rosmayati
Doctorate Program, Department of Agriculture, Faculty of Agriculture, North Sumatra University, Medan, North Sumatera, Indonesia

Fauziyah Harahap
Departement of Biology, FMIPA, State University of Medan, Jl. Willem Iskandar, Psr V, Medan, Sumatera Utara, Indonesia

Yashbir Singh Shivay
Division of Agronomy, Indian Agricultural Research Institute, New Delhi, India

Rajendra Prasad
Division of Agronomy, Indian Agricultural Research Institute, New Delhi, India

Rajiv Kumar Singh
Krishi Vigyan Kendra, Aligarh of Chander Shekhar Azad University of Agriculture & Technology, Kanpur, India Directorate of Seed Research, Village Kushmaur, Post Office, NBAIM, District Mau, Uttar Pradesh, India

Madan Pal
Division of Agronomy, Indian Agricultural Research Institute, New Delhi, India

Chi-I Chang
Department of Biological Science and Technology, National Pingtung University of Science and Technology, Pingtung, Taiwan

Chiy-Rong Chen
Department of Life Science, National Taitung University, Taitung, Taiwan

Yo-Chia Chen
Department of Biological Science and Technology, National Pingtung University of Science and Technology, Pingtung, Taiwan

Kuei-Wen Cheng
Department of Biological Science and Technology, National Pingtung University of Science and Technology, Pingtung, Taiwan

Bo-Wei Lin
Department of Biological Science and Technology, National Pingtung University of Science and Technology, Pingtung, Taiwan

Yun-Wen Liao
Department of Biological Science and Technology, National Pingtung University of Science and Technology, Pingtung, Taiwan

Wen-Ling Shih
Department of Biological Science and Technology, National Pingtung University of Science and Technology, Pingtung, Taiwan

Jhon Paul Mathews Delgado
Graduate program in Botany from the National Institute of Amazonian Research-INPA, Manaus, Brazil

Patrick Mathews Delgado
Department of Parasitology, Institute of Biology, University of Campinas, São Paulo, Brazil

Carlos Abanto Rodriguez
Peruvian Amazon Research Institute-IIAP, Pucallpa, Ucayali, Peru

Ricardo Manuel Bardales Lozano
Ret Bionorte (Multinsitutional programme of Amazon), Brazil

Ziad Barakat Al-Rawashdeh
Department of Agricultural Sciences, Al-Shoubak College, Al-Balqa' Applied University, Al-Shoubak, Jordan

Ezz Al-Dein Muhammed Al-Ramamneh
Department of Agricultural Sciences, Al-Shoubak College, Al-Balqa' Applied University, Al-Shoubak, Jordan

Muwaffaq Ramadan Karajeh
Department of Plant Protection & IPM, Faculty of Agriculture, Mu'tah University, Karak, Jordan

A. R. Khan
Department of Environmental Sciences, COMSATS
Institute of Information Technology, Abbottabad, Pakistan

S. M. Shah
Department of Environmental Sciences, COMSATS
Institute of Information Technology, Abbottabad, Pakistan

M. Irshad
Department of Environmental Sciences, COMSATS
Institute of Information Technology, Abbottabad, Pakistan

Vivian U. Oleforuh-Okoleh
Department of Animal Science, Ebonyi State University,
Abakaliki, Ebonyi State, Nigeria
Department of Animal Science, Rivers State University
of Science and Technology, Nkpolu-Oroworukwo, Port
Harcourt, Rivers State, Nigeria

Harriet M. Ndofor-Foleng
Department of Animal Science, University of Nigeria,
Nsukka, Enugu State, Nigeria

Solomon O. Olorunleke
Department of Animal Science, Ebonyi State University,
Abakaliki, Ebonyi State, Nigeria

Joesph O. Uguru
Department of Animal Science, Ebonyi State University,
Abakaliki, Ebonyi State, Nigeria

www.ingramcontent.com/pod-product-compliance
Lightning Source LLC
Chambersburg PA
CBHW050450200326
41458CB00014B/5125